Tasks for vegetation sciences 3

Series Editor

HELMUT LIETH

University of Osnabrück, F.R.G.

THE RADIATION REGIME AND ARCHITECTURE
OF PLANT STANDS

JUHAN ROSS

The radiation regime and architecture of plant stands

Dr W. Junk Publishers   The Hague-Boston-London 1981

Distributors:

for the United States and Canada

Kluwer Boston, Inc.
190 Old Derby Street
Hingham, MA 02043
USA

for all other countries

Kluwer Academic Publishers Group
Distribution Center
P.O. Box 322
3300 AH Dordrecht
The Netherlands

This volume is listed in the Library of Congress Cataloging in Publication Data

ISBN 978-90-6193-607-7 ISBN 978-94-009-8647-3 (eBook)
DOI 10.1007/978-94-009-8647-3

Preface by the T:VS Series editor

Dr Ross' book on 'Radiation Regime and Architecture of Plant Stands' is one of the first volumes of a new series of books concerned with the international development of vegetation science. The objective of research in vegetation science has shifted in recent decades from mainly descriptive to analytical and functional topics. The new tasks often require new methods, many of which are recent developments in chemistry and physics. The general use of these methods in vegetation science requires a broader understanding of the theory and the application potential as well. Many of these new approaches need more space for their presentation than journal papers can offer. It is therefore the specific aim of this series, Tasks for Vegetation Science (T:VS) to provide an outlet for such treatments.

The series is open to monographic treatments by individual authors or collective efforts during Symposia or other arrangements. Topics should be devoted to the analysis and interpretation of vegetation and plant distribution to structural features or to functional properties. While plants and/or vegetation will be the focal point, interactions between vegetation/soil, vegetation/atmosphere, vegetation/animals or vegetation/microbes will be included. Under the heading 'functions', we understand the functions of the vegetation that influence the environment as well as the functioning of the vegetation within given environmental conditions. Preference will be given to topics which open possibilities for utilizing new methods, techniques or concepts to understand and interpret vegetation patterns and vegetation functions. Descriptive treatments can only be included if the approach to the description is innovative, or significantly necessary for further work in the field.

In many cases new developments occur regionally. When the results are immediately useful for practical applications, they are published in the languages of the potential user. If this is not English, it takes a long time before such innovations cross the language barrier. It is for this reason that the series considers translations of books recently published, the content of which fits into our series.

Dr Ross' book is the first 'translation volume' in our series. Several more are in preparation. The original text of Dr Ross' book is printed in Russian. The book deals with biophysics of plants and vegetation pieces. This topic is not a neglected subject in western literature. The Russian approaches to experimentation and modelling are sometimes different, however, and therefore of great interest internationally.

We thank Dr Ross for consenting to include his book in our series. The content of the book is very important for future work in applied vegetation science and especially plant ecology. Noteworthy is the fact that the models and methods described in this book were applied in difficult environments and are therefore interesting for the comparison with models derived from moderate climates.

While this book appears on the international market, Dr Ross has coauthored a new book on 'Mathematical modelling of plant transpiration and photosynthesis under soil moisture stress', 223 p. This book, jointly written by S. Bichele, H. Moldau and J. Ross, and published in 1980 by Hydrometeoisdat in Leningrad expands on the modelling aspect of photosynthesis and the plant production process dealt with in several chapters of the book presented in this series. Though published in Russian, the book contains an English summary and figure captions in English. This makes it possible to use the book conveniently together with this T:VS volume, especially, since most mathematical notations are very similar in both books.

By including this volume into our series T:VS we hope to increase the discussion of scientific ideas in our field across one of the language barriers more difficult to pass.

In the next volumes to appear in this series we expect several more to deal with modelling aspects of vegetation. We hope that this will help to improve our insight into structural as well as functional phenomena of vegetation and ecosystems.

Osnabrück, February 1981 H. LIETH

Preface by I. A. Shulgin

In the last decade, plant stand and regularities of its development and activity have become an object of increasing interest for such seemingly remote sciences as botany, particularly plant physiology, ecology and morphology, geography, atmospheric physics (including satellite meteorology and atmospheric optics). Moreover, the scientific interests of specialists of various fields appear to be focused on plant stand whether it is a cultivated crop or a natural biocenosis.

As is known, vegetation is the primary source of human food, whereas the productivity of plants as autotrophic organisms, i.e. their very existence, depends chiefly, if not exclusively, upon photosynthesis, a phenomenon involving utilization of solar radiation in a complex, multistage process of the transformation of atmospheric carbon dioxide, water and the nutrients into various organic compounds of green plants.

It stands to reason, therefore, that radiation regime is of primary importance in the production process, both the input of radiation and its distribution in the stand being vitally important, the latter determining the 'light' nutrition of the leaves as well as of the other photosynthesizing organs, and thereby providing for the intensity and duration of their functioning, the quality and quantity of their products.

Moreover, radiant energy also plays an essential part in form development, determining the dimensions of the axial organs (i.e. stem internodes and the leafstalks), the tillering pattern, the shape and size of the leaves, their orientation, the time of the onset of differentiation and growth of the reproductive organs. In other words, along with the water-and-heat regime, solar radiant energy, or rather its physiologically active portion (380–760 nm), affects the habitus of the plant, its 'architecture', which, in turn, affects light, heat and CO_2 regimes in biocoenoses. This necessitates a detailed and strictly quantitative study of the radiation regimes, precisely of its time-spatial structure in relation to the architecture of plants, whose hereditary habitus largely determines the optical properties of plant stand, which, in turn, influence the architecture of plant organisms that compose the stand.

Properly speaking, investigation of the radiation regime in plant stand has an almost century-long history. However, partly owing to the lack of equipment, partly because of the complexity of measurements but, chiefly, because of the inexplicit statement of the principal aims of studies, the results have often been fragmentary and not always comparable. The central problem that united individual research efforts was the elaboration of a theory of photosynthetic production. It is not surprising, therefore, that the problem 'photosynthesis and utilization of solar radiation' has been included in the International Biological Programme as one of its most important items.

The radiation regime of plant stand may also be viewed in terms of atmospheric physics, for it is in the plant cover, this most tenuous stratum of life, that ultimate transformation of solar radiation occurs after selective attenuation in the atmosphere. In this sense plant stand acts as a peculiar transformer of solar radiation into other forms of energy which determine the regime of the earth surface and that of the air boundary layer.

The complexity and variability of plant stand and its components do not exclude application of the approach and methods of turbid-media optics. However, investigation of this aspect of the problem – radiation and plant stand – is in its initial stage.

The present monograph is the first of its kind in world literature to offer a systematic presentation of a scientific branch, which borders on geophysics, on one hand, and biology, on the other, and deals with solar radiation transfer in plant stand. The author, J. Ross, is the founder of the Tartu school of actinometry, which has gained recognition both within and outside this country.

An obvious asset of the monograph is that is clearly formulates the problem of the plant stand radiation regime as one pertaining to exact sciences, laying thereby the theoretical basis of phytoactinometry. In view of the fact that the radiation regime of plant stand is closely related to its architecture, the author found it necessary to accurately formulate methods of strict quantitative study of the structure of an individual plant as well as a plant stand as a whole.

The monograph gives a quantitative theory of the plant stand radiation regime, that is based, essentially, on the works of the author and on those of his colleagues.

The demand for a theory of this kind has become especially strong in recent years in view of the current development in agrometeorology of the 'weather-yield' problem, whose initial solution largely depends on process modelling. Subsequently, as more information becomes available, the solution will depend on the computation of the effect of human activity on the production process and, consequently, on the yield.

Another asset, not the least important of the book, is the systematic use of a vast body of experimental material obtained by the author, as well as by other workers. This is primarily true of the data on the architecture of a number of agricultural crop stands. Certain characteristics specified or newly introduced by the author, e.g. the inclination index of foliage area or the relative plant foliage area density as a function of relative height, enabled the author to systematize practically all the data available in the literature on the orientation and distribution of leaves in stands. A substantial body of these data were collected by the author and his associates in different regions of this country, including Estonia, Moldavia, Tadjikistan.

Owing to the high theoretical value of this work and the exhaustive coverage of material, available in the literature, the monograph can be justly regarded not only as an original research work by J. Ross, but also as a manual and a source of reference.

However, the physical treatment of the problem calls for the application of a mathematical approach, which will, obviously embarass the reader. To help the reader overcome this difficulty, the author has provided explicit verbal interpretation of the formulae and equations.

In summary, publication of this monograph is, without doubt, a notable contribution to the development of the actinometry of plant stand.

Author's preface to English translation

The Russian version of the book was written in 1973. Many new works both theoretical and experimental have appeared since then, which has made the present book a little out-of-date. The author, therefore, took this into account while preparing the English version, and has made a number of additions to the manuscript. Firstly, a new final chapter 'Net radiation of plant stand' has been added which places emphasis, like the whole of the book, on the theoretical aspect of the problem. Secondly, some new paragraphs have been written, i.e. 'Architecture, photosynthesis and productivity of plant stand', 'The method of hemispherical photographs', 'Monte-Carlo simulation models', and 'New theories of radiative transfer'. The paragraphs are concerned with recent achievements of theoretical phytoactinometry. And, finally, over a 100 new works have been reviewed and included in the list of references. These, again, are chiefly concerned with theoretical problems.

A specific feature of the present book is that it makes more reference to Russian literature. This stands to reason, because owing to language difficulties these works are little known or totally unfamiliar to Western authors.

The theoretical basis of both the present book and of all the principal works on theoretical phytoactinometry is the concept of the plate turbid medium and mathematical modelling of radiative transfer in this medium. From talks and discussions with biologists the author, who is a physicist, had the impression that assimilation of the concept of the plate turbid medium as a model of a plant stand and phytoelements as a model of leaves, stems and other aboveground parts involves certain difficulties. The problem is discussed in II.1.6 of the book. However, here again the author explains that a phytoelement is not an exact analogue of the entire leaf, stem, etc., as a plate turbid medium is not an exact analogue of a plant stand. It is no more than an arbitrary model. The basic difference between a real stand and a modelled stand is that the latter deals with small pieces of plant organs, not with their real size. These pieces of plant organs are substituted for the whole organs observing the equality of their respective areas and orientation.

Moreover, in the modelled plant stand natural arrangement of plants and the structure of their crowns are to a large extent eliminated. It is assumed, instead, that small phytoelements are randomly distributed and that the random distribution can be described by corresponding statistical laws. The modelled plant stand, (i.e. a plate turbid medium) is, therefore, a more simplified analogue of a real stand allowing, however, for the application of the quantitative methods of exact sciences. This simplification of complex biological objects in their mathematical modelling is common to science nowadays and reflects the current standard of our knowledge.

It is the author's sincere hope that, to a certain extent, the present book fills the gap that now exists, also in Western literature, in the systematic description of the current state of theoretical phytoactinometry.

In conclusion, the author wishes to thank B. K. Kapin for the tremendous work of translating the book.

Tartu, November 1979 JUHAN ROSS

Contents

Part two: Radiation regime in plant stand

List of main symbols

Symbol	Description	Unit
A_B	albedo (reflection coefficient) of the ground surface	–
A_k	albedo (reflection coefficient) of the plant stand	–
A'_k	albedo (reflection coefficient) of the dense plant stand ($L_0 \to \infty$)	–
	$k=S$ for direct solar radiation	
	$k=D$ for diffuse solar radiation	
	$k=Q$ for total solar radiation	
A_x	asymmetry of characteristic x*	–
$a_{km}(L)$	penetration function (relative irradiance) of solar radiation inside the stand averaged over the horizontal area at the depth L	–
	$k=S$ direct solar radiation flux density	
	$k=D$ diffuse sky radiation flux density	
	$k=D_z$ zonal sky radiation	
	$k=d$ diffuse sky radiation intensity	
	$k=Q$ total solar radiation flux density	
	$m=P$ photosynthetically active radiation (PAR)	
	$m=N$ near-infrared radiation (NIR)	
$a^H_{kl}(L)$	penetration function (relative irradiance) of complementary solar radiation inside the stand averaged over the horizontal area at the depth L	–
	$k=S$ direct solar radiation	
	$k=D$ diffuse solar radiation	
	$l=1$ downward radiation	
	$l=2$ upward radiation	
a_L	distance of leaf projection on the ground surface from the origin	cm
a_p	effective radius of plant basal area	cm

		Unit
$B(L)$	net radiation (radiation balance) inside the stand averaged over the horizontal area at the depth L	$W\,m^{-2}$
B_0	net radiation above the stand	$W\,m^{-2}$
$B_k(L)$	net solar radiation (shortwave radiation balance)	$W\,m^{-2}$
$B_J(L)$	net infrared radiation (longwave radiation balance)	$W\,m^{-2}$
b_B	ground surface brightness coefficient	sr^{-1}
b_k	stand brightness coefficient	sr^{-1}
	$k=S$ for direct solar radiation	
	$k=D$ for diffuse solar radiation	
	$k=Q$ for total solar radiation	
C_k	conversion factor from shortwave radiation to photosynthetically active radiation	–
	$k=S$ for direct solar radiation	
	$k=D$ for diffuse solar radiation	
	$k=Q$ for total solar radiation	
$\tilde{C}_S$	phytoelements distribution coefficient	
D_0	irradiance of the diffuse solar radiation above the stand	$W\,m^{-2}$
$D(z)$	irradiance of the diffuse solar radiation inside the stand averaged over the horizontal area at the height z	$W\,m^{-2}$
$D_{0z}(\theta)$	zonal brightness of the sky	$W\,m^{-2}\,rad^{-1}$
$D'_{0z}(\theta)$	relative zonal brightness of the sky	rad^{-1}
$d_0(\tilde{r})$	sky brightness at the direction r	$W\,m^{-2}\,sr^{-1}$
$d'_0(r)$	relative sky brightness at the direction r	sr^{-1}
$d(z, r)$	sky brightness inside the stand at the direction r averaged over the horizontal area at the height z	$W\,m^{-2}\,sr^{-1}$
d_{1j}	largest diameter	cm
d_{2j}	smallest diameter	cm
	$j=C$ stem	
	$j=P$ plant	
E_k	infrared radiation flux density	$W\,m^{-2}$
	$k=A$ from atmosphere	
	$k=B$ from ground surface	
	$k=L$ emitted by leaves	
$E_l(L)$	infrared radiation flux density inside the stand averaged over the horizontal area at the depth L	$W\,m^{-2}$
	$l=1$ downward flux	

Symbol	Description	Unit
	$l=2$ upward flux	
E_x	excess of the characteristic x*	–
$F_l(L)$	horizontal irradiance (flux density) of the total solar radiation inside the stand averaged over the horizontal area at the depth L	$\mathrm{W\,m^{-2}}$
	$l=1$ downward flux	
	$l=2$ upward flux	
$G(r)$	foliage area orientation function	–
$\frac{1}{2\pi} g_L(r_L)$	distribution function of foliage area orientation	–
$g'(\vartheta_L)$	leaf inclination distribution function	–
$\hat{g}_L(\vartheta_L) = g'_L \sin \vartheta_L$	leaf inclination distribution function	–
$\frac{1}{2\pi} g''_L(\varphi_L)$	leaf azimuth distribution function	–
$H(\eta)$	foliage thickening function	–
$H_{kl}(L)$	horizontal irradiance (flux density) of the complementary solar radiation averaged over the horizontal area at the depth L	$\mathrm{W\,m^{-2}}$
	$k=S$ complementary direct solar radiation	
	$k=D$ complementary diffuse solar radiation	
	$l=1$ downward flux	
	$l=2$ upward flux	
h_{jL}	leaf dimension	cm
	$j=1$ length	
	$j=2$ largest width	
	$j=3$ thickness	
$h_{\odot}$	solar elevation	rad
$J_{kl}(z, r_L)$	solar radiation irradiance (flux density) of the surface exposed inside the stand in the direction r_L averaged over the horizontal area at the height z	$\mathrm{W\,m^{-2}}$
	$k=S$ direct solar radiation	
	$k=D$ diffuse solar radiation	
	$k=Q$ total solar radiation	
	$l=1$ surface upperside	
	$l=2$ surface lowerside	
$i_\lambda(P, r)$	monochromatic radiation intensity	$\mathrm{W\,m^{-2}\,sr^{-1}}$
$i_{kl}(z, r)$	radiation intensity inside the stand averaged over the horizontal area at the height z	$\mathrm{W\,m^{-2}\,sr^{-1}}$

Symbol	Description	Unit
	$k=S$ complementary direct solar radiation	
	$k=D$ complementary diffuse solar radiation	
	$k=Q$ total solar radiation	
	$k=J$ infrared radiation	
	$l=1$ downward intensities	
	$l=2$ upward intensities	
$K_i(P,P',r)$	attenuation function of radiation field intensity	–
K_{xx}'	normalized correlation matrix	–
k_{LF}	leaf form-factor	–
k_k	area attenuation coefficient	–
	$k=M$ for plate medium	
	$k=S$ for direct solar radiation	
	$k=Q$ for total solar radiation	
$k^*_M = u_L k_M$	volume attenuation coefficient for plate medium	cm^{-1}
L_0	leaf area index	cm^2 leaf cm^{-2} ground
$L(z)$	downward cumulative leaf area index (relative stand depth)	cm^2 leaf cm^{-2} ground
L_{0j}	area index of the j-th plant organ	cm^2 organ cm^{-2} ground
$L_j(z)$	downward cumulative area index of the j-th plant organ	cm^2 organ cm^{-2} ground
	$j=C$ stem	
	$j=F$ reproductive organs	
	$j=R$ root	
m_j'	fresh weight	g
m_j	dry weight	g
	$j=L$ leaf	
	$j=C$ stem	
	$j=F$ reproductive organs	
	$j=R$ root	
	$j=P$ plant	
N_p	plant stand density	m^{-2} ground
N_X	adequate sampling number for characteristic x^*	–
n_L	total number of leaves per plant	–
n_{LA}	number of living leaves per plant	–
n_{LD}	number of dead leaves per plant	–

		Unit
$P=(x,y,z)$	point in the space	
$P_i(P,P',r)$	transmission function of radiation field intensity	
Q_0	total solar irradiance (flux density) above the stand	$W\,m^{-2}$
$Q(L)$	total solar irradiance inside the stand averaged over the horizontal area at the depth L	$W\,m^{-2}$
$Q^m(L)$	total solar irradiance inside the stand $m=O$ in gaps $m=T$ in shade $m=C$ on cloudy sky	$W\,m^{-2}$
$q_0(r)$	intensity of the total solar radiation above the stand	$W\,m^{-2}\,sr^{-1}$
R_{jl}	reflection coefficient of j-th plant organ $j=L$ leaf $j=C$ stem $j=F$ reproductive organs $l=1$ upper side $l=2$ lower side	–
$r=(\vartheta,\varphi)$	direction in space	–
$r_L=(\vartheta_L,\varphi_L)$	direction of leaf normal	–
$r_C=(\vartheta_C,\varphi_C)$	direction of stem	–
$r_S=(h_\odot,\varphi_\odot)$	direction of sunrays	
$\widehat{rr'}$	angle between directions r and r'	–
S_{00}	Solar Constant	$W\,m^{-2}$
$S_0(h_\odot)$	irradiance of direct solar radiation above the stand on a surface perpendicular to the sunrays	$W\,m^{-2}$
$S'_0(h_\odot)=$ $=S_0 \sin h_\odot$	horizontal irradiance of the direct solar radiation above the stand	$W\,m^{-2}$
$S'(L,r_S)$	horizontal irradiance (flux density) of the direct solar radiation inside the stand averaged over the horizontal area at the depth L	$W\,m^{-2}$
S_C	area of stem lateral surface	cm^2
S_{CC}	stem cross-sectional area	cm^2
S_{EC}	area of stem axial section	cm^2
S_P	plant lateral area	cm^2
S_{CP}	plant basal area	cm^2
S_L	leaf area	cm^2
$\hat{S}_L(z)$	vertical distribution of foliage area of an individual plant	cm^2 leaf cm^{-1}
$\hat{S}_{LP}$	foliage area of an individual plant	cm^2 leaf

		Unit
$\tilde{S}_L(\bar{z})$	relative plant foliage area density as a function of relative height $\bar{z}$	
s	pathlength	cm
T_{Ll}	leaf transmission coefficient	–
	$l=1$ upperside	
	$l=2$ lowerside	
t_L	leaf surface temperature	^0C
$u_j(z)$	area density of the j-th plant organ	cm^2 leaf cm^{-3}
$u_{mj}(z)$	dry weight density of the j-th plant organ	$g\ cm^{-3}$
$u'_{mj}(z)$	fresh weight density of the j-th plant organ	$g\ cm^{-3}$
	$j=L$ leaves	
	j=C stems	
	j=F reproductive organs	
	j=R roots	
V_j	volume	cm^3
	$j=L$ leaf	
	$j=C$ stem	
	$j=P$ plant	
V	variance	–
V'	relative variance	–
v_X	variation coefficient of the characteristic x*	–
W_j	water content of the j-th plant organ (characteristic as a random quantity)	–
z_L	leaf height above the ground	cm
z_{1L}	plant foliage uppermost level in its natural position	cm
z'_{1L}	plant foliage uppermost level	cm
z_{2L}	plant foliage lowest level	cm
z_{1C}	stem length	
z_{1P}	plant height in its natural position	cm
z_{0j}	uppermost level of the j-th plant organ in the stand	cm
z_{00j}	lowest level of the j-th plant organ in the stand	cm
$\bar{z}$	relative height	–
α_L	azimuth angle of the leaf projection on the ground surface	rad

		Unit
$\frac{1}{\pi}\Gamma_L^*(P,r',r)$	volume scattering phase function of the plate medium	cm^{-1}
$\frac{1}{\pi}\Gamma_L(P,r',r)$	area scattering phase function of the plate medium	–
$\frac{1}{4\pi}\gamma(P,r',r)$	scattering phase function	sr^{-1}
$\frac{1}{\pi}\gamma_L(r_L,r',r)$	leaf scattering phase function	sr^{-1}
$\gamma_B(r',r)$	bidirectional reflection function of the ground surface	–
$\delta(x-x')$	Dirac delta – function	–
δ_k	emissivity of the infrared radiation $k=L$ leaf $k=B$ ground surface	–
ε_k	intensity of the infrared radiation $k=A$ atmosphere $k=B$ ground surface	$W\,m^{-2}\,sr^{-1}$
η	overlapping index	–
η_L	leaf emission coefficient	$W\,m^{-2}\,sr^{-1}$
η_M	area emission coefficient for the plate medium	$W\,m^{-2}\,sr^{-1}$
$\eta_M^*=u_L\eta_M$	volume emission coefficient for the plate medium	$W\,m^{-3}\,sr^{-1}$
$\theta=\pi-\vartheta$	inclination angle	rad
ϑ	inclination angle	rad
ϑ_j	inclination angle of the j-th plant organ $j=L$ leaf j=C stem	rad
$\vartheta_\odot=\frac{\pi}{2}-h_\odot$	solar inclination	rad
κ_L	leaf absorption coefficient	–
κ_M	area absorption coefficient for the plate medium	–
$\kappa_M^*=U_L\kappa_M$	volume absorption coefficient for the plate medium	cm^{-1}
μ_j	part of dry weight of the j-th organ in the dry weight of the whole plant	–

		Unit
ν_L	leaf wetness	–
$\zeta = \arccos r_S r_L$	angle between direction of sunrays and leaf normal	rad
$\pi_{Ll}(z, r_L)$	leaf absorption function $l=1$ leaf upperside $l=2$ leaf lowerside $l=S$ for direct solar radiation $l=D$ for diffuse solar radiation $l=JS$ for complementary direct solar radiation $l=JD$ for complementary diffuse solar radiation	$W m^{-2}$
$\pi_k(z)$	foliage absorption function for solar radiation inside the stand $k=S$ direct solar radiation $k=JS$ complementary direct solar radiation $k=D$ diffuse solar radiation $k=JD$ complementary diffuse solar radiation	$W m^{-3}$
ρ_j	volumetric dry weight of the *j*-th plant organ	$g cm^{-3}$
ρ'_j	volumetric fresh weight of the *j*-the plant organ	$g cm^{-3}$
ρ_{wj}	volumetric water content of the *j*-th plant organ	$g cm^{-3}$
σ_L	specific leaf dry weight	$g cm^{-2}$
σ'_L	specific leaf fresh weight	$g cm^{-2}$
σ_{WL}	specific leaf water content	$g cm^{-2}$
σ_X	standard deviation of the characteristic x	
τ	time	s
$\tau(s,r)$	optical depth	cm^2 leaf cm^{-2} stand
φ	azimuth angle	rad
φ_C	stem azimuth	rad
φ_L	leaf azimuth	rad
$\varphi_\odot$	solar azimuth	rad
χ_L	foliage area inclination index	–
ψ_L	leaf main axis angle	rad

		Unit
Ω	solid angle	sr
Ω_1	solid angle 2π of the upper hemisphere	sr
Ω_2	solid angle 2π of the lower hemisphere	sr
Ω_1^L	solid angle 2π over the leaf upperside	sr
Ω_2^L	solid angle 2π over the leaf lowerside	
$d\Omega = \sin\vartheta d\vartheta d\varphi$	elemental solid angle	
		sr
ω_L	leaf scattering coefficient	–
ω_M	area scattering coefficient of the plate medium	–
$\omega_M^* = u_L \omega_M$	volume scattering coefficient of the plate medium	cm^{-1}

Abbreviations

PAR	photosynthetically active radiation
NIR	near infrared radiation
LWR	long-wave (infrared) radiation
SWR	short-wave radiation

Introduction

The solar radiant energy is in fact the only source of energy for the basic physical processes taking place in the atmosphere and on the earth's surface. When passing through the atmosphere and being reflected by the ground surface, solar radiation undergoes changes and conversions. Some of it is absorbed in the atmosphere and converted into other forms of energy, mainly into heat, and some is scattered by gases, by dust and by water vapour.

Because of absorption and scattering in the atmosphere, solar radiation is changed by the time it reaches the earth's surface. That part of it which arrives as a beam of parallel rays is referred to as *direct solar radiation*, and that which is scattered in the atmosphere and reaches the earth's surface from all directions of the sky is called *diffuse solar radiation*. Both of them are reflected back into the atmosphere when they reach the earth's surface, and this third type of radiation is defined as *reflected radiation*.

All of these radiations differ from solar radiation arriving at the upper level of the atmosphere in intensity as well as in spectral composition although they all fall within the spectral region of solar radiation. In atmospheric physics these types of radiation are known as *short-wave radiation* (SWR) as distinguished from *long-wave or infrared radiation* (LWR) emitted by the atmosphere and the earth's surface.

The study of transfer and conversion of solar radiation and of long-wave radiation from the atmosphere and earth's surface is the subject of two departments of geophysics, viz. actinometry and atmospheric optics.

Although studies of radiative transfer in the earth's atmosphere have been carried on on a wide scale, little has been done in a field bordering on geophysics and biology, viz. solar radiation transfer in plant stand.

Plant stand interested actinometrists chiefly as a special type of reflective surface on the earth, hence their studies were limited to calculations of the brightness coefficient and albedo for various types of stands (Krinov 1947; Shifrin 1953).

Apart from actinometrists, a similar interest in the subject was shown by micrometeorologists engaged in the study of physical processes observed in the atmospheric boundary layer and soil (Van Wijk 1968).

As compared with various types of ground surface, plant stand has a number of peculiar features, such as greater thickness which, in the case of a forest, may reach tens of metres.

A dense and well-developed stand is known to absorb from 60 to 90% of incident short wave radiation of the Sun and sky, reflect between 10 and 30% and allow only 5 to 10% to pass through and be absorbed by the ground. Plant stand also absorbs much of the carbon dioxide contained in the air giving off oxygen into the atmosphere. Of importance is the role of plant stand in creating the water-and-heat regime of ground surface (Budagovsky 1964). The above features make plant stand an active contributor to the climate formation having also an impact on the physical processes in the atmosphere, particularly, in its boundary layer.

But the role of solar radiation as a source of energy is not confined to the physical processes in the atmosphere and on the earth's surface alone; it is also the primary source of energy for the numerous biological processes taking place in plants.

The interaction between radiation and living plants can be divided into three distinct categories:

1. Thermal effects of radiation. More than 70% of solar radiation absorbed by plants is converted into heat and used as energy for transpiration, for maintaining plant temperature, etc. (Slatyer 1967; Gates 1968; Kleschnin 1960).

2. Photosynthetic effects of radiation. A part of solar radiation, up to 28% in terms of energy, absorbed in a spectral region of 0.38–0.71 μm (PAR) is used in photosynthesis and stored chemically in high-energy organic compounds (Rabinowitch 1945; Nichiporovich 1963).

3. Photomorphogenetic effects of radiation on growth and development (Shulgin 1973). Solar radiation plays an important role as a regulator and a controller in processes of growth and development. Several aspects of these interactions are not yet understood, although it has been established that the region of photomorphogenetically active radiation begins in the ultraviolet, extends over the whole visible region and ends in the near infra-red near 0.75μm. In addition to the spectral quality and energy content, the photomorphogenetic effects of radiation depend also on periodical variations of radition in time – alternation of light and dark periods, their duration, etc.

Absorption of radiation by green leaves depends on the wave-length and is characterized by the spectral absorption coefficient.

Since detailed data on spectral composition of narrow bands is very often difficult to obtain and is not always needed, use is made of rougher characteristics. From the point of view of biogeophysics it is convenient to divide the entire radiation spectrum of the Sun and sky into four spectral regions in the table below (Ross 1968).

The three aspects of solar radiation along with other environmental factors

Type of radiation	Spectral region (μm)	Percent of solar radiant energy	Effect of radiation on plant life		
			Thermal	Photo-synthetic	Photo-morphogenetic
Ultraviolet	0.29–0.38	0–4	Insignificant	Insignificant	Significant
Photosyn-thetically active radiation	0.38–0.71	21–46	Significant	Significant	Significant
Near infra-red radiation	0.71–4.0	50–79	Significant	Insignificant	Significant
Infra-red radiation	4.0–100	–	Significant	Insignificant	Insignificant

determine to a large extent the existence and development of stand in all its diversity.

The urgency for an all-round study of stand radiation regime and optical properties as factors of paramount importance in plant productivity and climate formation is evident. In terms of energy, solar radiation in plant stand is a highly complicated process dependent both on incident radiation and on the optical and geometrical properties of vegetation. Some short-wave radiation is transmitted by stand and reaches deeper layers of vegetation, some is absorbed, and some is scattered by leaves, stems, flowers, etc. and, after multiple scattering, either returns to the atmosphere or reaches the ground surface, or is absorbed by plant organs. Since scattering of radiation by plant organs is selective, the spectral quality of radiation is changed as it is being transferred. Because of the finite dimensions of the scattering and shading elements in a stand the radiation field is highly non-uniform horizontally. Fluctuation of plant organs due to wind results in large time variations of radiation. Taken together these features make radiation regime of a plant stand specific, calling for a special approach in its study. This is the reason why radiative transfer in stands has become the subject of a specialized branch of science known as phytoactinometry, which has shown a rapid development only in the last few decades.

By the late fifties when the author was just starting his investigations, there had been experimental works, predominantly dating from the prewar period, devoted to the study of an integral and spectral radiation regime, which dealt mainly with forest stands (Baumgartner 1953; Härtel 1954; Orth 1938; Rademacher 1950; Sauberer 1948; Sauberer & Trapp 1947; Sauberer & Härtel 1959; Sauberer 1937). Those investigations had been carried out with low accuracy instruments using incomparable techniques. For this reason most of the data obtained were of an illustrative character.

On the basis of ideas of Ivanov (Ivanov 1932; Ivanov 1946) who laid a solid foundation for regarding the light climate as the main factor in the life activity and productivity of forests, investigations of the radiation regime in plant stands were begun in the USSR (Galperin 1938; Kudryavtseva 1940; Yanishevsky 1940; Yanishevsky 1941).

In the post-war period the study of radiation regime in stands became topical in the USSR in connection with further development of the theory of photosynthetic activity of agricultural crops (Bell 1955; Bell 1956; Nichiporovich 1955; Nichiporovich 1958; Nichiporovich & Chmora 1958). Among actinometrists it was Lopukhin (1948a, 1948b, 1950, 1951a, 1951b) who devoted much of his time to studying the radiation regime of a cotton field. Then Shifrin (1953) put forward a fruitful idea of applying the theory of radiative transfer in a turbid medium for describing radiation fluxes in stands and proposed the first mathematical model for calculating vertical short-wave radiation in stands.

Things went a little better in the field of leaf optical properties research. Using fairly modern instrumentation based on an integrating sphere, measurements were made of the leaf transmission and reflection coefficients, both in the PAR region and in the NIR and long-wave radiation regions (Ilyina 1947; Billings & Morris 1951; Rabideau, French & Holt 1946; Sauberer 1948; Seybold 1933; Seybold & Weissweiler 1942).

But notwithstanding these achievements the study of optical properties of a plant stand was not yet posed as a physical problem, no general theory of radiative transfer in a stand was evolved, instrumentation was not yet developed to carry out measurements of the various parameters of the radiation regime in a stand, which was necessary for testing the developing transfer theory and for determination of the effects of optical properties in a quantitative theory of photosynthesis and productivity of a stand.

Beginning in 1957 the Department of Atmospheric Physics of the Institute of Astrophysics and Atmospheric Physics, Academy of Sciences of the Estonian SSR, joined in solving the above problems. Some of the results of that work form the basis of this book. Although the book touches on both theoretical and experimental problems, it is primarily concerned with the theoretical treatment of the problem, elaboration of calculation methods for determination of the geometrical structure characteristics and optical properties of a stand and creation of a mathematical model of radiation regime. For this reason such aspects as instrumentation, presentation of purely experimental data, comparison and analysis of numerous published phytoactinoclimatological data have been given less consideration in the book. The bulk of the book deals with the following problems of phytoactinometry.

1. Evolving a quantitative theory of the radiation regime in a stand based on the radiative transfer equation for a plate medium, which would make it possible to reveal the principal relations governing radiative transfer in a stand in dependence on various biogeophysical factors; calculation of the main characteristics of the optical properties of a stand.

2. Developing calculation and experimental methods for determination of statistical characteristics of geometrical structure of a stand, and also, obtaining experimental data on the architecture of a number of agricultural crops, which would allow the general relation governing stand structure to be revealed on the basis of the data obtained by the author and those available in the literature.

3. Obtaining reliable and comprehensive experimental data on the radiation regime of a number of agricultural crops and the experimental verification of the quantitative theory.

In dealing with the first problem a plant stand was treated as a turbid medium (Shifrin 1953). However, because of the large size of plant organs and the important role played by the geometrical structure of a stand in its radiation regime, it was impossible to automatically use the methods and mathematical approach evolved for the description of radiative transfer in a turbid medium in astrophysics and atmospheric physics. For this reason a more specified and generalized concept of a turbid medium was proposed, according to which a stand was treated as a plate turbid anisotropic medium homogeneous in a horizontal plane in terms of statistics (Ross 1964). According to this model, small parts of the surface of leaves, stems and other overground plant organs are distributed randomly in horizontal layers but their orientation is given by a statistical law. The vertical distribution of their area is also given. The main optical characteristics of and the radiative transfer equation for this type of medium were derived also (Ross 1964). A specific feature about a plate turbid medium is its anisotropy, i.e. dependence of the optical properties on direction. For this reason, the modified radiative transfer equation has, as will be seen, a more complicated form in which the structural geometrical characteristics of the optical medium itself play an important part. The importance of the structure is especially felt in the long-wave radiation region where there is no scattering, and in the PAR region where scattering is low. In the NIR region, where scattering is great, the optical properties of small parts of plant organs gain more prominence.

While the specific geometrical and optical properties of plant stands called for a variety of approaches to an approximated solution of the modified transfer equation, the complicated calculations demanded the use of a digital computer. However, this solution had to be analytical in form so as to facilitate assessment of the various factors.

Significance of the solution of the second problem becomes evident from the first problem, since architecture of a plant stand plays an important role in the radiation regime. Solution of the third problem is necessitated by the complexity of the first. Even at the time of formulating the problem and choosing a model of a plate turbid medium (for the description of a stand) the author was fully aware of the model being only a very rough approximation to such a multifarious and variable object as a plant stand. The author's quantitative theory of the stand radiation regime may not hold true for all types of vegetation and does not always yield satisfactory quantitative results.

A vast set of experimental data are needed for establishing the range of its applicability and for its verification, which will, in turn, serve as a basis for further improvement of the theory. Moreover, the mathematical intricacy of the theory has required simplified and semiempirical formulae with coefficients derived experimentally.

The above circumstances have called for an integrated approach to the study of the stand radiation regime, i.e. along with theoretical research work has also been undertaken on designing new instruments, developing new measuring techniques and a large number of measurements in various crop stands were carried out in Estonia, Moldavia and Tadjikistan (Mägi & Ross 1969a, b; Ross 1957; Ross & Vlasova 1967; Ross & Nilson 1965, 1967a, b; Ross & Ross 1969a, b; Ross & Tooming 1968, 1971b). For field experiments we have designed a special mobile phytoactinometric laboratory in a bus which allowed for the recording of the main radiation parameters both above and inside a stand.

The measuring techniques developed by our group as well as the devices we designed (Laisk 1964, 1965a, b, c; 1969; Reemann, 1963; Niilisk 1962, 1963, 1964, 1965) and calculating formulae are now in use in the Soviet Union (Abashina 1968; Abashina & Kozinets 1968; Gulyayev 1967; Gulyayev, Mitrofanov & Manuilsky 1969; Yefimova 1965, 1969), Japan (Udagawa, Uchijima & Horie 1969; Udagawa et al. 1968), Australia (Anderson 1969) and Bulgaria (Lingova & Stanev 1969). The above methods, devices and formulae were also used in compiling methodological instructions for the crop productivity studies (Tooming 1969; Tooming & Ross 1969).

It is gratifying to the author to express his sincere gratitude to his teacher of theoretical physics, academician A. Kipper, former director of the Institute of Astrophysics and Atmospheric Physics, Academy of Sciences, the Estonian SSR, for his support in the progress of phytoactinometry at his institute, to the corresponding member of the USSR Academy of Sciences A. A. Nichiporovich, Chairman of the Scientific Council on the problem 'Photosynthesis', for his assistance in the development of integrated work on biogeophysics, and to professor G. V. Rosenberg for fruitful discussions of the current problems of radiation transfer in non-homogeneous turbid media. The author is also indebted to his teacher of actinometry Dr. Yu. D. Yanishevsky and to his Moscow colleagues Dr. A. A. Budagovsky, Dr. S. N. Chmora and Dr. I. A. Shulgin for stimulating talks and discussions, also for assistance in the analysis of biological and biophysical problems.

Special thanks go to all the colleagues of the Department of Atmospheric Physics, and especially to Dr. T. Nilson, whose joint work and sustained effort laid the basis of this book.

PART ONE

Plant stand architecture

I.1 Role of phytometric investigations in the studies of plant stand architecture and radiation regime

Architecture herein is taken to mean a set of features delineating the shape, size, geometry and external structure of a plant. A quantitative study of the architecture of both a stand and an individual plant with its organs constitutes a fundamental task of phytometry.

Hence, plant stand phytometry, as part of biometry (Lakin 1968) is regarded here as a large-scale estimation of living plant qualitative and quantitative characteristics, statistical treatment, analysis and interrelation thereof. Phytometry of a stand also involves the development of new, more complex characteristics that are not readily amenable to direct measurements but calculated in terms of primary characteristics. Phytometrie measurements, therefore, involve the floristic analysis of stands, phenologic observations of phases of individual species development, quantitative description of the distribution of plants, as well as that of individual plant organs, within a given plot. Phytometric measurements also involve the estimation of the linear dimensions, area, volume, dry and wet weights of the individual components of a stand. As can be seen, they do not involve micrometeorological observations in a stand, nor do they include biophysical measurements or biochemical analyses, etc.

Plant stand phytometry as a science bordering upon mathematics and biology has certain specific features. Among these are: firstly, multitude of measurements, i.e. the phytometric methods are applied to a set of objects rather than to individual objects or to single results of measurements. Secondly, simplicity of stand phytometric measurements which do not require any sophisticated equipment, nor ãny high skill of the observer. Thirdly, a wide use of methods and techniques of mathematical statistics in carrying out and processing phytometric measurements.

The main purposes of phytometry can be stated as follows:

1. Assessment of stand phytometric features that allow for easy and fairly precise measurement in large quantities and are essential for the study of radiation and water-and-heat regimes, also for the study of productivity. The number and pattern of these features may vary depending on the task and types of a stand.

2. Elaboration of methods, instruments and appliances for the measurement and statistical treatment of the selected primary phytometric characteristics.

3. Statistical analysis of the characteristics.

4. Determination of the degree of independence of certain parameters or of their close interrelation at a given moment and their change during the growth period. In phytometry usually linear or root-mean-square regression is applied. Existence of statistical relation between two parameters makes it possible to measure only one of them, as the ofter one can be calculated.

The practical value of the calculation of a characteristic is particularly high where performance of numerous measurements is difficult or impossible. In forestry, for instance, calculation techniques are widely used in the assessment of stem phytomass, estimation of height, as well as of the phytomass of photosynthesizing organs, etc. from the stem diameter measured at breast height.

5. Finding of new, more sophisticated phytometric characteristics relating to the quantitative theory of radiation regime, mass exchange and stand productivity and not readily subject to direct measurement; also, development of methods and techniques for the calculation of those characteristics from primary directly measurable parameters.

6. Elaboration of new statistical methods for plant systematics and morphology.

7. Elucidation of the inherent laws of the architecture of both an individual plant and of a stand as a whole; discovery of interrelation between plant habitus and metabolism, between plant and environment, between geometric structure and photosynthetic activity (Loomis et al. 1967; Horn 1971); discovery of the intrinsic laws of growth and interrelation between the processes of growth and the geometric structure of a plant.

The latter two problems, whose solution is now only in the initial stage, lie outside the limits of conventional phytometry. We do feel, however, that this particular problem will subsequently become the main concern of stand phytometry when phytometry from a statistical and empirical science changes into a statistical and theoretical branch; development of the above problems underlies the progress of quantitative plant ecology.

In the quantitative theory of plant stand productivity the results of phytometric studies can be used, basically, along two lines: (1) quantitative approach to the architecture of both individual plants and a stand as a whole; (2) quantitative description of stand growth and development.

The emphasis in the former case is on the delineation of the stand structure at a given moment, the time herein being a certain parameter. And conversely, the latter case is concerned basically with the alteration of one or other stand characteristic during the growing season, time being regarded as the main variable.

It is to be noted that the stand architecture appears to be one of the most essential factors of the stand radiation regime as well as one of those most difficult to assess. The difficulty stems primarily from the multiformity and complexity of stand structure. To illustrate, let us consider the leaves that are the most important stand elements in view of the radiation regime and photosynthetic activity. The leaves vary in length and width, in their shape and number (compound leaves consist of a large number of single leaflets), etc. Moreover, individual plant species also vary in habitus, length, diameter, assimilating surface area, etc. Investigation and comparison of the above features is naturally the concern of plant morphology which currently limits itself to the descriptive level, while quantitative estimation of the geometric features of a plant and of its individual organs by mathematical methods, which is obviously the concern of phytometry, involves difficulties resulting from the complexity and great variability of the parameters of the object itself. The difficulties become all the greater when we proceed from an individual organ to a plant, from a plant to a pure stand, and from a pure stand to a plant community. And at this point, we have to observe that with regard to quantitative stand architecture only preliminary steps have so far been made although progress in phytometry and quantitative plant ecology has been rapid.

A statistical approach appears to be the most fruitful in view of the great interspecific differences and large intraspecific variations of the features relating to the geometric structure. On the other hand, the statistical approach divests the object under investigation of its diversity. The application of mathematical methods invariably leads to disregard of certain aspects of the structural features. We, therefore, have to be rather schematic in the phytometric delineation of the stand geometric structure and can capture only the most essential features of an object, thus ignoring many of its peculiarities.

From the methodological standpoint the phytometric description of a stand should be carried out on four levels, namely: individual organs of the plant, the whole plant, the pure stand and the plant community. Each higher level contains elements of the preceding ones, yet it does not represent a mere set of the lower level elements but has specific features of its own.

The problem of modelling a plant stand as a layer of turbid medium will be discussed separately.

I.2 Principal phytometric characteristics of stands

I.2.1 PHYTOMETRIC CHARACTERISTICS OF LEAF AND OTHER PLANT ORGANS

In the photosynthetic activity of a plant stand an individual leaf performs a number of important functions of which the following are of special interest.

1. *The leaf as a radiation receiver.* Of the total amount of solar radiation absorbed by a stand, leaves account for 80 to 90%, the rest being absorbed by stems, flowers and other overground organs. The following characteristics are used for the description of a leaf as a radiation receiver: (a) leaf location in a stand; (b) leaf location relative to the plant; (c) leaf spatial orientation; (d) leaf size and shape; (e) leaf optical properties.

2. *The leaf as a photosynthesizing organ.* Being a radiation receiver the leaf is at the same time the main photosynthesizing organ. It is known that of the total stand photosynthesis, leaves account for 60 to 95%, the rest belongs to stems, branches, spikelets and other green organs. The leaf as a photosynthesizing organ is characterized by the size of the photosynthetic apparatus, i.e. the leaf area, and rate of its activity dependent on the complex of leaf's anatomical, morphological and physiological properties.

3. *The leaf as a growing and developing plant organ* (in the ontogeny). Since any plant in its growth and development obeys its inherent laws, the leaf also follows a specific pattern in its growth and development during the growing season. In addition to more complicated morphological properties the following phytometric characteristics are used to describe leaf growth and development: (a) leaf ordinal number; (b) age; (c) leaf dry weight as a function of time; (d) leaf water supply characteristics, etc.

A similar list of quantitative characteristics can be prepared for other plant organs as well.

The enormous multiformity of leaves of various plant species makes one look for characteristics which are comparatively simple, yet universal enough to provide a phytometric description of the leaf. These include first of all h_{1L}, leaf's length in centimetres; h_{2L}, leaf largest width in centimetres; S_L, leaf area in sq. centimetres. Usually, only the area of the leaf upper side is meant by S_L.

A number of methods for measuring leaf area have been developed in phytometry to cope with the complicated shape of leaves. These will be dealt with in detail under I.3.1.

Of importance are also such phytometric characteristics of the leaf as m'_L and m_L, leaf fresh and dry weights in grammes, respectively.

Frequently, instead of the leaf fresh weight, leaf water content, W_L, is used defined as

$$W_L = \frac{m'_L - m_L}{m'_L}, \tag{I.2.1}$$

$(1 - W_L)100$ being leaf dry weight percentage, or leaf wetness v_L, defined as

$$v_L = \frac{m'_L - m_L}{m_L} = \frac{W_L}{1 - W_L}. \tag{I.2.2}$$

The leaf wetness determines the amount of water in grammes per g leaf dry weight.

The leaf thickness h_{3L} is seldom used as a phytometric characteristic as is also the leaf volume $V_L = h_{3L} S_L$. Instead, use is made of the specific leaf fresh weight

$$\sigma'_L = \frac{m'_L}{S_L} \tag{I.2.3}$$

and the specific leaf dry weight

$$\sigma_L = \frac{m_L}{S_L} \tag{I.2.4}$$

in addition to the specific leaf water content

$$\sigma_{WL} = \sigma'_L - \sigma_L = \frac{m'_L - m_L}{S_L}. \tag{I.2.5}$$

The unit of σ'_L, σ_L and σ_{WL} is g/cm^2.

The specific leaf weight is the amount of leaf mass per unit area. In general, this value increases with increasing leaf thickness and has a diurnal variation dependent on assimilation and transpiration.

In literature, one may come across σ'_L and σ_L replaced with their reciprocals that can be defined as specific leaf area.

The volumetric leaf fresh and dry weights

$$\rho'_L = \frac{m'_L}{V_L} \text{ and } \rho_L = \frac{m_L}{V_L} \tag{I.2.6}$$

are also rarely used as is the volumetric leaf water content

$$\rho_{WL} = \rho'_L - \rho_L = \frac{m'_L - m_L}{V_L}. \tag{I.2.7}$$

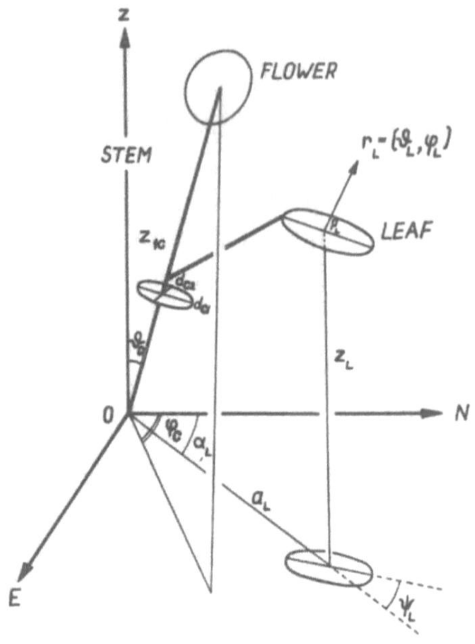

Fig. 1. Geometrical characteristics of plant.

From (I.2.4) and (I.2.6) it follows that $h_{3L}=\sigma_L/\rho_L$, ρ'_L and ρ_L being measured in g/cm^3.

To locate individual plant organs spatially, consider Fig. 1. The coordinate system $Oxyz$ is so chosen that its origin O coincides with the point of intersection of an inclined stem with the ground surface and the axis x is directed northwards.

The leaf middle point P_L is located in space by its coordinates z_L, a_L, α_L. Here, z_L is the leaf height above ground, a_L is the distance of leaf projection on the ground surface from the origin O, and α_L is the leaf azimuth counted off clockwise from north, with $0 \leqslant \alpha_L \leqslant 2\pi$.

Leaf orientation at the point $P_L=(z_L, a_L, \alpha_L)$ may be various and to determine it, use is made of the leaf normal r_L.

Horizontal leaves have $\vartheta_L=0$; vertical leaves facing south have $\vartheta_L=\pi/2$, $\varphi_L=\pi$, those facing west have $\vartheta_L=\pi/2$, $\varphi_L=\frac{3}{2}\pi$ and so on.

However, leaf location in space is not complete unless the three locating coordinates and the two orientation coordinates are supplemented with a leaf main axis angle ψ_L, i.e. an angle through which the projection of the leaf main axis (Fig. 1) on a horizontal plane must be turned clockwise in order to coincide with the projection of the vector OP_L on the same plane. This angle may vary between 0 and π.

In a similar way the characteristics of the stem are determined: z_{1C} for height, $r_C=(\vartheta_C, \varphi_C)$ for direction, d_{1C} and d_{2C} for largest and smallest diameters, respectively, $S_{CC}=(\pi/4)d_{1C}d_{2C}$ for cross-sectional area, $S_C=(\pi/2)z_{1C}(d_{1C}+d_{2C})$ for lateral surface, $V_C=S_{CC}z_{1C}$ for volume, m'_C for fresh weight, m_C for dry weight, W_C

$=(m'_C - m_C)/m'_C$ for water content, $\rho'_C = m'_C/V_C$ and $\rho_C = m_C/V_C$ for volumetric stem fresh and dry weights, respectively.

An important parameter in the study of the stand radiation regime is the area of stem axial section $S_{EC} = S_C/\pi$, but in dealing with photosynthesis, respiration and transpiration the entire lateral surface S_C participating in water and CO_2 exchange must be taken into account.

In some Australian works on phytometry (Warren Wilson 1965a) only half the lateral surface $S_C/2$ is used, which, in our opinion, has less ground.

Minor parts of a plant (twigs, leafstalks, etc.) are usually described phytometrically by integral characteristics, such as fresh and dry weights, alone, without considering the geometrical structure of the object.

In a similar way phytometric characteristics of reproductive organs and roots are determined.

I.2.2 PHYTOMETRIC CHARACTERISTICS OF AN INDIVIDUAL PLANT

The plant as a living organism is an integral and selfadjusting system undergoing a development cycle in its ontogeny. Characteristics of the plant are also its definite shape and geometric structure, i.e. its architecture that reflects the long-lasting evolutionary process and adaptation to environment.

From the point of view of architecture the plant can be regarded as a combination of individual organs arranged in space in a manner peculiar to a given species, the plant structure being determined both by the geometry of individual organs and that of their spatial arrangement. The external appearance of the plant, i.e. its habitus is due to hereditary properties and specific features of the organism. A plant's exterior is determined also by the environmental conditions which continually affect a living plant, by adaptation of plants to these. In addition, it is due to intra- and interspecific competition and to many other factors. Throughout a number of generations these permanently acting factors of the environment have brought about changes in vital activity of the plant and in its architecture. It may, therefore, be expected that the external form of the plant reflects its vital activity, peculiarities of its growth and development as well as the interaction between plants and their environment. Consequently, a plant's architecture as a quantified representation of its external form must have a functional interrelationship with other characteristics of plant activity.

An attempt by Paltridge (1973) to elaborate a quantitative theory in which the tree growth and shape depend on photosynthesis and distribution of stored reserves over the crown turned out to hold much promise. Smith and Scoullar (1975) proposed a computer model for simulation of coniferous tree crowns. The model consists of submodels of branch death and allocation of stored reserves. The geometrical parameters of growth, i.e. the branching angle and the branch length, are considered in a computer model of Honda (1972) describing the crown growth and crown shape.

The amount of solar radiation absorbed is a decisive factor determining the plant total weight. However, potential growth of the plant cannot be realized until the given amount of radiant energy is not matched by an adequate supply of water and nutrients in the soil. Apparently, the extent to which radiation regime is matched by water supply determines the relation between the assimilating area of the leaves and that of the roots. The less water at a given radiation, the less the leaf-to-root area ratio. The share contributed to the total weight by the stem, trunk, branches and other parts of the plant 'architectural frame' is largely dependent on competition for radiation.

In a forest, for example, trees have to spend the bulk of assimilates for the building and maintenance of this frame. But the heavier and larger the overground frame, the more firmly it has to be fixed under the ground by the supporting roots. To this must be added part of the assimilates used to build the reproductive organs which provide for the species survival. Such qualitative considerations indicate that plant structure is not a random phenomenon but is functionally related to the plant's entire activity.

Many of the plant species are symmetrical in their external shape with respect to a central, or axial line, although statistically it is achieved approximately (Sinnot 1960). This symmetry allows the modelling of individual plants as cylinders, spheres, cones and other simple geometric figures filled with plant elements.

The general shape of the plant is best described using linear characteristics pertaining to the geometric figure modelled. These characteristics include z^*_{1P} for plant height in its natural position, usually concurring with the plant foliage uppermost level z^*_{1L} (should the latter be lower, the height is determined separately), z^*_2, plant foliage lowest level, a^*_P, effective radius of plant basal area determined from its projection on the ground surface[1].

While geometric figures can be used to represent, with a certain degree of accuracy, the real shape of trees, bushes and some multifoliate plants, they are less suitable for representation, say, of cereals and legumes. Thus, for instance, the basal area of a maize plant is nearer in shape to an ellipse than to a circle, therefore it is better represented by two diameters of the plant, d^*_{1P} and d^*_{2P}, the plant's basal area being $S^*_{CP} = k_P d^*_{1P} d^*_{2P}$. In a more general case, where the shape of an axisymmetric plant is given by a generatrix equation $r = f(z)$, the plant volume V^*_P and lateral area S^*_P are given by the corresponding formulae

$$V^*_P = \pi \int_0^{z^*_{1P}} f^2(z)\mathrm{d}z \quad \text{and} \quad S^*_P = 2\pi \int_0^{z^*_{1P}} f(z)\mathrm{d}z. \tag{I.2.8}$$

For a more detailed description of a plant shape supplementary characteristics can be evolved, depending on the particular problem. Luzganov & Solodko

[1] Hereinafter phytometric values pertaining to an individual plant are marked with an asterisk (*). These values are statistically random when considering a stand as a community of plants. Values pertaining to a stand will have no asterisk.

(1968), for instance, study a drop-shaped crown of a tree described by a pair of parabolas.

In agrometeorology plant growth is characterized by the uppermost level of plant foliage z'^*_{1L}, by which is meant the distance from the ground surface to the uppermost point of an upwardly stretched leaf. This characteristic is inapplicable, say, for the study of a plant radiation regime. Therefore, in dealing with the energy and mass exchange in stands, use is made of the quantity z^*_{1L}, the uppermost level of plant foliage in its natural position, by which is meant the distance from the ground surface to the uppermost point of plant foliage in its natural position. The height z^*_{1L} is somewhat dependent on such factors as wind velocity, stem inclination, upper leaf bending, etc., and in general, $z^*_{1L} \leqslant z'^*_{1L}$. Thus, during the growing season of a maize plant there is a simple correlation $z'^*_{1L} = z^*_{1L} + 20$ (heights measured in cm) (Ross & Vlasova 1967) between z'^*_{1L} and z^*_{1L}. This means that the height of the plant in its natural position is, on the average, 20 cm less, due to the bending of upper leaves. Similarly, z^*_{1L} is the lowest level for the plant foliage in its natural position.

Very often phenological properties of a plant are used as highly distinctive, easily obtainable phytometric characteristics giving a general idea of the plant's condition, development phase and external shape (Instructions to Hydrometeorological Stations and Observation Posts 1963; Schnelle 1955).

Among the phenological characteristics also are n^*_L, the total number of leaves per plant, n^*_{LA}, number of living leaves per plant or, in the general case, n^*_j, number of plant j-th organs with $j = L$ for leaves, $j = C$ for stems, $j = R$ for roots, and $j = F$ for reproductive organs.

The phytometric characteristics pertaining to plant productivity and growth include m^*_j and m'^*_j, plant j-th organ dry and fresh weights, respectively,

$$m^*_P = \sum_j m^*_j \tag{I.2.9}$$

whole plant dry weight,

$$\mu^*_j = \frac{m^*_j}{m^*_P} \tag{I.2.10}$$

share of dry weight of the j-th organ in the dry weight of the whole plant.

However, to describe the geometrical structure of a plant within its volume, proceeding from the shape and location of each individual element appears to be too complicated in some cases. Therefore, in a quantitative description of the structure of an individual plant, rougher and averaged values and also used, while individual characteristics of elements of single plant are replaced with values averaged within the plant over a certain volume.

Consider a volume $V^*(P)$ around point $P = (z, a, \alpha)$ within a plant, containing n leaves, with each leaf given by its geometrical parameters. If S_{Lk} is the area of the k-th leaf, then $\sum_{k=1}^{n} S_{Lk}$ is the total leaf area inside the volume, and the foliage area density as a function of point P is equal to

$$u_L^*(P) = u_L^*(z, a, \alpha) = \frac{1}{V^*(P)} \sum_{k=1}^{n} S_{Lk}. \tag{I.2.11}$$

Since leaf distribution inside the entire plant volume is not uniform and individual leaf area varies, the foliage area density inside the plant volume is a function of point P coordinates.

The function $u_L^*(z, a, \alpha)$ is an important geometrical characteristic for both an individual plant and a stand; it is referred to as *foliage area density* because it defines the amount of leaf area per unit plant volume; it is measured in cm^2/cm^3.

Summing-up the foliage area density over all points in the plant volume gives us the total leaf area per plant

$$\hat{S}_{LP}^* = \iiint_{(V_P^*)} u_L^*(P)dP = \int_{z_{2L}^*}^{z_{1L}^*} dz \int_0^{2\pi} d\alpha \int_0^{a_P^*} u_L^*(z, a, \alpha)a da. \tag{I.2.12}$$

If integration is performed only over the height between level z and foliage uppermost level z_{1L}^*, we obtain a quantity $L^*(z, a, \alpha)$ equal to the foliage area within a cylinder of unit cross-section constructed about a vertical with coordinates (a, α) in the plant upper part between levels z and z_{1L}^*

$$L^*(z, a, \alpha) = \int_z^{z_{1L}^*} u_L^*(z', a, \alpha)dz'. \tag{I.2.13}$$

Let us call the quantity $L^*(z, a, \alpha)$ the *downward cumulative leaf area index* at the point (a, α) and in the layer $z_{1L}^* - z$.

If integration is performed over the entire foliage height, we obtain a quantity

$$L_0^*(a, \alpha) = L^*(z_{2L}^*, a, \alpha) = \int_{z_{2L}^*}^{z_{1L}^*} u_L(z', a, \alpha)dz' \tag{I.2.14}$$

defined as the *leaf area index* at the point (a, α). The L^* and L_0^* are non-dimensional quantities showing by how many times the area of the leaves exceeds the basal area of the vertical cylinder within which these leaves are located.

In many cases a plant is assumed to be axisymmetrical and the azimuthal dependence can be neglected.

If the function $u_L^*(P)$ is integrated over a and α, we obtain the leaf area within a horizontal layer of unit thickness at level z:

$$\hat{S}_L^*(z) = \iint_{(S_{1P(z)}^*)} u_L^*(P)dS = \int_0^{2\pi} d\alpha \int_0^{a_P^*} u_L(z, a, \alpha)a da \tag{I.2.12a}$$

with

$$\hat{S}_L^*(z_{2L}^*) = \hat{S}_L^*(z_{1L}^*) = 0. \tag{I.2.12b}$$

The function $\hat{S}_L^*(z)$ characterizes the vertical distribution of foliage area of an individual plant.

The total plant leaf area will be

$$\hat{S}_{LP}^* = \int_{z_{2L}^*}^{z_{1L}^*} \hat{S}_L^*(z)dz. \tag{I.2.12c}$$

In a similar way the area density, downward cumulative leaf area index and vertical distribution of area can be defined for other plant organs.

The orientation of leaves within a volume $V(P)$ is characterized by the distribution function of leaf normals. To find it, divide the surface of a unit sphere into a number of equal portions each having a solid angle of $\Delta\Omega = \sin\vartheta_L \Delta\vartheta_L \Delta\varphi_L$, and determine what number of leaves within the volume $V(P)$ have their normals oriented within the i-th solid angle $\Delta\Omega_{Li}$ located in the direction r_{Li}. Let n_i denote the number of leaves and the area of the j-th leaf having its normal oriented within $\Delta\Omega_{Li}$, and let n be the number of leaves in $V(P)$. Then the total area of leaves having their normals oriented within the i-th solid angle $\Delta\Omega_{Li}$ is equal to $\Delta S^*_{Li} = \sum_{j=1}^{ni} S^*_{Lij}$, and the area of leaves or parts thereof, whose normals fall within the unit solid angle around direction r_{Li} equals $\Delta S^*_{Li}/\Delta\Omega_{Li}$. This area is a part of the total foliage area $\sum_i S^*_{Li}$ inside the volume $V(P)$ and is equal to

$$\frac{1}{2\pi} g_L(P, r_{Li}) = \frac{1}{\Delta\Omega_{Li}} \frac{\Delta S^*_{Li}}{\sum_i S^*_{Li}} \tag{I.2.15}$$

and $\sum_i n_i = n$ and $\sum_i \Delta\Omega_{Li} = 2\pi$.

From (1.2.15) it follows that

$$\frac{1}{2\pi} \sum_i g_L(P, r_{Li}) \Delta\Omega_{Li} = 1$$

or, in terms of integrals,

$$\frac{1}{2\pi} \int_{\Omega_1} g_L(P, r_L) d\Omega_L \equiv 1, \tag{I.2.16}$$

where $d\Omega_L = \sin\vartheta_L d\vartheta_L d\varphi_L$.

Let us call $(1/2\pi)g_L(P,r_L)$ *the distribution function of foliage area orientation or leaf normal distribution function.* It determines the part of the foliage area oriented inside some volume around point P so that the leaf normals are within the unit solid angle around direction r_L.

In its general form the $(1/2\pi)g_L(P,r_L)$ function was introduced and analysed in detail by the author together with Nilson (Ross & Nilson 1967b). It is a non-dimensional function determining the probability of foliage area being oriented in the direction r_L 'inside' a plant.

In principle, leaves can be oriented in all directions and their normals can be in the upper as well as in the lower hemisphere, although actual phytometric measurements show that there are very few leaves on a plant that have their upper side facing down. This means that practically all leaf normals are in the upper hemisphere and in the lower hemisphere $(1/2\pi)g_L \approx 0$. If there is no preferred azimuth orientation, as is assumed in most analyses, then by denoting.

$$\hat{g}_L(P, \vartheta_L) = \frac{\sin\vartheta_L}{2\pi} \int_0^{2\pi} g_L(P, \vartheta_L, \varphi_L) d\varphi_L \tag{I.2.17}$$

we obtain the following expression from (I.2.16)

$$\int_0^{\pi/2} \hat{g}_L(P, \vartheta_L)\mathrm{d}\vartheta_L = 1. \tag{I.2.16a}$$

In this case the function $\hat{g}_L(P, \vartheta_L)$ defines the probability of leaf normals falling within a unit interval of inclinaton angle ϑ_L.

If ϑ_L and φ_L are independent quantities in the two-dimension distribution function $(1/2\pi)g_L(\vartheta_L, \varphi_L)$ then

$$\frac{1}{2\pi} g_L(z, \vartheta_L, \varphi_L) = g'_L(z, \vartheta_L)\frac{g''_L(z, \varphi_L)}{2\pi},$$

where $g'_L(z, \vartheta_L)$ and $g''_L(z, \varphi_L)/2\pi$

are the distribution functions of leaf inclination and azimuth, respectively. Taking into account (I.2.16) and (I.2.17) we obtain

$$\hat{g}_L(z, \vartheta_L) = g'_L(z, \vartheta_L)\sin\vartheta_L. \tag{I.2.17a}$$

The other distribution function of leaf inclination $\tilde{g}_L = u_L\hat{g}_L$ with the normalizing condition

$$\int_0^{\pi/2} \tilde{g}_L(\vartheta_L)\mathrm{d}\vartheta_L = u_L$$

was introduced by Philip (1965a). It should be noted, that the normalizing condition he uses lacks the lucidity of condition (I.2.16a) and makes comparison of $\tilde{g}_L$ function of between various species and between various foliage layers difficult. Probability interpretation of the function $\tilde{g}_L$ is also impeded in this case.

De Wit (1965) used an integral distribution function of leaf inclination

$$G_W(\vartheta_L) = \int_0^{\vartheta_L} g'_L(\vartheta'_L)\sin\vartheta'_L\mathrm{d}\vartheta'_L.$$

This function as compared with the differential function g'_L is less sensitive to changes of orientation, which makes it less suitable for analysis of plant and stand architecture.

Analogous distribution function can be defined for stems $(1/2\pi)g_C(P, r'_C)$ (*Philip* 1966a), flowers and other plant organs. In the case of stems, $r'_C = (\vartheta'_C, \varphi'_C)$ denotes the normal of the stem area element.

G-function. In mathematical modelling of the radiation regime in a stand the $G(z, r)$-function (Ross & Nilson 1965) is very often considered instead of $(1/2\pi)g_L(z, r_L)$, defined by

$$G(z, r) = \int_{\Omega_1} \frac{g_L(z, r_L)}{2\pi}|\cos\widehat{r_L r}|\mathrm{d}\Omega_L, \tag{I.2.18}$$

where

$$\cos\widehat{r_L r} = \cos\vartheta\cos\vartheta_L + \sin\vartheta\sin\vartheta_L\cos(\varphi - \varphi_L). \tag{I.2.19}$$

Geometrically the G-function can be interpreted as follows: $(g_L/2\pi)|\cos\widehat{r_L r}|\mathrm{d}\Omega_L$ is the projection of the part of the area of leaves with normals within the solid angle $\mathrm{d}\Omega_L$ in the direction r, and G is the projection of a unit foliage area in the same direction r. In other words, G is the cosine of the angle between a given direction r and the effective leaf normal $r_{LE}=(\vartheta_{LE}, \varphi_{LE})$ at the height z, i.e.

$$G(z,r)=\frac{1}{2\pi}\int_{\Omega_1} g_L|\cos\hat{r}r_L|\mathrm{d}\Omega_L\approx$$

$$\approx|\cos\hat{r}r_{LE}|\int_{\Omega_1}\frac{g_L}{2\pi}\mathrm{d}\Omega_L=|\cos\hat{r}r_{LE}|. \tag{I.2.18a}$$

Hence, the expression

$$u_{LE}(z,\mathrm{r})=u_L(z)G(z,r) \tag{I.2.20}$$

defines the projection of foliage area in a unit volume at the height z in the direction r, i.e. $u_{LE}(z,r)$ is the effective leaf area density in the direction r and at the height z, and

$$L^*_{OE}(r)=\int_{z^*_{2L}}^{z^*_{1L}} u^*_L(z)G(z,r)\mathrm{d}z\approx G(r)L^*_0 \tag{I.2.21}$$

is the effective leaf area index in the direction r.

Thus, the G-function characterizes the dependence of the effective leaf area on both the beam direction and on leaf orientation.

Consider a few specific cases.

1. *Uniform orientation.* Foliage area is equally oriented in all directions:

$$g_L(z,r_L)=1.$$

It can be shown that

$$G=\frac{1}{2\pi}\int_{\Omega_1}|\cos\hat{r}r_L|\mathrm{d}\Omega_L=\tfrac{1}{2}. \tag{I.2.18b}$$

The spherical distribution of leaves (Nichiporovich 1961) and the uniform distribution are one and the same.

2. *Horizontal orientation*, i.e. leaf normals are vertically oriented:

$$g_L(z,r_L)=\delta(\vartheta_L-0),$$

where δ is the Dirac's delta-function and

$$G=\cos\vartheta\int_0^{\pi/2}\delta(\vartheta_L-0)\cos\vartheta_L\sin\vartheta_L\mathrm{d}\vartheta_L=\cos\vartheta \tag{I.2.18c}$$

because

$$\int_0^{\pi/2} f(\vartheta_L)\delta(\vartheta_L-\vartheta)\sin\vartheta_L\mathrm{d}\vartheta_L=f(\vartheta)$$

3. *Vertical orientation with uniform azimuthal distribution:*

$$g_L(z, r_L) = \delta\left(\vartheta_L - \frac{\pi}{2}\right)$$

and

$$G = \frac{\sin\vartheta}{2\pi}\int_0^{2\pi}\cos(\varphi - \varphi_L)\mathrm{d}\varphi_L = \frac{2}{\pi}\sin\vartheta. \tag{I.2.18d}$$

This particular case also covers vertical round stems. However, in the case of stems $G_C = (1/\pi)\sin\vartheta$ since half of the stem lateral area is shaded.

4. *All leaves have an inclination ϑ_0 and uniform azimuthal distribution:*

$$g_L(z, r_L) = \delta(\vartheta_L - \vartheta_0).$$

It is possible to show that

$$G = A(\vartheta, \vartheta_0) = \begin{cases} \cos\vartheta\cos\vartheta_0, \quad \vartheta + \vartheta_0 \leqslant \frac{\pi}{2} \\ \cos\vartheta\cos\vartheta_0 + \arcsin(\cot\vartheta\cot\vartheta_0) + \\ +\frac{2}{\pi}\sin\vartheta\sin\vartheta_0(1 - \cot^2\vartheta\cot^2\vartheta_0)^{\frac{1}{2}}, \\ \vartheta + \vartheta_0 > \frac{\pi}{2} \end{cases} \tag{I.2.22}$$

This particular case was first brought up in the work by Monsi and Saeki (1953) and then by Reeve (Supplement to (Warren Wilson 1960)) and has been given a detailed analysis in the works by Anderson (1966a) and Warren Wilson (1967a).

5. *Uniform azimuthal distribution of leaf orientation:*

$$g_L(z, r_L) = g'_L(z, \vartheta_L)$$

and

$$G(z, \vartheta) = \int_0^{\pi/2} g'_L(z, \vartheta_L) A(\vartheta, \vartheta_L)\sin\vartheta_L\mathrm{d}\vartheta_L. \tag{I.2.23}$$

In particular cases when $\vartheta = 0$ and $\vartheta = \pi/2$, we obtain from (I.2.23):

$$G_1 = G\left(z, \frac{\pi}{2}\right) = \frac{2}{\pi}\int_0^{\pi/2} g'(z, \vartheta_L)\sin^2\vartheta_L\mathrm{d}\vartheta_L \approx$$

$$\approx \frac{2}{\pi}\sin\vartheta_{LE}\int_0^{\pi/2} g'\sin\vartheta_L\mathrm{d}\vartheta_L = \frac{2}{\pi}\sin\vartheta_{LE}. \tag{I.2.23a}$$

$$G_2 = G(z, 0) = \int_0^{\pi/2} g'(z, \vartheta_L)\cos\vartheta_L\sin\vartheta_L\mathrm{d}\vartheta \approx$$

$$\approx \cos\vartheta_{LE} \tag{I.2.23b}$$

and

$$\tan \vartheta_{LE} = \frac{\pi}{2} \frac{G_1}{G_2}. \tag{I.2.24}$$

6. *Uniform distribution of leaf inclination*

$$g_L(z, r_L) = g''_L(z, \varphi_L)$$

and

$$G(z, \vartheta, \varphi) = \int_0^{2\pi} g_L(z, \varphi_L) B(\vartheta, \varphi - \varphi_L) \mathrm{d}\varphi_L,$$

where

$$B(\vartheta, \varphi - \varphi_L) = \begin{cases} \dfrac{\cos \vartheta}{2} + \dfrac{\pi}{4} \sin \vartheta \cos(\varphi - \varphi_L), \dfrac{\pi}{2} \leqslant \varphi - \varphi_L \leqslant \tfrac{3}{4}\pi \\[2ex] \dfrac{\cos \vartheta}{2} + \dfrac{\pi}{4} \sin \vartheta \cos(\varphi - \varphi_L) + \\[2ex] + \sin \vartheta \cos(\varphi - \varphi_L) \arctan\left(\dfrac{-1}{\tan \vartheta \cos(\varphi - \varphi_L)} \right), \\[2ex] -\dfrac{\pi}{2} \leqslant \varphi - \varphi_L \leqslant \dfrac{\pi}{2} \end{cases}$$

By measuring $G(z, r)$ in some way the inverse problem can be solved. The expression (I.2.18) is then an integral equation for the determination of

$$(1/2\pi) g_L(z, r_L).$$

This subject will be given a more detailed consideration under (I.3.3) in connection with the inclined point quadrats. As can be seen from (II.1.4), the function $G(z, L)$ is the main characteristic of leaf orientation in mathematical modelling of the radiation regime in stands. More complicated is the case where a leaf of graminae is twisted, i.e. the leaf shows curvature and there is a rotation of the lamina along the midrib as one proceeds from one end of the leaf to the other. The theory of a shadow projected by a twisted leaf is presented by Ledent (1976). As pointed out by Ledent, the effect of twisting is similar to that of an increased erectness of the canopy. It is greatest at high sun altitudes and with droopy canopies.

All geometric characteristics of an individual plant vary, variability being inherent in biological objects, therefore, they should be treated as random values.

From this it follows that one has to perform as many measurements on individual plants as are necessary for statistical treatment in order to obtain reliable averaged data on a stand.

The structural parameters of an individual plant are essential in the study of its growth and productivity, in comparing plants of a given species growing in different climates, in the comparison and classification of different species, etc. but they are less important in the case of dense stands where considerable mutual shading and overlapping of plant organs exists.

I.2.3 PHYTOMETRIC CHARACTERISTICS OF A PURE STAND

By a pure stand we will designate a combination of plants of the same species that are arranged over the land area in a particular fashion.

Under natural conditions pure stands are rare (reeds, homogeneous boreal forests, bamboo plantings), while as cultivated stands they are predominant (agricultural crop stands, plantations, afforestations).

Description of the geometry of a pure stand requires, apart from the introduction of the individual plant structure parameters, new concepts and characteristics which allow a quantitative estimation of an object's new features, primarily, plant density in the stand, the pattern of spacing as well as the degree of overlapping and shading.

Let $N_P(P')$ be the plant canopy density at point P', i.e. the number of plants per unit ground area around point P'. N_P is measured in plants/m^2. If measurements are taken over a small plot around point P', the result will be uneven and varies stepwise.

Such density must be treated as a statistically random value. As is common with quantitative ecology, the density N_P can be used to characterize plant distribution over a sampling area (see, for example, Greig-Smith 1964) and much attention is paid to its study.

Pure stands fall into four types, according to pattern of plant dispersion:

1. *Regular dispersion* (e.g. square sown crop, plantation).
2. *Semi-regular dispersion*, tending from regular towards random (e.g. row crop).
3. *Random dispersion* (e.g. broadcasting).
4. *Clumped dispersion*, tending from random towards crowding (e.g. natural stands).

Fig. 2 illustrates the above dispersion types with the same mean density $N_P = 24$. Geobotanic studies have established that a regular dispersion does not occur in natural communities. Predominant in these is the fourth type, where there is tendency towards crowding (see, for example, Greig-Smith 1967).

Characteristic of cultivated stands are the first three types of plant dispersion, the fourth being of rare occurrence. Of the four types the least difficult for a mathematic treatment is the third type, while description of the intermediate types (the second and the fourth) involves great difficulties.

Let us now consider the types of dispersion in more detail.

Regular dispersion. Plant dispersion is viewed as regular when plants are found at the vertices of a parallelogram with given lengths x_1 and y_1 of its sides and an angle α_k between them (Figure 2). In a particular case, when $\alpha_k = 90°$, dispersion is

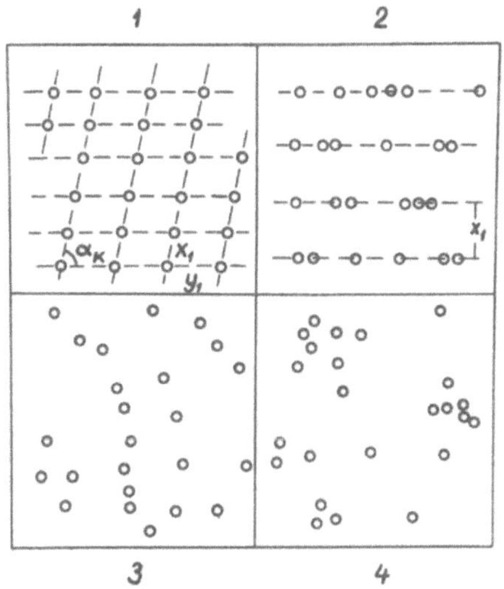

Fig. 2. Four types of plant dispersion with equal density $N_P = 24$.
1 – regular dispersion (square sown crop); 2 – semi-regular, tending from regular to random (row crop); 3 – random (broadcasting); 4 – clumped, tending from random towards crowding (natural stands).

square sown. In another case, when $x_1 = y_1$ and $\alpha_k = 60°$, we will have a dispersion in which adjacent plants form equilateral triangles. In the absence of closure, dispersion of this kind provides the maximum possible coverage[1] which is equal to 0.91, while the coverage at $x_1 = y_1$ and $\alpha_k = 90°$ equals 0.79 (assuming that the basal radius is $\alpha_P = x_1/2$).

Random dispersion. Plant dispersion is treated as random when occurrence of a plant in a given place does not diminish or augment the probability of another plant occurring elsewhere, in other words, the probability of plant occurrence at a particular point of the plot is the same for all the points. In the case of regular dispersion this probability decreases since the occurrence of a plant at a particular point means that adjacent plants are to be found at a definite distance from it. The probability increases in the case of a clumped dispersion.

The random dispersion is given by the binomial formula

$$p(m) = C_N^m \left(\frac{N_P}{N}\right)^m \left(1 - \frac{N_P}{N}\right)^{N-m}, \tag{I.2.25}$$

where N is the maximum possible canopy density, $p(m)$ is the probability for canopy density m, with $m = 0, 1, 2, \ldots, N$ and

$$C_N^m = \frac{N(N-1)\ldots(N-m+1)}{m!}.$$

[1] Coverage is that part of the ground surface which is covered by the overground parts of plants when viewed vertically.

In particular cases

$$p(0)=\left(1-\frac{N_P}{N}\right)^N$$

and

$$p(1)=N_P\left(1-\frac{N_P}{N}\right)^{N-1}$$

are the probabilities of occurrence of plant-free plots and plots with unit density, respectively.

If the real plant density is much lower than the maximum possible density, but isn't an infinitesimal quantity, i.e. if $N_P/N \to 0$ and N_P isn't an infinitesimal quantity, then from (I.2.25) we obtain

$$\lim_{N_P/N \to 0} p(m)=P(m)=\frac{N_P^m}{m!}\exp(-N_P). \tag{I.2.26}$$

Expression (I.2.26) defines the Poisson distribution.

In particular cases

$$P(0)=\exp(-N_P)$$

and

$$P(1)=N_P\exp(-N_P)$$

are the probabilities of the occurrence of plant-free plots and plots with unit density, respectively, with $\lim_{m\to\infty} P(m)=0$.

It must be stressed that the Poisson distribution does not always show a random dispersion of plants, but only in the case when probability of plant occurrence is low (as compared with what theoretically might be the case), but not equal to zero at a fairly vast sampling. The maximum canopy density is a theoretical quantity which is related to the plant basal area and morphology, yet cannot be determined.

The mean values of binomial and Poisson distributions are the same and equal to N_P. The variance V of binomial distribution equals $N_P(1-N_P/N)$, and in the case of the Poisson distribution it is equal to the mean value. The relative variance $V'=V/N_P$ thus is unity for the Poisson distribution and $V'=1-N_P/N<1$ for the binomial distribution. The Poisson distribution is central not only in the description of plant dispersion within a canopy but also applies to the spatial distribution of foliage of any kind. This law is at present the cornerstone in the description of radiant energy attenuation within a stand, for statistical measurements often require the knowledge of whether or not the Poisson distribution is present. A number of criteria have been developed (see, for example, (Greig-Smith 1964)), among which the above relative variance is one of the simplest.

According to the relative variance criterion, the types of plant distribution are classified as shown in Table I.2.1.

Table I.2.1 *Dispersion of plants in relation to variance*

Type of dispersion	Regular	Semi-regular	Random	Clumped
Relative variance V'	$V'=0$	$0<V'<1$	$V'=1$	$V'>1$

Mention should be made, however, that at relative variance values slightly lower than unity and at very high density $N_P \lesssim N$ distinction between the semi-regular and random dispersions becomes impossible. Further data are needed to differentiate between the above types.

Semi-regular dispersion, which is an intermediate type between regular and random dispersions is frequent in cultivated stands. Thus, in a drill sown crop the dispersion is regular across the rows and is determined by the row width, whereas along the rows the dispersion is random or clumped, the relative variance being close to zero across the rows and is $V' \geqslant 1$ along them.

Another type of semi-regular dispersion occurs in an approximately square-sown crop. The mean values of the square determine the method of sowing, while deviations in individual plant dispersion from the mean values follow close to normal dispersion, the relative variance herein approximating zero, yet not equalling it.

Clumped dispersion is the main type of plant distribution in natural communities. It stands to reason, therefore, that quantitative ecology lays a particular stress on its study. However, no satisfactory and fairly simple method for the mathematical description of clumped dispersions has yet been found owing to the complexity of the problem.

In general, elucidation of the patterns of plant spacing involves a great deal of difficulty. It is not surprising, therefore, that though dispersion in a stand is an important characteristic of plant stand structure, its adequate quantitative description can only be attained in relation to extremal cases, i.e. when plants are dispersed either regularly or randomly.

When studying dispersion within a stand we designate plants as dots or small circles at a certain distance from one another.

However, at some moment of growth when density is sufficiently high, plants assume such dimensions that their marginal parts overlap thus leading to closure of the stand.

We will now determine a new characteristic, *the overlapping index*, from the density N_P and the plant's basal area S^*_{CPi}

$$\eta = \sum_{i=1}^{N_P} S^*_{CPi} \qquad \text{(I.2.27)}$$

If there exists no plants on the sampling plot, then $\eta=0$.

If $0<\eta<1$, the plot is sparsely filled with plants, no closure takes place; the equality $\eta=1$ marks the instant of closure, while the values of $\eta>1$ indicate its extent. Obviously, to the maximum value of N_P corresponds the maximum possible plant canopy density η.

It should be noted that $\eta = 1$ clearly marks the initial closure only in the case of regular or nearly regular dispersion. If the dispersion deviates from regular, neighbouring plants overlap earlier but the stand closure occurs at $\eta > 1$ and is dependent on the pattern of dispersion. This question will be further discussed in Section II.4.2.

In detailed studies of the vertical structure of a canopy the overlapping index η can be treated as a function of the height z, closure thereby can be estimated for individual layers. The function $\eta(z)$ can also be treated as random, which will then allow the estimation of other statistical characteristics of η in addition to its mean value.

From the expression (I.2.27) it follows that the overlapping index η depends on the density and the basal area of plants.

Fig. 3, based on the results of expeditions to Tadjikistan and Estonia (Jõgeva) shows variation of the basal area S_{CP} of an average maize plant during the growing season.

The maize plant basal area was modelled by an ellipse, whose area was calculated from the formula $S_{CP} = \pi a_{1P} a_{2P}$. The time τ was counted off from the emergence of seedlings. Figure 3 provides a graphic illustration of the fact that an increase in the stand density inhibits the increase in the plant basal area S_{CP}, especially after closure.

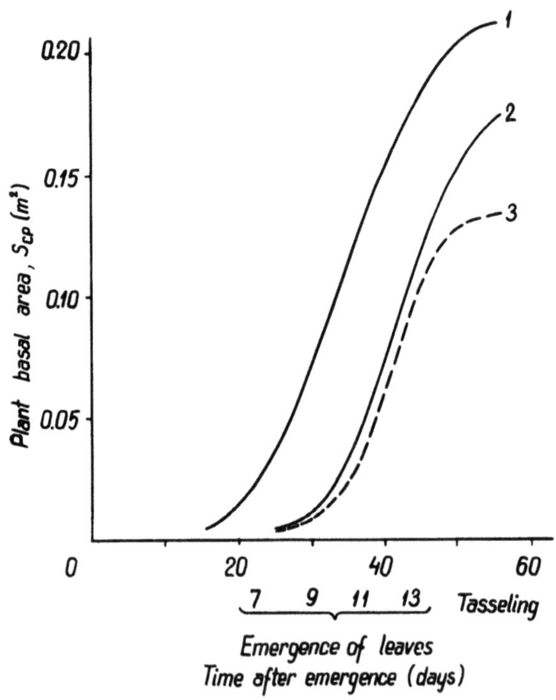

Fig. 3. Seasonal change of maize plant basal area S_{CP}.
1 – maize stand, experiment No. 14, Tadjikistan, 1965, plant density $N_P = 4.5$ plants/m^2; 2 – maize stand, experiment No. 19, Estonia, 1966, $N_P = 27.8$ plants/m^2; 3 – maize stand, experiment No. 20, Estonia, 1966, $N_P = 44$ plants/m^2.

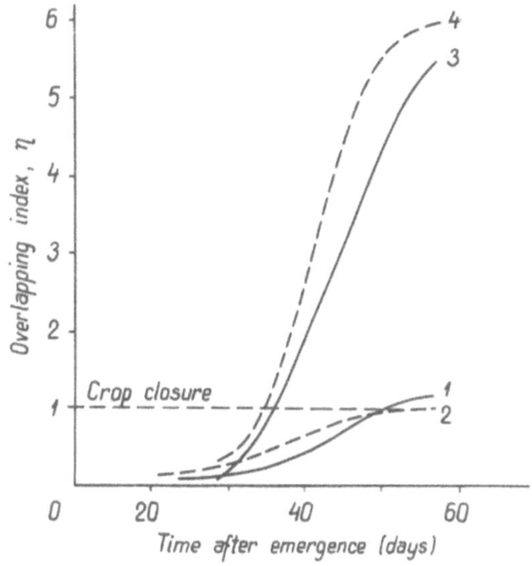

Fig. 4. Seasonal change of overlapping index η.
1 – hybrid swede stand, experiment No. 21, Estonia, 1966, $N_P = 2.8$ plants/m^2; 2 – maize stand, experiment No. 14, Tadjikistan, $N_P = 4.4$ plants/m^2; 3 – same, experiment No. 19, Estonia, 1966, $N_P = 27.8$ plants/m^2; 4 – same, experiment No. 20, Estonia, 1966, $N_P = 44$ plants/m^2.

Fig. 4 represents the variation of the overlapping index η during the growing season obtained from the same data. In addition, it shows the variation of η for hybrid swede (curve 1). As can be seen, in maize stands in Tadjikistan and in swede stands in Estonia, closure occurs on the 50th day upon emergence of seedlings and towards the end of vegetation, η only slightly exceeds unity, i.e. closure in these stands is relatively low. However, in dense silage maize stands in Estonia, closure is observed as early as the 35th day upon emergence of seedlings and towards the end of the summer, η exceeds the value 6, closure herein is unusually high.

Of theoretical interest are the maximum values of the overlapping index η in relation to individual species under diverse environmental conditions. As follows from the formula (I.2.27) the maximum possible value of the overlapping index η_{max} is proportional to the maximum possible density N, i.e. $\eta_{max} = NS_{CP}$. In natural canopies the density N_P, highest during sprouting, decreases constantly with time; the basal area, increasing at the initial stage, subsequently remains more or less constant. This indicates that during sprouting η is equal to zero and then increases. After attaining its maximum magnitude it eventually declines following the death of some individual plants.

In cultivated stands the point at which $\eta = 1$ is attained is of theoretical value from the point of view of the stand geometric structure modelling.

If $\eta < 1$, i.e. the stand is open, it should be regarded as an aggregate of distantly spaced geometric formations (plants), specifically arranged on the ground surface, rather than a layer. In terms of the radiation regime a stand of the kind can be considered as a poorly reflecting surface with turbid bodies on it which absorb and scatter radiation. The problem then is to determine the amounts of absorbed radiation for the surface and the turbid bodies (i.e. plants), separately.

If $\eta > 1$, i.e. the stand is closed, it is reasonable to regard it as a layer of foliage elements. In this situation the geometry of individual plants and peculiarities of their dispersion are of no particular importance, the plot filled with plant elements is thus regarded as a horizontal layer.

Studies of the structure of plant communities are the concern of quantitative ecology. In view of the extreme complexity of the object, quantitative description is confined to floristic composition, i.e. the occurrence of individual species in a community, elaboration of plant distribution criteria, particularly for clumped dispersion. The studies also include interrelations between individual species, as well as the causes of the emergence and development of a community.

The present book is, however, not intended to discuss these questions in detail.

I.2.4 PLANT STAND AS A HORIZONTAL LAYER

Many biogeophysical and agrometeorological works treat a closed stand as a horizontal layer filled with plant elements more or less uniformly. Studies of the geometry of this layer concentrate on characteristics relating to the layer as a whole or to a unit of the layer volume rather than on the distribution and geometry of individual plants in the stand.

To make the quantitative approach to the description of a stand geometry possible, individual plant organs – leaves, flowers, stems, etc. are conventionally divided into a certain number of smaller metameric elements, whose geometric parameters, such as distribution, orientation, etc., come under examination. In so doing we ostensibly 'destroy' part of the stand geometry, thus simplifying the object and ignoring certain structural features. But, on the other hand, such treatment allows the application of quantitative methods for the description of the interrelationship between a stand and the environmental factors, i.e. the methods developed in physics for turbid media.

In physics a simple and extensively used model of a medium is a horizontally homogeneous, optically turbid medium. We will further treat a plant stand as a turbid medium and the aerial organs of plants as scattering elements of the medium. The turbidity of a medium such as a stand will be determined not only by the distribution and orientation of the elements (leaves, stems, etc.) but also by their optical properties.

In accordance with this model it will be assumed that the so-called 'elementary' volume of the medium contains a sufficient number of elements to describe its properties with a fair degree of precision, using statistical laws. Underlying here is the notion that the medium elements are dispersed randomly in all directions, i.e. following the Poisson distribution. Horizontal homogeneity in the statistical sense designates that the mean value, variance and higher moments of statistical characteristics of the turbid medium components are independent of the coordinates x and y, being a function of the vertical coordinate z alone.

The question as to what extent the model of a horizontally homogeneous turbid medium will be realized in a stand is a challenging one and remains open until now. The answer to it depends on the degree of randomness of stand phytoelement dispersion as well as on the way the 'elementary' volume is determined.

The foregoing section dealt with different types of plant dispersion within a stand. A similar question can be raised with regard to the dispersion of phytoelements in a stand separately for leaves, stems and other organs. Even qualitative ideas indicate that there are no strong reasons to assume a random dispersion of phytoelements as invariably present within a stand. Direct measurements (Nilson 1971b, Warren & Wilson 1961, 1965a) also bear witness to the assumption that in reality the dispersion of phytoelements deviates from random to regular as well as to clumped. Moreover, the pattern of dispersion also depends on locality and direction. We can therefore presume that treatment of a stand as a horizontally homogeneous turbid medium is not always possible. With reference to real stands this model seems best suited for small-leaved, dense stands or for those of lesser density ($\eta \approx 1$), yet composed of species with uniform density of vegetation inside the individual plant volume. But on the other hand, departure from the concept of a random dispersion of phytoelements in a stand leads to enormous mathematic complications. It stands to reason, therefore, that for the purpose of quantitative description of the stand structure we must use the model of a horizontally homogeneous turbid medium as the first approximation, mindful though, of its limitations.

Let us now proceed to the determination of the mean values of stand structure characteristics bearing in mind that the term 'a plant stand' is here applied to a pure stand modelled as a horizontally homogeneous turbid layer.

As in the case of an individual plant, $u_j(z)$ designates the area density of the j-th component in the layer. On general consideration $u_j(z)$ is a random quantity assuming different values in a horizontal layer of the stand at the height z. However, we will further limit ourselves to its mathematical expectation.

Transition from characteristics of the plant to those of the layer is achieved as follows below.

Let σ_0 be some area of a stand, which carries n plants. Further, let $\hat{S}^*_{Lk}(z)$ be the vertical distribution function of leaf area for the k-th plant. Then, the mean leaf area density in the stand $u_L(z)$ is determined, with (I.2.12a) taken into account, by the expression

$$u_L(z) = \frac{1}{\sigma_0} \sum_{k=1}^{n} \hat{S}^*_{Lk}(z) = N_P \bar{S}_L(z), \qquad \text{(I.2.28)}$$

where $N_P = n/\sigma_0$ is the stand density and

$$\bar{S}_L(z) = \frac{1}{n} \sum_{k=1}^{n} \hat{S}^*_{Lk}(z) \qquad \text{(I.2.29)}$$

is the leaf area in the unit layer of the 'mean' plant as a function of z.

Taking (I.2.27) into account, from (I.2.28) we obtain

$$u_L(z) = \eta(z) u_{LP}(z), \tag{I.2.28a}$$

where

$$u_{LP}(z) = \sum_{k=1}^{n} S^*_{Lk}(z) / \sum_{k=1}^{n} S^*_{CPk}(z)$$

is the foliage area density inside the volume of the 'mean' plant.

Similar formulae can be obtained for other plant organs.

A formula of the type (I.2.28a) relates plant phytometric characteristics to those of the stand as a layer. In the case of sparse stands $\eta < 1$ and $u_L < u_{LP}$. But in dense stands, $\eta > 1$ and $u_L > u_{LP}$. As η is also height dependent, the vertical profiles of u_L and u_{LP} will slightly differ in the general case.

In the case of a row-crop or a square sown crop, when the two- or three-dimensional foliage area density distribution has to be determined, the test area of plant canopy is selected in accordance with the canopy geometry, individual plant geometry being disregarded. For instance, in the case of a row-crop with the distance between rows x_1, the area σ_0 is a rectangle of nx_1 length, where n is an integer, and of Δy width with $\Delta y > a_P$, where a_P is the effective radius of the plant basal area.

This distance x_1 is divided into a whole number of lengths equal to Δx_i, in dependence on particular conditions, and foliage area is measured for every length, i.e. foliage area is determined in volumes $\Delta V_i = \Delta y \Delta z \Delta x_i$. The averaged two-dimensional foliage area density $u_L(z, x)$ and other characteristics are calculated similarly.

The definition of 'plant stand height' or, more accurately, 'the uppermost and lowest levels of the j-th plant organ in canopy' requires a detailed consideration. It would be natural to relate these notions to the foliage area density $u_j(z)$ and determine them from the conditions

$$\left.\begin{aligned} u_j(z_{0j}) &= 0, \\ u_j(z_{00j}) &= 0, \end{aligned}\right\} \tag{I.2.30}$$

where z_{0j} and z_{00j} are the uppermost and lowest levels, respectively, of the j-th plant organ in canopy.

How are the levels z_{0j} and z_{00j} determined in practice?

In the case of the uppermost level, visual estimation of z_{rj} is often used, this being a mean level of the j-th plant organ in the layer. However, this method is seldom applicable, as there are always plants whose height exceeds the mean level and therefore the condition (I.2.30) is not satisfied.

On the other hand a stand is composed of individual plants each of which has its own uppermost and lowest level of the j-th plant organ, z^*_{1j} and z^*_{2j}. It would, therefore, be reasonable to regard z_{1j} and z_{2j} of individual plants as statistically random quantities and to determine z_{0j} and z_{00j} of a stand through statistical characteristics of these random quantities.

Experimental data (Kupchenko 1968; Ross & Vlasova 1967; Ross & Ross 1969a) show that the distribution of height z^*_{1P} is nearly normal. Similar results are obtained for the plant foliage lowest level. In the general case, the uppermost and lowest levels z_{0j} and z_{00j} are determined for every organ in the layer.

It is known from the law of probability that with a normal distribution of a random quantity z^* the interval $(z-2\sigma, z+2\sigma)$, where z and σ are the mathematical expectation and standard deviation, respectively, covers 95.5% of all cases. In view of this fact it is further assumed that

$$\left.\begin{aligned} z_{0j} &= z_{1j} + 2\sigma_{z_{1j}} \\ z_{00j} &= z_{2j} - 2\sigma_{z_{2j}} \end{aligned}\right\} \qquad \text{(I.2.31)}$$

This implies that the uppermost level of the j-th plant organ in a canopy equals the mean height of the j-th plant organ of individual plants making up the stand, plus a two-fold standard deviation of the height z_{1j}. In the case of the lowest level, the latter is taken with a minus sign. In actual measurements conditions (I.2.31) are satisfied but approximately, depending on plant sampling number.

The foliage area density $u_j(z)$ characterizes the vertical profile of the area of the j-th plant organ in a canopy. But it is better sometimes to replace it with an integral characteristic $L_j(z)$. The downward cumulative phytoarea index $L_j(z)$ and the phytoarea index L_{0j} for the j-th plant organ in a canopy are determined in the same way as (I.2.13) and (I.2.14) and are given by the expressions

$$L_j(z) = \int_z^{z_{0j}} u_j(z')\mathrm{d}z' \qquad \text{(I.2.32)}$$

and

$$L_{0j} = \int_{z_{00j}}^{z_{0j}} u_j(z')\mathrm{d}z'. \qquad \text{(I.2.33)}$$

The latter quantity denotes the area of phytoelements of the j-th plant organ in a canopy, located over a unit surface area.

For stems z_{00C} is always equal to zero.

If u_j, L_j and L_{0j} functions are summed over all the plant organs, i.e. over j, this will give us

$$u(z) = \sum_j u_j(z),\ L(z) = \sum_j L_j(z),\ L_0 = \sum_j L_{0j}. \qquad \text{(I.2.34)}$$

When dealing with canopies in which leaves are the major plant organs, height z, being an independent variable, is often replaced with the quantity L which is taken here to characterize the canopy depth counted off, in accordance with formula (I.2.29), from the uppermost level of the stand and measured in relative units of foliage area. Accordingly, the vertical distribution of the stand phytometric characteristics is given with reference to depth L and not to height z.

Other phytometric characteristics of the stand are determined in much the same way as the vertical distribution of relative area of individual components in a stand $u_j(z)$.

Thus, for instance, the vertical distribution of the absolute dry phytomass density of plant's j-th organ as a function of height, $u_{mj}(z)$ is determined by the formula

$$u_{mj}(z)=\frac{1}{\sigma_0}\sum_{k=1}^{n} m_{jk}^*(z)=N_P\bar{m}_j^*(z), \tag{I.2.35}$$

where $m_{jk}^*(z)$ and $\bar{m}_j^*(z)=(1/n)\sum_{k=1}^{n} m_{jk}^*(z)$ are the vertical distribution of the absolute dry phytomass density of the j-th plant organ for the k-th and 'mean' plants, respectively.

The fresh weight density of the j-th plant organ as a function of height, $u'_{mj}(z)$ is determined similarly.

The distribution function of the j-th plant organ area orientation $(1/2\pi)g_j(z, r_L)$ for the plant stand layer is determined in the same way as for an individual plant (I.2.15). According to the model of a horizontally homogeneous layer, the function $(1/2\pi)g_j$ does not vary horizontally and is only dependent on height.

While phytometric measurements provide a means for direct determination of mean characteristics of a stand, indirect, calculation methods can be used for the determination of more complex phytometric characteristics. These will be treated in detail under I.3.6.

Our concept of a square-sown or row crop as of a horizontally homogeneous vegetation layer is, in fact, a very rough approximation, especially where it concerns canopies that are not closed. It would, therefore, be more correct to consider the function u_j as dependent also on the horizontal coordinates x and y, where x is the direction of rows.

In the case of a row crop, dependence on the coordinate x can be neglected and dependence of u_j on y be considered as a periodical function with the period equal to the distance between rows, y_1.

In the case of a square sown crop, u_j depends also on x and is a periodical function with the period x_1, i.e. $u_j=u_j(z, x, y)$. For the sake of mathematical simplification separation of the variables should be allowed

$$u_j(z, x, y)=u_j(z)\bar{u}_{j2}(y)\bar{u}_{j3}(x), \tag{I.2.36}$$

$u_j(z)$ being a non-periodical function and $\bar{u}_{j2}(y)$ and $\bar{u}_{j3}(x)$, periodical functions with the periods y_1 and x_1, respectively.

The functions $\bar{u}_{j2}(y)$ and $\bar{u}_{j3}(x)$ should be normalized for the conditions

$$\left.\begin{aligned} \frac{1}{y_1}\int_0^{y_1} \bar{u}_{j2}(y)\mathrm{d}y=1 \\ \frac{1}{x_1}\int_0^{x_1} \bar{u}_{j3}(x)\mathrm{d}x=1 \end{aligned}\right\}. \tag{I.2.37}$$

Then $u_j(z)$ characterizes the area density of the j-th plant organ at the height z, averaged over the horizontal area, while $\bar{u}_{j2}(y)$ and $\bar{u}_{j3}(x)$, are relative deviations from the mean value in the directions y and x, respectively.

In the case of a row crop $\bar{u}_{j3}(x)=1$.

Some experimental data on the functions $u_j(z, x, y)$ and $\bar{u}_{j2}(y)$ are given under I.5.3.

Strictly speaking, the separation of the variables in accordance with the formula (I.2.36) is not quite correct, especially when the degree of closure varies with height and the form of the functions $\bar{u}_2$ and $\bar{u}_3$ changes with height.

In our treatment of a stand as a horizontally homogeneous turbid medium we have to deal with the mean values of phytometric characteristics averaged over horizontal layers. This model, as has already been stated, is far from being a true representation of the highly complicated geometrical structure such as a stand, and utterly ignores the major fact that in reality a stand is always a community of individual plants, each being an organism by itself with a geometrical structure characteristic of the given species, and liable to changing both under the influence of the environment and under the mutual influence of the neighbouring plants.

Therefore, if statistical phytometry is to develop further, the model of a horizontally homogeneous turbid medium must be abandoned and a plant stand treated statistically as a population of individual plants.

I.3 Phytometrical methods

I.3.1 DETERMINATION OF LEAF AREA

Much literature has been published by now relating to the methodology, instrumentation and the results of leaf area measurements both in stands and in laboratory conditions. Some reviews of the methods are given in books by Milthorpe (1956) and Nichiporovich (1961). A more recent review is contained in the work by Marshall (1968). The proposed methods and techniques which vary in precision, efficiency and difficulty of technical performance, can be divided into five groups:

1. Direct methods.
2. Method of standards using simple shapes.
3. Calculation methods.
4. Indirect methods (photoelectric planimeters).
5. Methods based on the leaf area to leaf weight relationship.

1. *Direct methods.* The procedure is basically as follows. The contour of a leaf is drawn on paper or related to a reference grid and its area is thus measured. Historically, it is the earliest method for leaf area determination, and it has found wide application. Its obvious advantages are high precision (the error, as claimed by Rozhnyatovsky (1954), is of the order of 1%), simplicity, and possibility of measuring the area without damage to the leaf. The major drawback of the method is its low efficiency. Moreover, it is practically inapplicable to crimpy leaves as well as to small and compound leaves such as carrot leaves. At present direct methods are used for reference to prove other methods.

2. *Method of standards.* Several dozens of standard leaf shapes of different areas are made. A close-fitting shape is chosen for each leaf, so that both coincide in the area. The error, according to estimations by different authors, ranges from 1.5 to 8.0%. High efficiency is, without doubt, the chief advantage of this method, while the relatively low precision and a certain subjectivity in the selection of shapes are its obvious limitations. Added to these is also the necessity of preparing a complete set of chapes for each plant species.

3. *Calculation methods* are based on modelling the leaf as a simple geometric figure as near as possible to the actual leaf shape. The area of the figure is determined by its linear dimensions, i.e. length h_{1L} and width h_{2L}. The leaf area is then determined by the formula

$$S_L = k_{LF} f_L(h_{1L}, h_{2L}), \qquad \text{(I.3.1)},$$

where the function f_L is the area of the geometric figure used as model for the leaf, and k_{LF} is the leaf form-factor, a certain mean correction factor taking account of the deviation of the actual leaf area from that of the elementary geometric figure. The closer the coincidence of the shape of the leaf with that of the geometric model, the nearer is k_{LF} to unit.

Using the calculation methods one more often confines oneself to the simple formula

$$S_L = k_{LF} h_{1L} h_{2L}, \qquad \text{(I.3.1a)},$$

where k_{LF} is obtained from a correlation chart.

Sometimes only one linear dimension is used and in the general case

$$S_L = k'_{LF} h^n_{1L} \text{ or } S_L = k''_{LF} h^n_{2L}, \qquad \text{(I.3.1b)},$$

where usually $n=1$ and 2 (Anderson 1967; Fulga 1965; Jain & Misra 1966; McKee 1964).

The simple forms (I.3.1a) and (I.3.1b) of the formula (I.3.1), undoubtedly, reduce the accuracy of calculation but simplify the calculation procedure, which is essential in mass scale phytometric measurements. Actually, a dependence like (I.3.1a) is not strictly linear, besides, the factors k_{LF}, k'_{LF} and k''_{LF} may vary during the growing season and depend on the leaf index number (Ondok 1968). In our measurements for barley (Mägi & Ross 1969a), for example, $S_L = 0.33 h^{1.32}_{1L}$ for the first leaf and $S_L = 0.13 h^{1.55}_{1L}$ for the remaining leaves.

Among the calculation formulae of the type (I.3.1) the most suitable one is (I.3.1a). The factor k_{LF} of this formula, as obtained for the same species by different authors, may vary considerably, and apparently depends on cultivar peculiarities, development stage, conditions of growing, accuracy of the reference method for the determination of S_L, etc. This necessitates repeated determinations of the factor k_{LF} for each stand de novo during the growing season.

For many species $k_{LF} = 0.667$ (Dolph 1977)

The attractive features of the calculation methods are their simplicity, relatively high efficiency, the possibility of leaf area estimation without damage to the leaf; the low precision and the necessity of preliminary drawing of a correlation chart are, however, the shortcomings of these methods.

4. *Indirect methods* require the use of special instruments, i.e. automatic planimeters which can be divided into the following types:

(a) photoplanimeters with a parallel beam;
(b) photoplanimeters with an integrating sphere;
(c) photoplanimeters with a scanning beam;
(d) air flux planimeter.

The photoplanimeter with an integrating hemisphere, which is an improved version of the photoplanimeter by Gavrilov & Yeremenko (1959), was devised by Laisk (1965a) at our laboratory. The instrument allows leaf area mensuration, with an error that does not exceed 3%. This planimeter was extensively used in our phytometric measurements in 1963–1966.

The high efficiency of planimeters is their chief attraction. But they have a number of drawbacks: most photoplanimeters are operated on cut-off leaves under laboratory conditions. Moreover, there are certain design deficiencies leading to both random and systematic errors.

The main sources of errors in the use of photoplanimeters:

(a) since some radiation is transmitted by the leaf itself, its area is overestimated by the instrument; transmission is due to the optical nature of the leaf and varies depending on species, cultivar and biological age, which, in turn, necessitates in each individual case, the introduction of a correction factor that takes account of the optical properties of the leaf in a particular experiment;

(b) the working surface of the photoplanimeter is of non-uniform sensitivity, which drawback is minimized by special correcting appliances in more advanced types of photoplanimeters (Laisk 1965a);

(c) the photocell used in photoplanimeters is a non-linear receiver of unstable sensitivity, and thereby needs frequent calibration;

(d) supply voltage fluctuations cause an instrumental zero drift.

To diminish the errors resulting from the latter two sources, two identical photocells are recommended for use (Rvachev & Yatsenko 1964; Maggs 1956; Rodskjer & Kornher 1974) Generally, the average error in measuring the leaf area by current photoplanimeters does not exceed 2–4%, these instruments, therefore, are increasingly used in biology nowadays.

5. *Methods based on the leaf area to leaf weight relationship.* These are based on the formula $S_L = \sigma'_L m'_L$ (I. 2.3) and require sufficiently constant mean values for the specific leaf weight σ'_L (Alekseyenko 1965a; Hozumi et al. 1962; Watson 1937).

If the number of samples is small, entire leaves are dried and weighed, with subsequent determination of σ_L, σ_{WL} W_L nad v_L from entire leaves. However, desiccation and weighing become very laborious and time-consuming in situations involving a vast scope of work, such as occur in measuring a complete set of phytometric parameters of a crop stand.

Therefore, only a certain number of disks are cut out of a leaf, which are exposed to further manipulations. But here an important methodological question arises: where, how many and of what area should cut-out disks be in order to obtain the data that would approximate the mean values for the entire leaf? Special methodological studies (Ross 1967) were initiated in our laboratory to answer the above question. These showed that there are definite diurnal dynamics and variability of parameters σ_L, W_L and v_L over the leaf.

Figure 5 shows variations of these parameters along the midrib and lateral part of a maize leaf. The distance $\bar{x} = h/h_{1L}$ from the base to the top of the leaf is expressed in relative units. There is an appreciable decrease in leaf water content

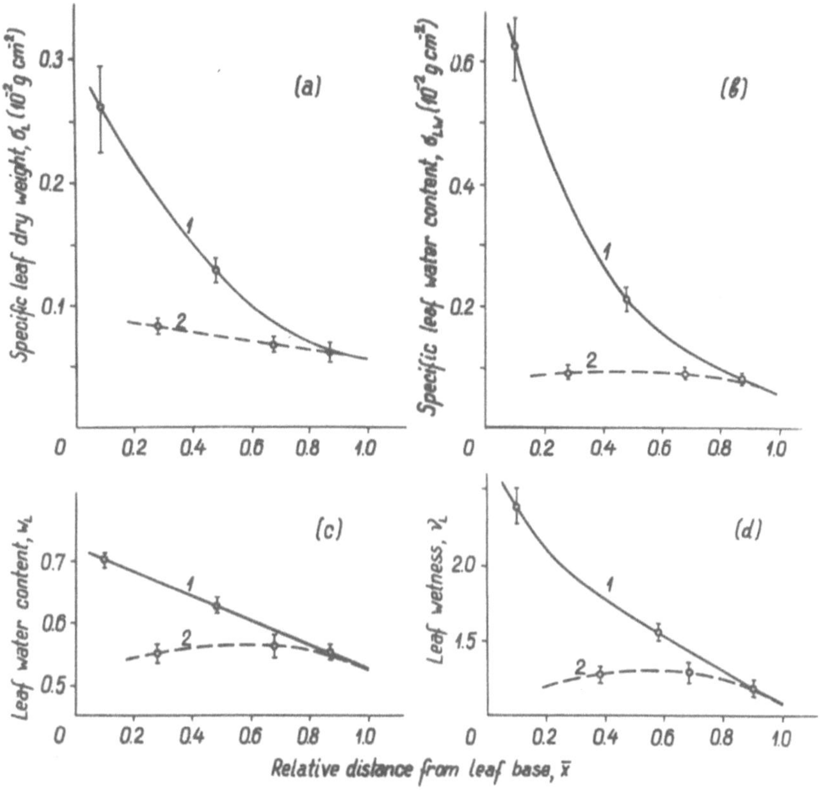

Fig. 5. Variation of phytometric characteristics along mid-rib (1) and lateral part (2) of a maize leaf. Experiment No. 14, Tadjikistan, August 11, 1965, 01:00 p.m.

and wetness, as well as in specific leaf dry weight along the length of the mid-rib, whereas in the lateral part the changes are slight, the varying thickness of the maize leaf in its different parts being presumably the chief cause of the above topography.

The leaf parameters W_L, σ_L and σ_{WL} also show diurnal variations. In conditions of Tadjikistan the leaf water content is at its minimum after midday, it reaches its maximum after midnight, the amplitude of diurnal fluctuations reaching 15%. The diurnal variation of water content is the same for any point of the leaf. Similar diurnal variations occur also in the specific leaf water content σ_{WL}, while the specific leaf dry weight σ_L shows little, if any, variation during the twenty-four hours (within the limits of accuracy achievable when weighing in field conditions).

A somewhat different pattern is observed in Estonia for horse beans and potatoes. Here the diurnal variation of water content is less marked, the amplitude not exceeding 10%. The minimum water content is observed at 5:00–6:00 p.m., the maximum at 3:00–4:00 a.m. Unlike maize stands in Tadjikistan, σ_L has the diurnal variation with the maximum in the evening and minimum in the morning, the amplitude of the variation of σ_L reaching 40%. Such dramatic differences are obviously due to the diurnal dynamics of photosynthesis and the assimilate transfer. Among the plants under study hybrid swede showed the highest water content (86–90%).

Diurnal variations of the above parameters must be considered in the performance of phytometric measurements and comparison of the results.

In view of the data presented in Figure 5, it can roughly be assumed that

$$\sigma_L = \sigma_{L1} \tag{I.3.2}$$

for the lateral part,

$$\sigma_L = \sigma_{L2}(\bar{x})$$

for the leaf mid-rib.

One disk from each leaf at a point $\bar{x}_E$, where $\sigma_L(\bar{x}_E) = \bar{\sigma}_L$ would seem most appropriate.

As has been shown experimentally, this point on the midrib can be found. For green leaves with the disk area $S_{LB} = 2\,\text{cm}^2$, $\bar{x}_E = 0.63 \pm 0.01$ for maize, 0.69 ± 0.01 for sorghum, 0.81 ± 0.005 for sunflower, and 0.89 ± 0.025 for hybrid swede. In cotton with $S_{LB} = 0.815\,\text{cm}^2$ $\bar{x}_E = 0.92 \pm 0.005$.

Effective distance $\bar{x}_E$ is dependent on disk area S_{LB}, diminishing with the growth of the latter. Optimal disk area is chosen in relation to the leaf size and also regarding conveniences of weighing. The experiments performed also demonstrated that effective distance is a value characteristic for a given plant species, varying little, if at all, from leaf to leaf, also during the growing season. A good way to determine it is as follows: a point $\bar{x}_E$ is marked on a rubber cord of the length slightly less than the shortest leaf. The cord is then stretched to cover the whole length of a leaf, the mark $\bar{x}_E$ moves accordingly showing the right spot on the leaf for cutting out the disk.

The entire procedure for the determination of leaf area by disk weighing elaborated by V. Ross consists in the following. The leaf under investigation is cut off the plant, its fresh weight m'_L being determined by weighing. Effective point $\bar{x}_E$ is then found on the leaf mid-rib using the rubber cord, at which a disk is cut out with a leaf bore. Bore diameters should be available providing for disks of 0.5; 1; $2\,\text{cm}^2$ cross section depending on the size of leaf under investigation. By further weighing disk fresh weight m'_{LB} is determined and after drying, disk absolute dry weight m_{LB} is obtained. Leaf water content and wetness are calculated according to the formulae (I.2.1) and (I. 2.2), repectively, while specific leaf dry weight is obtained using the formula (I. 2.4). Leaf area is found following the formula

$$S_L = \frac{1 - W_L}{\sigma_L} m'_L. \tag{I.3.3}$$

Precision of the methods based on the leaf area-to-weight relationship depend on the extent to which specific leaf dry weight σ_L and water content W_L for the effective point of the leaf $\bar{x}_E$ characterize the mean values for the leaf and for the entire stand. Methodological studies showed dependence of σ_L, σ_{WL}, W_L and v_L on leaf height. Moreover, diurnal changes and variation of the above parameters during the growing season have also been observed. These questions will be further discussed in more detail under I.5.2.

The error of the disk weighing method can be reduced to 3% by suitably taking into account dynamics of variation of σ_L and W_L in the stand. When leaf area alone is measured, determination of fresh weight would be sufficient: the leaf area S_L can be found from the correlation $S_L = m'_L/\sigma'_L$, where $\sigma'_L = m'_{LB}/S_{LB}$. This makes the disks drying and weighing procedure needless.

Huxley (1971) points out that between the leaf volume on the one hand, and the leaf dry weight or area, on the other, there exists proportionality. Using a volumeter for leaf volume determination S_L and m_L can also be found.

Of all the methods for measuring leaf area the one based on weighing the disks is most efficient, its efficiency being that by a set of simple procedures we succeed in obtaining the leaf main phytometric parameters, such as area S_L, absolute dry weight m_L, water content W_L, wetness v_L, etc.

Insufficiently high precision and the need for preliminary work on finding and checking for stability the effective point $\bar{x}_E$ on the leaf are deficiencies of this method.

I.3.2 THE MEASUREMENT OF LEAF ORIENTATION

Certain instruments are in current use for direct measurement of leaf orientation. Nichiporovich (1961) describes a simple device for the determination of the inclination angle ϑ_L. Shiman (1967) proposed a device allowing the measurement of the angle ξ, between leaf area and direction of the sunrays, i.e. $\xi = \pi/2 - \widehat{r_s r_L}$.

Květ et al. (1967) designed a simple installation for measuring leaf inclination in plants with long non-horizontal leaves (maize, cattail, etc.).

Laisk (1965a) of our laboratory designed an instrument allowing the determination of both coordinates, ϑ_L and φ_L.

In Laisk's device for determining the distribution function of foliage area orientation $(1/2\pi)g_L(r_L)$ the scale for the inclination angles ϑ_L from 0 to 90° was divided into six divisions spaced 15° apart, and the azimuth scale φ_L on the compass from 0 to 360° was divided into eight divisions 45° apart (Ross & Nilson 1967b). Consequently, the upper hemisphere towards which all leaf normals are directed, was divided into 48 sections, with six graduations for the inclination angle and eight graduations for the azimuth. A box with 48 sections was designed for cut-off leaves. With the above device a given leaf or its part was related to one or another of the 48 sections. After measuring, the leaf was cut off and placed in the appropriate section. After collecting sufficient quantity of leaf material, the leaf area for each section was determined and the value of $(1/2\pi)g_L(r_L)$ was calculated following the method described in the work of Ross & Nilson (1966b).

A technically perfect instrument for measurement of Cartesian coordinates in three dimensions is constructed by Lang (1973). The device consists of a linked set of four 40-cm long arms pivoted so that they actually move in a single plane, while angles between the arms are measured by potentiometers. The instrument is connected to a computer, which increases the speed of data collecting to about 20 leaves per minute.

The instrument allows determination of both, the leaf area and leaf orientation $r_L = (\vartheta_L, \varphi_L)$. The leaf surface is divided into a set of contiguous triangles and coordinates of the apices of each triangle are measured by means of the instrument. Then the area and orientation of each triangle are calculated automatically from the Cartesian coordinates of its apices. For cotton leaves, for instance, the division of the leaf into eight triangles and the measurement of 15 detached leaves of the triangles were found to give the total leaf area within 5% of the accuracy obtained with planimetry.

Indirect methods for the estimation of $(1/2\pi) g_L(r_L)$ have also been developed in addition to direct measurement of leaf orientation. These are discussed below.

I.3.3 INCLINED POINT QUADRATS METHOD

The inclined point quadrats method was elaborated in Australia in the 1930's (Levy & Madden 1933; Tinney et al. 1937). It was originally applied for determining the coverage of ground surface in grass stands. The method was further developed by Warren Wilson in whose works (1959a, b, 1960, 1963b, 1965a, b) it was theoretically substantiated and used for the measurement of vertical distribution of phytoelements with regard to individual components of the stand $u_j(z)$ along with the mean inclination of phytoarea in each layer $\bar{\vartheta}_j(z)$. Warren Wilson

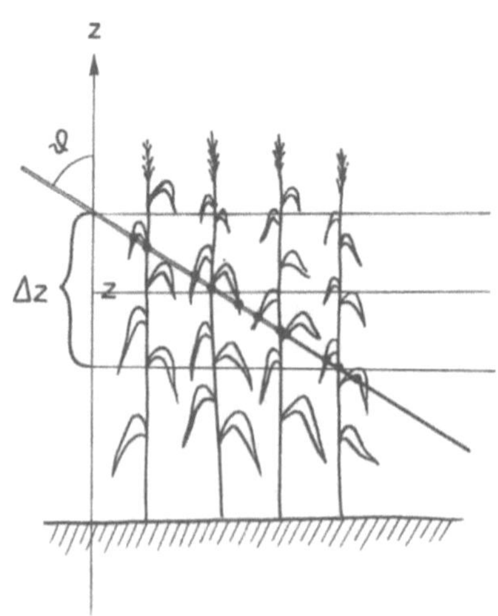

Fig. 6. The inclined point quadrats principle.

(1963a) also analysed and evaluated the errors due to the finite value of the point diameter. A comprehensive mathematical analysis of the method, evaluation of its applicability and accuracy as well as its further generalization for the estimation not only of the mean inclinatio angle of phytoarea $\bar{\vartheta}_j(z)$ but also the distribution function of leaf inclination angle $(1/2\pi)g_j(z, \vartheta_j)$ are provided in the works of Philip (1965a, b, 1966a, b).

The method consists in the following (Figure 6): the stand is pierced with a long point quadrat in the direction $r=(\vartheta, \varphi)$, and the number of contacts of the point quadrat with plant parts is counted. In Figure 6 the contacts are marked by dots, their number being seven for the leaves and four for the stems.

Let us consider in a plant stand at the height z a volume ΔV with a basal area σ_0 and height Δz, i.e. $\Delta V=\sigma_0\Delta z$. Let us now divide the basal area σ_0 into a large number of squares and insert the point quadrat n times so that it would pass through the centre of each small square. Let ΔN_L be the total number of contacts with the foliage in the volume ΔV at n insertions of the point quadrat. As can be seen from Figure 6, the pathlength of one insertion is equal to $\Delta z/\cos\vartheta$, and the total pathlength of all insertions is $\Delta s=n\Delta z/\cos\vartheta$. Then

$$v_L(z, r)=\frac{\Delta N_L}{\Delta s}=\frac{\Delta N_L}{n}\frac{\cos\vartheta}{\Delta z} \qquad \text{(I.3.4)}$$

is the contact frequency function, i.e. the number of contacts with foliage per unit length in volume ΔV when the point quadrat is inserted in direction $r=(\vartheta, \varphi)$. If the number n of insertions is sufficiently large, then v_L does not depend on the shape and size of the leaves, but is dependent exclusively on the total leaf area in the volume ΔV, and leaf orientation.

In this case, considering (I.2.20),

$$v_L(z, r)=u_{LE}(z, r)=u_L G=\frac{u_L(z)}{2\pi}\int_\Omega g_L|\cos\widehat{rr_L}|d\Omega_L \qquad \text{(1.3.4a)}$$

i.e. the number of contacts v_L is equal to the effective leaf area density u_{LE}.

All the above holds true for other plant parts.

Ignoring the azimuthal dependence of foliage area orientation, i.e.

$$g_L(z, r_L)=g'_L(z, \vartheta_L),$$

and considering (I.2.23), we obtain

$$v_L(z, \vartheta)=u_L(z)\int_0^{\pi/2} A(\vartheta, \vartheta_L)g'_L(z, \vartheta_L)\sin\vartheta_L d\vartheta_L, \qquad \text{(I.3.4b)}$$

where $A(\vartheta, \vartheta_L)$ is provided by the formula (I.2.22).

Integrating $v_L(z, \vartheta)$ over ϑ in the interval from 0 to $\pi/2$ and considering (I.2.16) also that

$$\int_0^{\pi/2} A(\vartheta, \vartheta_L)d\vartheta_L=1$$

we obtain

$$u_L(z)=\int_0^{\pi/2} v_L(z, \vartheta)\mathrm{d}\vartheta. \tag{I.3.5}$$

The formulae (I.3.4b) and (I.3.5) are the basic ones in the investigation of the foliage geometry by the inclined point quadrat method and were first derived by Philip (1965a). Their essence is in the following. A plant stand is pierced by the point quadrat at different angles ϑ, the number of contacts in separate horizontal layers is counted and a two-dimensional function $v_L(z, \vartheta)$ computed using the formula (I.3.4). Integrating it over ϑ for each layer by (I.3.5), we determined the vertical distribution of foliage area $u_L(z)$. Furthermore, since v_L and u_L are known, the formula (I.3.4b) can be treated as an integral equation for the determination of the foliage orientation function $g'_L(z, \vartheta_L)$. The kernel of the integral equation $A(\vartheta, \vartheta_L)$ is also known for it contains trigonometric functions alone.

The inclined point quadrats method has been predominantly used in Australia and New Zealand, basically in studying the geometry of grass stands. As was noted by Philip himself (1966b), the number of insertions had to be great to ensure reliable results. Thus, for example, in the case of random distribution of phytomass components within a stand, as many as 10000 point quadrat contacts with the leaves are needed to determine the function $v_L(\vartheta)$ to 1% precision. Moreover, slight inadequacies in the computation of v_L may involve gross errors in the determination of $g'_L(\vartheta_L)$ due to the specificity of the integral equation (I.3.4b). In practice, therefore, the procedure is normally restricted to the determination of the effective leaf inclination angle $\bar{\vartheta}_{LE}(z)$ at various heights instead of the function $g'_L(\vartheta_L)$. The large amount of work in the field and complexity of mathematical calculations in obtaining parameters of a stand geometry appear to be the major disadvantages of the point quadrat method.

Another limitation of the method is that in view of the technical difficulties the point quadrat cannot be made sufficiently long, the method, therefore, is applicable only to stands whose height does not exceed 100–150 cm.

An undeniable asset of the point quadrats method is that it does not involve the cutting and destruction of plants, thus allowing for measurements in the same place during the entire growing season to be carried out repeatedly. Its chief value, however, lies in the estimation of detailed statistical characteristics of a stand radiation intensity rather than in the determination of the characteristics of the stand geometry.

Passing the point quadrat through vegetation many times, we can, in principle, establish the location of each contact with the plant parts and from the data can compute the statistical function of distance distribution between individual components of foliage area in relation to height and direction of the point quadrat; in other words, this allows us to determine the probability of the free path length l in a plant stand, $f_l(l, z, r)$ as a function of the height z and direction of the point quadrat insertion r. The latter quantity is a detailed characteristic of the stand radiation regime and, simultaneously, provides information on the stand geometry. The first steps in this direction were made by Warren Wilson (1965a).

Ledent (1974) stresses that data obtained by the inclined point quadrats method fail to provide us the only way of interpreting the geometrical structure of a plant stand. At $V' > 1$ (negative binomial distribution), for instance, the results can be interpreted either as a stand in which the leaves change their inclination from layer to layer, or as a stand with the clumped leaf distribution.

I.3.4 STRATIFYING CLIP METHOD

Stratifying clipping suggested by Monsi & Saeki (1953) is a considerable step forward in the development of the methods of stand phytometry.

In Japan the method immediately found wide application in investigations of the vertical structure of agricultural crop stands, as well as that of natural stands and is being used everywhere at present.

The stratifying clip method consists in the following. The plants under investigation or a stand study plot is divided into a certain number of horizontal layers, after which all plant elements in a given layer are clipped successively starting from the top. The clipped elements are subsequently classified according to plant parts and mean phytometric characteristics $u'_{mj}(z)$, $u_{mj}(z)$, $u_j(z)$, $\sigma_j(z)$, $\sigma_{Wj}(z)$ and $W_j(z)$ of each layer are then found for each class.

Most works by Japanese scientists subdivide the aerial parts into two classes only, i.e. photosynthesizing plant organs and non-photosynthesizing parts. Moreover, they provide no information bearing on an essential methodological problem, namely, the problem of sampling, as to how the sampling of experimental plants or study plots should be effected in order to obtain adequate mean values over a given stand.

Along with the use of the stratifying clip method in our laboratory in 1962–1964 we also found the function of leaf area orientation $(1/2\pi)g_L(z, r_L)$ (Ross & Nilson 1967a, b). To achieve this goal, we established the direction of the leaf normal in each layer using the Laisk device prior to clipping, and classified all the leaves in the layer according to the direction of their normals. In the case of compound or large leaves (e.g. maize leaf) these are separated into a definite number of parts, each one being classified separately, following the orientation.

Although the stratifying clip method allows, in principle, for a highly detailed and highly precise measurement of all the basic phytometric parameters of a stand, its application in practice involves certain difficulties that diminish its value.

Firstly, by clipping we destroy the experimental plants and cannot, therefore, carry out further measurements on the same plot. Above all, we thereby cause disturbances in the regime and condition of the adjacent plots of the stand. Therefore, when studying the dynamics of a stand development, we have to attempt sampling from different plots, while unification of measurements made for different plots and at different times is not always easy to achieve.

Secondly, the field work is very laborious. While measuring, clipping, repeated weighing and drying do not demand high skill of the observer, the whole series of field work required for a single phytometric survey of the stand has to be carried out within a short space of time in order to avoid errors which are brought about by disturbances in the water supply and in the diurnal dynamics of the plants characteristics, particularly, in a hot and dry climate.

Above all, for technical reasons, both the stratifying clip method and the inclined point quadrat method cannot be extensively used in forests. These require indirect calculation methods which have yet to be elaborated.

I.3.5 THE METHODS OF STATISTICAL MEASUREMENTS

Limitations of the methods discussed in the foregoing sections, necessitate a search for indirect ways of ensuring reliable and effective determination of the vertical distribution of foliage area and of other phytometric characteristics.

At out laboratory we initiated a search for new calculation methods of determining the main phytometric parameters of crop stands, being guided by the following considerations:

(a) a plant stand representing a combination of variable objects, adequate estimation of averaged characteristics requires a great number of measurements of parameters, providing initial data and serving as the basis for further computation;

(b) in view of the variability of the object and a large bulk of initial information, the methodological approach to the task must be mathematico-statistical;

(c) damage to plants in the stand or to their individual parts resulting from the measurements of initial statistical characteristics should be minimized;

(d) computational schemes must comply with requirements for machine computation in view of the statistical character of the data;

(e) apart from measurements for the entire stand, the method must also make possible, without damage to the plants, determination of parameters on limited areas, where measurements such as biogeophysical ones may be in progress.

In view of the above we proceeded from the following phytometric parameters as the basis for the initial statistical information about a plant stand:

z_{1P}^* – plant height in its natural position,
z_{1L}^* – uppermost level of plant foliage in its natural position,
$z_{1L}'^*$ – uppermost level of plant foliage,
z_{2L}^* – plant foliage lower level,
z_C^* – stem height,
d_{1C}^* and d_{2C}^* – large and small stem diameters, respectively, measured in situ max d_{1C}^*,
n_L^* – total number of leaves per plant,
n_{LA}^* – number of living leaves per plant at the given moment,
n_{LD}^* – number of dead leaves per plant at the given moment.

Measurement of the above parameters is a simple matter, requiring little time and no special appliances.

For storage and computer processing of the obtained statistical data V. Ross elaborated a dual marked card prototype which is completed in the field with a soft pencil through special templates. There is a punched card for every plant, providing an initial record of measurement.

For the statistical analysis of the data the punched cards are fed into a computer, and statistical parameters are calculated according to a preset programme for each characteristic X*:

X – mean value,
σ_X – standard deviation,
A_X – asymmetry,
E_x – excess,
$v_X = (\sigma_X / X)$ – variation coefficient.
N_X – adequate sampling number corresponding to a confidence probability of 0.99 and a maximum admissible error of 0.05,
$K_{XX'}$ – normalized correlation matrix, i.e. a set of coefficients of correlation between all the basic parameters.

The obtained parameters of the initial phytometric characteristics and their analysis provide a first idea of the phytometric properties of stands, as well as lay the basis for further investigation aimed at revealing more intimate relationships between the stand phytometric features and its photosynthetic activity, also at elucidating certain special questions.

Some actual data on the phytometric statistical characteristics of stands are presented in chapter I.4.

In order to obtain the above statistical characteristics, plants are randomly selected following the routine procedure of random sampling. The number of plants N_X varies from 30 to 300 and is dependent on the stage and homogeneity of the stand, as well as on the phytometric characteristic.

Depending on the objective, other parameters easily amenable to measurement can also be taken for the initial statistical phytometric characteristics.

I.3.6 NUMERICAL METHODS FOR OF DETERMINATION FOLIAGE AREA VERTICAL DISTRIBUTION

Results of statistical phytometric measurements and experimental data on the foliage area vertical distribution have enabled us to elaborate two numerical methods for the determination of the latter. These are the method of statistical sampling and the triangle method.

The method of statistical sampling was developed by the author in collaboration with V. Ross (Ross & Ross 1969a) and is based on height and foliage area measurements with randomly selected plants in a stand.

As was noted in I.2.3 and shown by direct measurements, the vertical distribution function of leaf area for a randomly selected plant $\hat{S}_L^*(z)$ is a fairly variable quantity, a similar variability is characteristic of the height z_{1P}^* and the total foliage area $\hat{S}_{LP}^*$ of an individual plant.

Since the uppermost level of plant foliage z_{1L}^* is a random quantity with an approximately normal distribution, it then follows from (I.2.12c) that the upper limit of the argument of the function $\hat{S}_{1L}^*(z)$ is also a random quantity. Moreover, experimental data indicate that the total leaf area per plant $\hat{S}_{LP}^*$ correlated with $\hat{S}_L^*(z)$ by the ratio (I.2.12c) appears to be random too, while the function $\hat{S}_L^*(z)$ with the values of z_{1L}^* and $\hat{S}_{LP}^*$ unchanges varies within certain limits.

In terms of mathematics the function $\hat{S}_L^*(z)$ should therefore be treated as statistically random and possessing the properties (I.2.12b) and (I.2.12c).

Let us now express $\hat{S}_L^*(z)$ through statistically random quantities z_{1P}^* and $\hat{S}_{LP}^*$, also through a new function $\tilde{S}_L^*(\bar{z})$, i.e. *the relative plant foliage area density as a function of relative height*, using the formula

$$\tilde{S}_L^*(\bar{z}) = \frac{z_{1L}^*}{\hat{S}_{LP}^*} \hat{S}_L^*(z), \tag{I.3.6}$$

where $\bar{z} = z/z_{1L}^*$.

The relative function $\tilde{S}_L^*(\bar{z})$ is also a statistically random one; from the properties (I.2.12b) and (I.2.12c) we obtain the following conditions for $\tilde{S}_L^*$:

$$\tilde{S}_L^*(1) = 0 \tag{I.3.7}$$

$$\int_0^1 \tilde{S}_L^*(\bar{z}) d\bar{z} = 1. \tag{I.3.8}$$

The relative function $\tilde{S}_L^*(\bar{z})$ thus characterizes the relative plant foliage area density as a function of relative height, the height being expressed in relative units, $\bar{z}$, i.e. in foliage level fractions.

A somewhat different, and, in our opinion, less fitting, normalized function $u_L'(\bar{z})$ is proposed by Udagawa et al. (1968), which also makes use of the relative height $\bar{z}$, but the function u_L' is normalized for the mean foliage area density, i.e.

$$\bar{u}_L = \frac{L_0}{z_{1L}} = \int_0^1 u_L'(\bar{z}) d\bar{z}$$

Since $\bar{u}_L$ varies with time, normalization of the above kind does not permit comparison of the vertical foliage area distribution curves within a stand at different values of L_0/z_{1L}.

The method of statistical sampling is based on formula (I.3.6) and consists in calculating $\hat{S}_L^*(z)$ and $u_L(z)$ using statistical parameters z_{1L}^*, $\hat{S}_{LP}^*$ and $\tilde{S}_L^*(\bar{z})$ and proceeding from experimentally obtained relations between these quantities.

To elucidate the properties of the random quantities z_{1L}^*, $\hat{S}_{LP}^*$ and $\tilde{S}_L^*(\bar{z})$ and their relationships, V. Ross undertook special phytometric measurements, basically in maize stands, during the expeditions of 1964–67. Statistical treatment and

Table I.3.1 *Characteristics of lineair regression $\hat{S}_{LP}=a_1(z_{1L}-a_2)$ between plant foliage area $\hat{S}_{LP}$ and uppermost level z_{1_L}*

Plant	Region	Date	Phenological stage	a_1	a_2
Maize	Moldavia	July 28-Aug. 14, 1964	Wax ripeness	2.3	1.3
Ditto	Ditto	Aug. 17–23	Ditto	2.3	2.2
Ditto	Tadjikistan	June 6, 1965	9th leaf	3.6	1.3
Ditto	Ditto	June 11	10–11th leaf	3.6	1.5
Ditto	Ditto	June 23	14th leaf	5.4	4.2
Ditto	Ditto	July 10	Tassel flowering	5.7	4.8
Ditto	Ditto	July 23	Ear flowering	7.7	17.0
Ditto	Estonia	July 27-Aug. 4, 1966	14th leaf	4.1	5.4
Sorghum	Tadjikistan	July 29–30, 1965	Flowering	2.5	6.5
Potatoes	Estonia	Aug. 16, 1966		4.1	1.8
Sunflower	Tadjikistan	June 9, 1965	16th leaf	1.6	0
		June 27	28th leaf	5.5	5
		July 2–13	Flowering	7	10
		Aug. 14	Ripening	9	16

analyses of the data (Ross & Ross 1969a) allow the following conclusions which, in our opinion, hold also for other agricultural crops.

1. The uppermost level of plant foliage z_{1L}^*, as has already been noted, in its first approximation obeys to the normal distribution and is, therefore, specified by the mean value z_{1L} and standard deviation $\sigma_{z_{1L}}$.

2. A linear correlation exists between the uppermost level of plant foliage z_{1L}^* and plant foliage area $\hat{S}_{LP}^*$ of a given species at a given point in time. This correlation was obtained for maize stands in Estonia, Moldavia and Tadjikistan, also for sunflower and sorghum stands in Tadjikistan.

Examination of Table I.3.1, which shows the parameters of regression lines given by the formula

$$\hat{S}_{LP}=a_1(z_{1L}-a_2) \tag{I.3.9}$$

demonstrates that the coefficient a_1 is dependent on phenophase. For maize it is lowest during the time of ear flowering and milk ripeness.

It is noteworthy that certain crops in which the stems are lacking show no correlation between the uppermost level of foliage and the foliage area.

3. The relative plant foliage area density as a function of relative height $\tilde{S}_L^*(\bar{z})$ is a normal random function, i.e. its sections at a fixed $\bar{z}$ follow normal distribution. Estimates also indicate that the variance of the random function $\tilde{S}_L^*(\bar{z})$ is roughly constant.

4. Rough estimates show that the form of the averaged relative plant foliage area density as a function of relative height $\tilde{S}_L(\bar{z})$ is practically independent of the plant mean height z_{1L} and the average total plant foliage area $\hat{S}_{LP}$.

The mean value of the relative plant foliage density as a function of relative height $\tilde{S}_L(\bar{z})$ was determined by the following method. Depending on the stage of stand development 20–80 plants were selected to represent the whole range of heights z_{1L}^*. All plants thus selected were arranged in groups according to the height intervals Δz_{1L}, the number of plants in groups being different and roughly proportional to the distribution function of the foliage uppermost levels $f_{z_{1L}}$. In each group estimates were made of the mean foliage area per plant $\hat{S}_{LP}$ and the relative plant foliage density $\tilde{S}_L(\bar{z})$, the latter being estimated for the stand by averaging $\tilde{S}_L^*$ in all the groups since it did not vary with the mean values of z_{1P} and $\hat{S}_{LP}$ at various heights.

A more detailed discussion of the averaged relative plant foliage area densities as functions of relative height $\tilde{S}_L(\bar{z})$ will follow under I.5.2 (Figures 10–14).

A comprehensive representation of an individual plant vertical foliage area distribution function $\hat{S}_L^*(z)$ as a random function of three random variables using the formula (I.3.6) is a laborious task, the formulation of which in a general form is hardly possible at this stage.

The vertical distribution function of foliage area density $u_L(z)$ for the entire stand would prove sufficient for the purpose. In any case, $u_L(z)$ can be supplemented with the variance function.

Considering (I.2.28), (I.3.6) and (I.3.9) and also the fact that the plant height z_{1L}^*, as a random quantity, follows normal distribution and that the form of $\tilde{S}_L(\bar{z})$ is independent of z_{1L} and $\tilde{S}_{LP}$, we obtain an approximate formula for the function $u_L(z)$ of the vertical foliage area distribution within a stand. This is

$$u_L(z) = N_P \bar{S}_L(z) =$$

$$= N_P a_1 \int_{z_{1L} - 2\sigma_{z_{1L}}}^{z_{1L} + 2\sigma_{z_{1L}}} \left(1 - \frac{a_2}{h}\right) \tilde{S}_L\left(\frac{z}{h}\right) f_{z_{1L}}(h) \mathrm{d}h \qquad \text{(I.3.10)}$$

with

$$\tilde{S}_L\left(\frac{z}{h}\right) = \begin{cases} \tilde{S}_L\left(\frac{z}{h}\right), & z \leqslant h \\ 0, & z > h \end{cases} \qquad \text{(I.3.11)}$$

and

$$f_{z_{1L}} = \frac{1}{\sigma_{z_{1L}} \sqrt{2\pi}} \exp\left[-\frac{(z_{1L} - h)^2}{2\sigma_{z_{1L}}^2}\right]. \qquad \text{(I.3.12)}$$

In the general case, the distribution function of the foliage uppermost level $f_{z_{1L}}$ in the formula (I.3.10) may not be normal and may have any form determined from direct measurements.

Thus, estimation of the foliage area vertical distribution in a stand $u_L(z)$ by statistical sampling requires the following initial data:

(a) distribution function of the foliage uppermost level for individual plants $f_{z_{1L}}(h)$;

(b) line of regression $\hat{S}_{LP} = a_1(z_{1L} - a_2)$ between the foliage area S^*_{LP} and the uppermost level of foliage z^*_{1L} for individual plants;

(c) the mean relative plant foliage area density as a function of relative height $\tilde{S}_L(\bar{z})$. If the above data are known from direct phytometric measurements, $u_L(z)$ is calculated using (I.3.10).

The greatest difficulty is the determination of the mean relative density $\tilde{S}_L(\bar{z})$. Applicability of the statistical sampling method depends largely on the stability of $\tilde{S}_L$ in relation to weather conditions, stand density, soil factors, and its variability during the growing season. If the regression line and $\tilde{S}_L(z)$ are known, then determination of $u_L(z)$ for any plot of the stand can be achieved through merely determining $f_{z_{1L}}$ for the given plot, a procedure that does not entail damage to the plants. And this is the basic advantage of statistical sampling.

The difficulty of determining the mean relative density $\tilde{S}_L(\bar{z})$, not invariably reliable correlation between $\hat{S}^*_{LP}$ and z^*_{1L}, and the painstaking calculation of the integral (I.3.10) are the drawbacks of this method. For example, for oat plants the correlation between the plant effective phytoarea $\hat{S}^*_{EP}$ and z^*_{1L} has the form $\hat{S}^*_{EP} = az^{*n}_{1P}$, where a and n are empirical constants (Samarina 1969).

A partially modified method of statistical sampling was developed apart from our own by the American authors Kinerson and Fritschen (1971) for coniferous forests, *Pseudotsuga menziesii* (Mirb.) Franko, in particular.

The triangle method was developed by the author in collaboration with Mägi Ross & Mägi 1971) for barley crops. Judging from preliminary estimates it may as well be applied to other plant species, predominantly to cereal crops. The triangle method permits estimation of the variation of the stand foliage area and its vertical distribution during the entire growing season. It is not as much labour-consuming as statistical sampling, although less accurate.

Model of Saeki and Kuroiwa. The model delineated in the Japanese work by Saeki and Kuroiwa (1959) takes account of the effect of the upper part of the individual plant geometry on foliage area vertical distribution $u_L(z)$ within a stand. The model is, however, practically inapplicable in view of considerable simplifications.

Numerical method by Udagawa et al. (1968). This method characterizes the vertical distribution of foliage area in a maize stand, by means of a normalized function $u'_L(\bar{z})$, represented as a cubic parabola

$$u'_L(\bar{z}) = a + b\bar{z} + c\bar{z}^2 + d\bar{z}^3 \tag{I.3.13}$$

and satisfying the conditions

$$u'_L(1) \equiv 0 \quad \text{and} \quad \int_0^1 u'_L(\bar{z})\mathrm{d}\bar{z} = \bar{u}_L = \frac{L_0}{z_{0L}}, \tag{I.3.14}$$

where $\bar{z}$ is the relative height.

The normalized function u'_L and the function u_L are related by the formula

$$u_L(z) = u'_L\left(\frac{z}{z_{oL}}\right). \tag{I.3.15}$$

Owing to the conditions (I.3.14) two of the four empirical coefficients a, b, c and d, are independent, while the other two are substituted by L_0 and z_{0L}.

The expression (I.3.13) for $u'_L(\bar{z})$, in which the empirical constants are obtained experimentally, is a fairly adequate representation of the actual course of vertical distribution.

However, the above method needs further verification. As can be seen, the empirical constants are dependent on plant density in a stand and vary non-uniformly over a comparatively wide range during the growing season, which becomes obvious from the work of the authors. Moreover, the calculation of the four empirical constants from experimental data is a timeconsuming procedure.

Horie formula. Horie (1969) proposed an empirical formula for determining the function $L(z)$:

$$L(z) = \frac{L_0}{1 + a\left(\frac{z - z_{00}}{z_0 - z}\right)^r}, \tag{I.3.16}$$

where a and r are empirical constants.

Differentiation of $L(z)$, following (I.2.32) yields for $u(z)$ the expression

$$u(z) = \frac{\frac{L_0 ar}{z_0 - z} \frac{z_0 + z_{00}}{z - z_{00}}}{\left(\frac{z_0 - z}{z - z_{00}}\right)^r + a^2\left(\frac{z - z_{00}}{z_0 - z}\right)^r}. \tag{I.3.17}$$

The formula (I.3.16) gives a fairly adequate description of the vertical distribution of foliage area in a sunflower stand although the constants a and r vary during the growing season over the ranges 0.33–1.36 and 1.18–2.42, respectively.

The model of Stephens. The experiments made by Stephens (1969) in red pine tree stands indicate that foliage dry weight of individual trees (from 9 to 23 m tall) is normally distributed along the length of the crown and the distribution is little affected by age or size of tree or density of stand. The relative mean varies between 0.33 and 0.55 and the relative standard deviation $\sigma_{z_{1L}}$, between 0.17 and 0.26.

The method of Jordan. This method is based on the different attenuation of the red and near-infrared radiation in a plant stand (Jordan 1969) and allows determining L_0 in forest stands from a correlation of spectral fluxes at wavelengths 800 and 675 μm (Q800/Q675), respectively. It was found that for a range of L_0 from 1 to 10 there is a linear relationship between L_0 and $\log(Q800/Q675)$ approximated by the formula $\log(Q800/Q675) = 0.3813 + 0.0989\, L_0$. The method cannot be applied at a cloudy sky and its accuracy is dependent on solar elevation.

In summary, the statistical nature of stand structure must be taken into account if reliable estimation of foliage area vertical distribution of an entire stand or of its individual parts is our objective. On the other hand, the large volume of statistical phytometric measurements calls for the initial statistical parameters to be simple and easily measured. Again, phytometric measurements should be carried out in a manner that would minimize the number of damaged and totally destroyed plants. Mathematical modelling of plant stand structure, therefore, will, in our opinion, be fruitful if adequate correlations are found between the simple, initial plant stand characteristics and the final phytometric quantities used in the mathematical models of photosynthetic productivity. Unless purely empirical, this search for the above correlation must be directed towards revealing the underlying morphological phenomena that determined the plant habitus.

I.3.7 A RATIONAL METHOD FOR DETERMINING PHYTOMETRIC CHARACTERISTICS OF STAND ARCHITECTURE AND PRODUCTIVITY

Combination of the above various methods and techniques led us to the elaboration (Ross & Ross, 1969) of a rational procedure for determining the main phytometric parameters characterizing the architecture, and productivity of a number of stand types, primarily agricultural crop stands. The procedure we elaborated is as follows:

1. Species, cultivar, phenophase, stage, density and other general characteristics of the stand are determined.

2. Primary statistical characteristics, such as the plant height z^*_{1P}, the uppermost and the lowermost foliage levels z^{1*}_{1L} and z^*_{2L}, stem height z^*_{1C} and diameter d^*_C, total number of leaves n^*_L and the number of living leaves n^*_{LA} are measured with randomly selected plants. In view of the task and the stand type the list of characteristics subject to investigation can be supplemented or substituted by other characteristics, the number of the plants for measurements being 150–300. A punched card is prepared for each plant.

3. By mathematical treatment of the data obtained the first and second moments are found for each characteristic, describing the 'mean' plant of a given stand at a given time moment. A histogram of plant height distribution is then constructed.

4. Using the histogram and following the stand homogeneity, 15–30 plants are chosen to represent all the classes. In the classes of heights approximating the height of the mean plant, the experimental plants are chosen so that the value of their remaining statistical characteristics correspond to the mean values of the other selected characteristics.

5. A comprehensive phytometric analysis of the experimental plants is made following the order below:

(a) a phenological description of the plants is given;
(b) primary statistical characteristics are determined;
(c) the plant is subdivided into a certain number of horizontal layers;
(d) the inclination angle and azimuth of leaves or their parts in the layer are

measured using the Laisk device; the leaves are cut off and grouped according to their orientation;

(e) immediately after this procedure an estimate is made of the fresh weight of the grouped leaves, the total fresh weight equals to the fresh weight of the whole layer;

(f) leaf disks are punched out, subsequently weighed and subjected to drying;

(g) area and fresh weight of other parts of the plant in the given layer are determined; the other parts are also dried;

(h) after the drying the dry weight of the disks as well as that of the other parts of the plant is estimated;

(i) from the results of weighing, using the formulae (I.2.1)–(I.2.7) values of water and dry matter content σ'_L, ρ'_j and ρ_j are found for individual organs in the layer;

(j) from the formula $S_L^* = m_L'^*/\sigma_L'^*$ the plant foliage area $\hat{S}_L^*(z_i)$ in the i-th layer is determined, and the formula $m_L = m'_L(1 - W_L)$ is used to determine the dry weight $m_L(z_j)$;

(k) estimation is made of the foliage area $\hat{S}_{LP}^* = \sum_j \hat{S}_L^*(z_j)$ of each plant and a correlation chart is drawn for the values of $\hat{S}_{LP}^*$ and z_{1P}^*;

(m) relative plant foliage area density as a function of relative height $\tilde{S}_L^*(\bar{z})$ is determined from the formula (I.3.6), also the mean function $\tilde{S}_L(\bar{z})$, averaged over individual plants.

6. Using the formula (I.3.10) the stand foliage area vertical distribution $u_L(z)$ is calculated and from the formulae $u'_{mL}(z) = \sigma'_L(z)u_L(z)$ and $u_{mL}(z) = \sigma_L(z)u_L(z)$ the vertical distribution of the leaf dry and fresh weights, respectively, are determined.

7. From the fresh weight of the orientation-grouped leaves and from the formula $S_L = m'_L/\sigma'_L$ the foliage area for each group is determined, and the formula (I.2.15) serves to calculate the distribution function of foliage area orientation $(1/2\pi)g_L(z, r_L)$ for each layer.

The above method involving integrated phytometric measurements allows for the determination of all the essential characteristics of the architecture and productivity of both an individual plant and a plant stand as a whole. These, in turn, can be used for the calculation of other characteristics used in the quantitative theory of a stand radiation regime and productivity, namely $L_j(z)$, $G(z, r)$ and $u_j(z)$.

I.4 Statistical characteristics of a stand

I.4.1 PARAMETERS OF STATISTICAL CHARACTERISTICS

As was noted in I.3.5, the primary statistical characteristics of a stand do not merely describe its preliminary phytometric condition. They serve also as initial information for the calculation of more sophisticated characteristics incorporated in the theory of the photosynthetic activity of a stand.

The present chapter discusses, by way of example, the statistical characteristics of the following three plant species with a totally different habitus and stand geometry: (1) maize – experiment No. 14, Tadjikistan, 1965, also experiment No. 19, Estonia, 1966; (2) sunflower – experiment No. 16, Tadjikistan, 1965; hybrid swede – experiment No. 21, Estonia, 1966.

Statistical measurements in the stand were performed, on the average, once a week. The sample included 200 plants in Tadjikistan and 120 plants in Estonia.

In an earlier work (Ross & Vlasova 1967) we reported measurements of a maize stand of 1963 in Estonia.

Table I.4.1 shows, by way of example, the statistical parameters of a number or primary phytometric characteristics for a maize stand and their dynamics during the growing season.

Analysis of the data obtained permits a number of general conclusions.

Throughout the growing season all the above stands show a comparatively smooth variation of the mean values and standard deviation of statistical characteristics. Increase in the standard deviation is slow as distinct from that of the mean value. Consequently, the coefficient of variation v_X, as a rule, decreases during the growing season, as well as the adequate sampling number N_X, which, in turn, indicates that statistically the homogeneity of the crops increases with time. The only exception is the leaf number n_L^* of plants, in which case the coefficient of variation v_{n_L} has no seasonal variability.

The standard deviation σ_X and coefficient of variation v_X can be useful in comparing the accuracy in measurements of some characteristics. The number of living leaves per plant n_{LA}^* proved to be one of the most dependable statistical

Table I.4.1 *Phytometric characteristics and their statistical parameters. A stand of maize. Experiment No. 14, Tadjikistan, 1965*

	Uppermost level of plant foliage in its natural position z^*_{1L} (cm)						Uppermost level of plant foliage z'^*_{1L} (cm)					
Date	x	σ_x	V_x	N_x	A_x	E_x	x	σ_x	V_x	N_x	A_x	E_x
May 31	24.4	6.4	0.26	180	0.03	0.18	37.9	10.1	0.27	194	−0.29	−0.45
June 4	33.6	8.7	0.26	180	0.07	−0.51	50.5	13.0	0.26	180	−0.26	−0.71
June 11	44.8	12.0	0.27	194	0.19	0.10	69.5	16.1	0.23	141	−0.18	−0.39
June 17	66.9	16.5	0.25	166	0.02	−0.07	88.0	18.8	0.21	118	−0.14	0.12
June 25	96.4	26.9	0.28	208	0.24	−0.23	122.3	30.5	0.25	166	−0.15	−0.72
July 2	134.7	24.8	0.18	87	−0.15	0.52	165.1	26.4	0.16	68	−0.66	0.38
July 16	198.9	33.2	0.17	77	−0.71	0.58	217.1	33.3	0.15	60	−0.67	0.13
July 23	224.7	36.3	0.16	68	−0.56	0.41	246.8	36.5	0.15	60	−0.58	0.39
July 30	231.6	35.2	0.15	60	−0.76	0.44	257.6	35.0	0.14	53	−0.73	0.20
Aug. 6	232.4	33.4	0.14	53	−0.17	2.07	255.4	31.9	0.12	39	0.06	−0.02
Aug. 13	225.0	39.7	0.18	87	−0.55	2.37	256.1	32.5	0.13	45	−0.14	−1.35
Aug. 19	237.2	27.0	0.18	33	−0.38	−0.08	263.0	29.2	0.11	33	0.00	0.03

Date	Lowest level of plant foliage z^*_{2L} (cm)						Stem height z^*_C (cm)					
	x	σ_x	V_x	N_x	A_x	E_x	x	σ_x	V_x	N_x	A_x	E_x
May 31	1.6	0.8	0.50	665	1.46	1.71	6.6	2.6	0.39	405	0.15	−0.52
June 4	1.7	0.9	0.54	700	1.34	1.64	10.1	3.5	0.35	326	0.10	−0.58
June 11	2.4	1.2	0.51	690	1.24	1.63	15.7	5.4	0.35	326	0.31	0.14
June 17	3.2	4.9	1.50	700	4.09	16.39	23.8	11.8			6.23	
June 25	5.5	5.1	0.93	700	2.20	7.14	47.8	20.0	0.42	470	0.24	−0.76
July 2	6.1	3.7	0.61	700	1.28	1.53	83.1	22.8	0.27	194	−0.19	0.31
July 16	22.2	8.3	0.37	365	0.01	−0.47	161.4	39.4	0.24	153	−0.44	0.71
July 23	24.3	12.2	0.50	665	0.72	0.93	208.3	44.9	0.22	129	−1.00	1.33
July 30	37.5	14.5	0.39	405	0.57	−0.23	224.5	40.7	0.18	87	−1.23	2.18
Aug. 6	38.6	16.3	0.42	470	0.20	−0.53	223.9	32.0	0.14	53	−0.45	−1.04
Aug. 13	28.4	19.2	0.68	700	0.80	0.32	224.0	35.9	0.16	68	0.27	−1.22
Aug. 19	45.0	22.5	0.50	665	0.50	−0.76	234.5	26.9	0.11	93	−0.33	−0.15

Table I.4.1 (*Continued*)

Date	Large stem diameter d^*_{1C} (cm)						Small stem diameter d^*_{2C} (cm)					
	x	σ_x	V_x	N_x	A_x	E_x	x	σ_x	V_x	N_x	A_x	E_x
May 31												
June 4	1.4	0.4	0.30	240	−0.21	−0.86	0.9	0.3	0.31	256	−0.02	−0.13
June 11	2.0	0.5	0.25	166	−0.23	−0.06	1.2	0.4	0.30	240	0.40	0.04
June 17	2.5	0.6	0.24	153	−0.37	−0.33	1.8	0.6	0.32	273	0.01	−0.85
June 25	3.0	0.7	0.24	153	−0.11	−0.25	2.4	0.6	0.26	180	−0.55	−0.52
July 2	3.2	0.6	0.17	77	−0.36	−0.05	2.7	0.4	0.17	77	−0.61	0.63
July 16	3.0	0.5	0.17	77	−0.38	1.17	2.5	0.4	0.17	77	−0.52	0.46
July 23	3.0	0.5	0.18	87	−0.67	0.45	2.5	0.5	0.18	87	−0.75	0.68
July 30	2.8	0.6	0.20	107	−0.49	0.06	2.4	0.4	0.18	87	−0.75	0.75
Aug. 6	2.8	0.5	0.19	97	−0.40	−0.16	2.4	0.4	0.18	87	−0.44	0.15
Aug. 13	2.9	0.5	0.17	77	−0.06	−0.45	2.6	0.5	0.18	87	−0.20	−1.04
Aug. 19	2.8	0.4	0.15	60	0.22	0.70	2.4	0.3	0.13	45	0.06	−0.15

	Total number of leaves n_L^*						Number of living leaves n_{LA}^*					
Date	x	σ_x	V_x	N_x	A_x	E_x	x	σ_x	V_x	N_x	A_x	E_x
May 31	7.4	1.2	0.16	68	−0.74	0.75						
June 4	9.2	1.4	0.15	60	−0.57	−0.31						
June 11	9.8	1.3	0.13	45	−0.66	−0.16						
June 17	11.9	2.1	0.17	77	−0.16	−0.82						
June 25	14.4	3.3	0.23	141	5.51							
July 2	16.6	2.1	0.13	45	−0.16	0.22						
July 16	17.3	1.6	0.09	22	−0.17	2.67	14.2	1.6	0.11	33	−0.20	−0.14
July 23	16.8	1.7	0.10	27	0.10	−0.45	14.0	1.6	0.11	33	0.37	−0.12
June 30	16.9	1.7	0.10	27	−0.67	0.53	13.3	1.7	0.13	45	−0.28	−0.27
Aug. 6	16.7	1.4	0.08	17	−0.02	−0.24	12.8	1.7	0.13	45	−0.01	−0.51
Aug. 13	18.7	1.8	0.10	27	−0.04	−1.17	14.0	1.3	0.10	27	−0.33	1.39
Aug. 19	17.2	1.5	0.08	17	0.15	−0.25	12.2	1.7	0.14	53	0.44	−0.50

characteristics for all the crops considered. The total number of leaves per plant did not lend itself so easily to determination because it included both living and dead leaves and the latter were difficult to count towards the end of the growing season. The uppermost level of plant foliage z'^{*}_{1L} is another dependable characteristic. Variation of the uppermost foliage level of the plant in its natural position z^{*}_{1L} is greatest, being dependent on wind velocity as well as on the age of the uppermost leaf.

In the early period of maize vegetation stem height z^{*}_{C} measurements are of much lower accuracy, which is due to difficulties in determining the stem top end.

Of all the primary phytometric characteristics considered the lowest level of plant foliage z^{*}_{2L} is most difficult to determine. First, because the location of the lowermost living leaf is highly variable, second, because it is difficult to tell a living leaf from a dead one towards the end of crop growth.

Summing up, we may say that out of the primary phytometric characteristics we have just discussed the easiest to determine and at the same time least variable are the uppermost level of the plant (foliage) in its stretched position and the number of living leaves. The uppermost foliage level of the plant in its natural position, the stem diameter and the total number of leaves are less reliable.

The statistical variability of the characteristic x determines the corresponding value N_x.

For a maize crop, the sampling number N_x at the beginning of the growing season is between 100 and 200 and in the middle of the growing season it is between 30 and 80.

The stem height shows less uniformity: in the middle of the growing season N_x is between 40–150. The lowest level of plant foliage z^{*}_{2L} is subject to highest of variation, with the values of N_x ranging from 200 to 700. A crop of hybrid swede is comparable to that of maize in respect of statistical uniformity; a sunflower crop shows less uniformity. For example, the sampling number for in a sunflower crop varies between 50 and 300.

All the above-said leads to the conclusion that the sampling number equal to 200 in Tadjikistan and 120 in Estonia exceeds in most cases the adequate sampling number N_x.

According to data by Kupchenko (1968) much more uniformity is observed in winter wheat and rye stands. The sampling number for z'^{*}_{L} is 5 to 7 in the case of winter wheat and 10 to 15 in the case of winter rye. However, the data obtained by Timoshenko (1966) for a winter wheat stand in the Ukraine indicate that the adequate sampling number should be between 60 and 80.

The asymmetry and excess of primary characteristics may have both positive and negative values changing irregularly during the growing season. The plant height, its uppermost foliage level, stem height and diameter of the crops considered are, as a rule, distributed normally, though often enough this law is not obeyed.

The law of normal distribution of primary phytometric characteristics is best obeyed by a hybrid swede stand and worst by a sunflower stand. According to the data by Kupchenko (1968), the law of normal distribution of primary phytometric characteristics is also well obeyed by stands of winter wheat and rye, although the data obtained by Timosheko (1966) indicate frequent variations from normal distribution, predominantly, toward a negative symmetry.

But on the other hand, data exist (Hozumi et al. 1955; Koyama & Kira 1956; Obeid et al. 1975, Smurov 1975) which indicate that early in the growing season the weight of the aerial organs and the reproductive organ characteristics of plants adhere to normal distribution, but as the plant grows further, they gradually turn from normal to normal logarithmic distribution. In view of interspecific competition, the denser the stand, the quicker the change from one type of distribution to the other.

Furthermore, the interplant and interleaf distance distributions fail to follow the normal distribution law. Therefore, it cannot be maintained that all phytometric characteristics, as statistically random quantities, adhere to normal distribution.

I.4.2 CORRELATION BETWEEN STATISTICAL CHARACTERISTICS

In developing a rational and efficient method for performing statistical phytometric measurements in stands one should know exactly which of the primary phytometric characteristics selected are independent and which are correlated with other characteristics and what type of correlation exists between them. In search of correlations between phytometric characteristics one can confine oneself to linear relations if the characteristics sought are measurable during a short period of time, say, a day. The reason for it is the narrower range of variation in the value of a primary characteristic at a given time as compared with the range of variation in mean value of the same characteristic during the whole growing season. Furthermore, since the primary statistical characteristics are distributed normally in most cases, it could well be assumed that at a given moment the differences between the plants are random. For this reason linear relationships characterized by the normalized correlation matrix $K_{XX'}$ and linear regression equations $x^* = a_o + b_{XX'}X'^*$ were taken as the basis for the correlation analysis of primary characteristics.

Table I.4.2 presents, by way of example, correlation matrices $K_{XX'}$ for a maize stand at definite times. Examination of the table and analysis of all the material allows the following conclusions to be drawn.

1. The correlation coefficients can be divided according to their value and time variation into three groups:

(a) high correlation coefficients ($K_{XX'} > 0.7$) showing relatively little variation during the growing season;

(b) varying correlation coefficients showing, as a rule, a high value in the growing season and low values in the reproduction season;

(c) low correlation coefficients ($K_{XX'} < 0.5$) showing, as a rule, irregular variations.

Table I.4.2 *Normalized correlation matrices of stand primary statistical characteristics. A stand of maize. Experiment No. 14. Tadjikistan*

July 16, 1965, tassel flowering		z^*_{1L}	z'^*_{1L}	z^*_C	z^*_{2L}	d^*_{1C}	d^*_{2C}	n^*_L	n^*_{LA}	n^*_{LD}
Height of plant in its natural position	z^*_{1P}	0.936	0.954	0.924	0.350	0.652	0.674	0.416	0.399	0.031
Uppermost level of plant foliage in its natural position	z^*_{1L}		0.939	0.865	0.343	0.634	0.650	0.479	0.459	0.030
Uppermost level of plant foliage	z'^*_{1L}			0.923	0.334	0.670	0.700	0.474	0.447	0.043
Stem height	z^*_C				0.314	0.606	0.633	0.444	0.433	0.020
Plant foliage lowest level	z^*_{2L}					0.085	0.083	0.169	−0.006	0.228
Large stem diameter	d^*_{1C}						0.831	0.509	0.470	0.067
Small stem diameter	d^*_{2C}							0.464	0.433	0.063
Total number of leaves per plant	n^*_L								0.790	0.311
Number of living leaves per plant	n^*_{LA}									−0.333

Aug. 19, 1965, wax ripeness		z_{1L}^*	$z_{1L}'^*$	z_C^*	z_{2L}^*	d_{1C}^*	d_{2C}^*	n_L^*	n_{LA}^*	n_{LD}^*
Height of plant in its natural position	z_{1P}^*	0.918	0.861	0.927	0.225	−0.082	0.108	0.226	0.173	0.018
Uppermost level of plant foliage in its natural position	z_{1L}^*		0.872	0.989	0.155	−0.061	0.068	0.348	0.189	0.103
Uppermost level of plant foliage	$z_{1L}'^*$			0.874	−0.030	−0.072	0.151	0.173	0.319	−0.154
Stem height	z_C^*				0.185	−0.084	0.069	0.340	0.189	0.093
Plant foliage lowest level	z_{2L}^*					−0.195	−0.149	0.280	−0.345	0.457
Large stem diameter	d_{1C}^*						0.726	0.263	0.325	−0.075
Small stem diameter	d_{2C}^*							0.181	0.457	−0.251
Total number of leaves per plant	n_L^*								0.307	0.492
Number of living leaves per plant	n_{LA}^*									−0.658

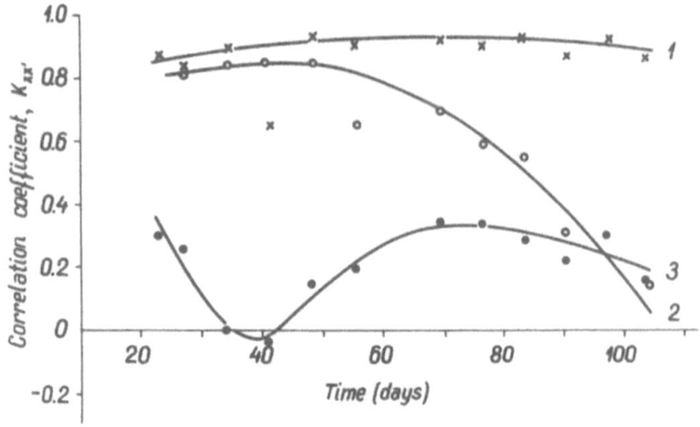

Fig. 7. Vegetational dynamics of correlation coefficient K_{XX}' between primary statistical characteristics of plants. Maize stand, experiment No. 14, Tadjikistan, 1965.
1 – high correlation between foliage level z'^*_{1L} and stalk height z^*_C; 2 – highly variable correlation between foliage level z^*_{1L} and stalk diameter d^*_{2C}; 3 – low correlation between uppermost level z'^*_{1L} and lowest level z^*_{2L} of foliage.

To illustrate this, Figure 7 presents the variation of correlation coefficients for a maize stand. Curve 1 belongs to the first group and shows correlation between the uppermost level of plant foliage in its stretched position z'^*_{1L} and the stem height d^*_{2C}. The correlation coefficient in this example varies between 0.83 and 0.93 throughout the growing season.

Curve 2 which represents the variation of correlation between the uppermost foliage level z^*_{1L} and the smallest stem diameter d^*_{2C} runs more or less gradually from the high correlation values in the growing season to the low correlation values in the ripening period. This is due to the specific features of maize growth when both the stem height and diameter increase during the first half of the vegetation season and then after tasseling the stem diameter not only stops growing but diminishes owing to natural drying and loss of leaf vagina, while the stem height is still increasing.

Curve 3 depicts a low correlation between the foliage uppermost and lowest levels z^*_{1L} and z^*_{2L}.

2. A maize stand. All heights z^*_{1P}, z^*_{1L}, z'^*_{1L} and z^*_C, as well as stem diameters d^*_{1C} and d^*_{2C} show good correlation. A varying correlation is observed between the heights z^*_1 and diameters d^*_C, also between the total number of leaves n^*_L and the number of living leaves n^*_{LA}. Correlation was very low between heights and leaf number. The lowest foliage level is an independent quantity having no correlation with other phytometric characteristics of the stand.

3. A sunflower stand. Correlation is good between heights z^*_{1P}, z'^*_{1L} and z^*_C and between n^*_L and n^*_{LA}. A varying correlation is observed between heights z^*_i, stem diameter d^*_C and total number of leaves n^*_L; correlation steeply diminishes in the second half of the growing season. The plant foliage lowest level correlates moderately with the heights and total number of leaves ($K_{XX'} \approx 0.5$).

4. A hybrid swede stand. Heights z_{1L}^* and $z_{1L}'^*$ correlate comparatively well. The high correlation between n_L^* and n_{LA}^* in the initial growth period decreases monotonously towards the end of the growth period. Correlation between n_L^* and n_{LD}^* shows an inverse variation. The hypocotil characteristics correlate poorly with one another and do not correlate at all with the leaf characteristics.

5. In general, the $K_{XX'}(\tau)$ curves run smooth. A few points outside the curves on some days must be due to measurement errors.

Measuring and processing the primary statistical characteristics is labour consuming and tedious work, but this work is necessary as a preliminary step on the way of developing in future a quantitative theory of plant structure and functioning. It is useful for determining which of the primary characteristics are independent and easiest to measure, and to what extent certain species in a given locality can change due to both environmental factors and mutability of the species.

I.5 Spatial distribution of phytoelements in stands

I.5.1 GENERAL

A systematic quantitative study of the vertical distribution of phytomass and phytoarea was started in the 1950's as a result of the fundamental work by Monsi & Saeki (1953) and the stratifying clip method suggested in the same work. The two authors also obtained data on the vertical distribution of the fresh phytomass $u'_m(z)$ separately for the synthesizing and non-synthesizing organs of a number of pasture plants (*Chenopodium album, Pennisetum japonicum, Phragmites communis, Thalictrum simplex var. affine*, etc.). The stratifying clip method was further extensively used by Japanese workers in studies of the vertical structure, radiation regime and productivity of buckwheat and horse beans (Ivaki 1959), Jerusalem artichokes (Kasanaga & Monsi 1954; Hogetsu et al. 1960), rice (Ichimura 1954; Tsunoda 1959), *Sasa nipponica* (Saeki & Kuroiwa 1959), *Miscanthus sacchariflorus* (Saeki 1960a), *Fagopyrum esculentum* and communities of *Cirsium* (Midorikawa 1957), sorghum, trees of birch, spruce and *Triadica sebifera* (Shinozaki et al. 1964); the method was also applied by the Chinese scientists (Yin et al. 1959).

Australian workers made use of the inclined point quadrat method in studies of foliage area vertical distribution for perennial clover, white clover, *Achillea millefolium* and *Taraxacum officinale* (Warren Wilson 1959b), horse beans (Warren Wilson 1965b) and *Atripex nummularia* (Philip 1965b).

Saeki (1960b, 1961) made an interesting attempt to define the function $u_L(z)$ theoretically using a mathematical model based on the analysis of plant growth patterns.

Similar phytometric studies in the USSR were started somewhat later except for a comprehensive work of Rabotnov (1950) on the meadow grass stand structure. Abundant experimental data have been amassed and analyzed in a number of works by Alekseyenko (1958a, b; 1960, 1964, 1965b, 1967) relating to the phytomass and leaf area vertical structure of perennial grasses, such as timothy grass, fescue, orchard grass, red clover, blue alfalfa, and foxtailgrass, as well as their mixtures. Data on phytometry and vertical phytomass distribution of forests have been published by Rauner (1968, 1972) and Utkin & Dylis (1966) and of forest steppe, by Grin et al. (1970).

Similar studies of the vertical structure of agricultural crop stands were performed at out laboratory during 1961–1967. The basic results obtained in these studies are discussed below.

I.5.2 SPACE-TIME VARIABILITY OF TRANSITION FUNCTIONS

As follows from formulae (I.2.1)–(I.2.5) the characteristics of phytomass m'_K, m_K, u'_{mK}, u_{mK} and phytoarea S_K and u_K are interconnected by the transition functions W_K, v_K, σ'_K, σ_K and σ_{WK} which vary with time and space. These functions, apart from providing a means for transition from phytomass to phytoarea, have a physiological significance of their own, since they are characteristics of the plant (or stand) water supply and photosynthetic activity. Their diurnal dynamics and leaf length-wise variation were discussed in Chapter I.3.1. Below we consider their variation with height and during the growing season.

Maize leaves in Estonia have 20% more water content than in Moldavia and Tadjikistan, although the plot in Tadjikistan was irrigated.

In Estonia there are 3 to 8 g of water to a gramme of dry weight, which is 1.5–3.0 times as much as in Moldavia and Tadjikistan.

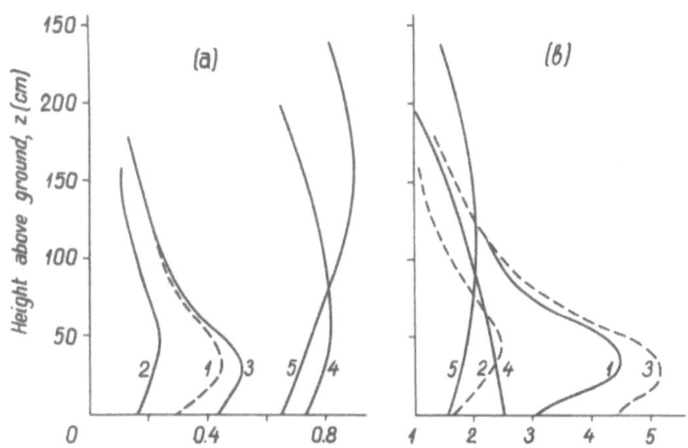

Fig. 8. Specific leaf dry weight (a) and specific leaf water content (b) as functions of height, in maize stands.

1 – experiment No. 4, Estonia, August 3, 1963 at the stage of 10th leaf emergence; 2 – experiment No. 4, Estonia, August 13, 1963, emergence of 12th leaf; 3 – experiment No. 4, Estonia, August 20, 1963, tasseling; 4 – experiment No. 11, Moldavia, August 1964, milk ripeness; 5 – experiment No. 14, Tadjikistan, August 13, 1965, milk ripeness.

As seen from Figure 8 the vertical variation of specific leaf water content in maize differs from that of water content. The highest values of σ_{WL} in Estonia are observed at heights between 30–40 cm, a far too weaker maximum in Tadjikistan is associated with heights between 100–120 cm; and in Moldavia, the specific leaf water content σ_{WL} is constantly diminishing with height as is also the water content W_L.

The curves for the specific leaf water content σ_{WL} and specific leaf dry weight σ_L are similar – as σ_L decreases σ_{WL} also decreases, but σ_{WL} diminishes at a faster rate. Depending on weather conditions σ_L and σ_{WL} undergo a twofold variation in Estonia. Values for σ_L are 3 to 5 times higher in Tadjikistan than in Estonia.

The phytometric characteristics W_L, σ_L, σ_{WL}, etc. also change throughout the growing season influenced by both ontogeny and environmental conditions. Thus, for instance, for a young birch tree, $1/\sigma_L$ of leaves in spring is two times more than that in summer and autumn (Araki 1972). Besides, with increasing relative illumination σ_L grows for the birch (Araki 1972) as well as for many other plants (Tooming 1977). Figure 9 shows the course of W_L, σ_{WL} and σ_L in stands of Estonia and Tadjikistan during the period of growth.

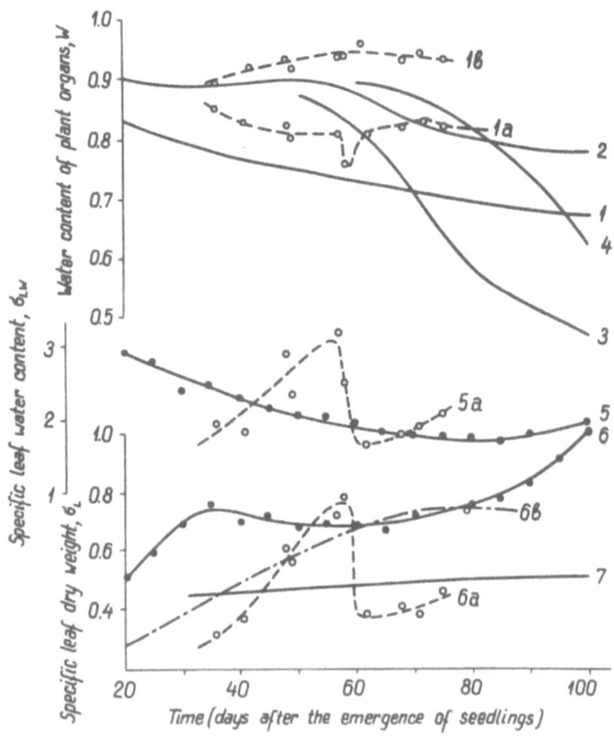

Fig. 9. Seasonal change of plant organ water content (curves 1–4), specific leaf water content (5 and 5a) and specific leaf dry weight (6–7) in a maize stand.
1–6 – experiment No. 14, Tadjikistan, 1965: 1 – leaves, 2 – stalks, 3 – tassels, 4 – ears, 1a, 5a, 6a – experiment No. 4, Estonia, 1962; 1b and 6b – experiment No. 19, Estonia, 1966; curve 7 – hybrid swede, experiment No. 21, Estonia, 1966.

The weather in Tadjikistan in 1965 was free of severe variations, so it may be assumed that it was ontogeny alone that determined the course of phytometric characteristics. Water content generally diminshes in individual plant organs (as illustrated by curves 1–4, Figure 9), particularly in the tassel and ears.

The specific leaf water content σ_{WL} (curve 5) diminishes until the milk ripeness time, then increases again. The specific leaf dry weight σ_L (curve 6) increases initially reaching its maximum in the 11th day phase. At the tassel flowering time σ_L falls down, then, as the ear starts flowering, it grows monotonously reaching twice its size during the observation period.

Quite a different picture was observed in 1963 in Estonia (see also Ross & Vlasova 1967). The end of July (40th–55th day of vegetation) was dry and partly cool. A heavy rain occurred on July 29 and made conditions for the development of the crops early in August (57th–62nd day of vegetation) highly favourable, with clear sky and daytime air temperature between 25–30 °C. Curves 1a, 5a and 6a in Figure 9, representing the dynamics of W_L, σ_{WL} and σ_L, respectively, can obviously be interpreted as follows: the drought brought about a decrease in the water content but not to the extent to cause leaf withering. Combined with low air temperature the drought retarded the growth and outflow of assimilates from the leaves notably. But the leaves were still capable of synthesizing causing assimilates to stay there and bringing about a sharp increase in the specific leaf dry weight and water content. Growth was remarkably good on favourable days, as the crop height increased by 15 cm during August 3, 4. This was accompanied by an intensive outflow of assimilates from the leaves, abrupt drops of σ_L, σ_{WL} and some reduction in water content, then σ_L and σ_{WL} slowly began to increase.

In 1966 the weather in Estonia was free of sharp changes, and curves 1b and 6b in Figure 9, representing the course of W_L and σ_L, respectively, in a maize crop show a smooth change. Curve 7 in the same Figure refers to hybrid swede and shows a slight increase of σ_L during the summer.

The above examples show that unless affected by severe weather changes, the plant phytometric characteristics interrelating the phytomass and phytoarea of leaves and other plant organs have a smooth shape during the growing season being determined by general laws of plant growth and development. However, severe weather changes are capable of causing significant alterations in curve smoothness. It is, therefore, very important when describing stand development phytometrically, to have a thorough knowledge of the above characteristics and their variation in time and space.

I.5.3 VERTICAL DISTRIBUTION OF PHYTOMASS AND PHYTOAREA

The vertical structure of foliage area in a stand can be described with one of the three functions: $u_L(z)$, $u'_{mL}(z)$ or $u_{mL}(z)$. These functions are related by formulae (I.2.2)–(I.2.4). Because of the low dependence of functions σ'_L, σ_L, v_L and W_L on height z, the curves of $u_L(z)$, $u'_{mL}(z)$ and $u_{mL}(z)$ slightly differ in shape although they preserve the main common qualities. In the study of relationship between radiation regime and stand architecture $u_L(z)$ will be undoubtedly the best-grounded characteristic; in a productivity analysis all the three functions can be used.

Similar functions for an individual plant will be $\tilde{S}_L(z)$, $m'_L(z)$ and $m_L(z)$, respectively. These are also interrelated through formulae (I.2.2)–(I.2.4).

Here we should point to the difference between the vertical distribution for a 'mean' plant of a stand $\hat{S}_L(z)$ and stand vertical distribution calculated for an individual plant of a stand, $\bar{S}_L(z)=u_L(z)/N_P$. The latter is determined in the interval (0, z_{0L}), while $\hat{S}_L(z)$ is determined in the interval (0, z_{1L}) using formula (I.2.12c), with $z^*_{1L}=z_{1L}$ and $\hat{S}^*_{LP}=S_{LP}$. The extent of the difference depends mainly on the form of the distribution function of foliage uppermost level $f_{z_{1L}}$ and on how much vacant space is left between plants. In considering the curves one should be conscious of whether they refer to a 'mean' plant or to the whole stand, and to act correspondingly depending on the objective set, i.e. use either the phytometric quantities of the 'mean' plant or the quantities of a horizontally averaged stand.

We now turn to the discussion of the experimental data and the mean relative plant foliage area density as a function of relative height, $\tilde{S}_L(\bar{z})$, the calculation method for which was dealt with in I.3.6.

The most detailed data (Figures 10 and 11) were obtained for maize during our expeditions to Estonia, Moldavia and Tadjikistan. A study of these figures and an analysis of the experimental data indicate that during the growing season $\tilde{S}_L(\bar{z})$ changes according to a definite pattern which is different for the sparse (Moldavia, Tadjikistan) and dense (Estonia) stands after their closure.

In the early stages of vegetation the densiest foliage is found in the lowest layers. But as the plant develops, the relative height $\bar{z}_m$ of the densiest foliage $\tilde{S}_L(\bar{z})$ tends to increase reaching its maximum value of 0.7–0.8 at the time of tassel emergence and flowering. In a sparse stand leaves droop a little; with maturity their inclination increases and max $\tilde{S}_L(\bar{z})$ falls to a relative height of 0.4–0.5. Further on, as leaves turn yellow, $\tilde{S}_L(z)$ of living leaves rises again, as most yellowing occurs in the lower layers. The foliage area density in the thickest horizontal layer is 1.5 times greater than that of uniform vertical distribution and does not vary significantly during the growing season, despite relative height variations of the thickest layer. Subsequent investigations by Goisa and Rogachenko (1969) on maize crops in the Ukraine confirmed our observations. The same authors also disclosed that the height of the thickest layer $\bar{z}_m$ increases with increasing plant density (within 2.4 to 8.2 plants/m^2 density).

Upon closure the vertical distribution of foliage area in dense maize stands (with high Np) becomes less uniform (see Figure 10) as compared with that of a sparse stand. The amount of leaf area in the lower strata gradually diminishes with a corresponding increase of density at a relative height of 0.6–0.9. Maximum values for $\tilde{S}_L(\bar{z})$ are in a range of 2.3–2.5. The period of tasseling and tassel flowering is associated with a mass yellowing of the leaves, which causes the $\tilde{S}_L(z)$ curve lower end to rise from zero to a relative height of 0.2 and the upper part of

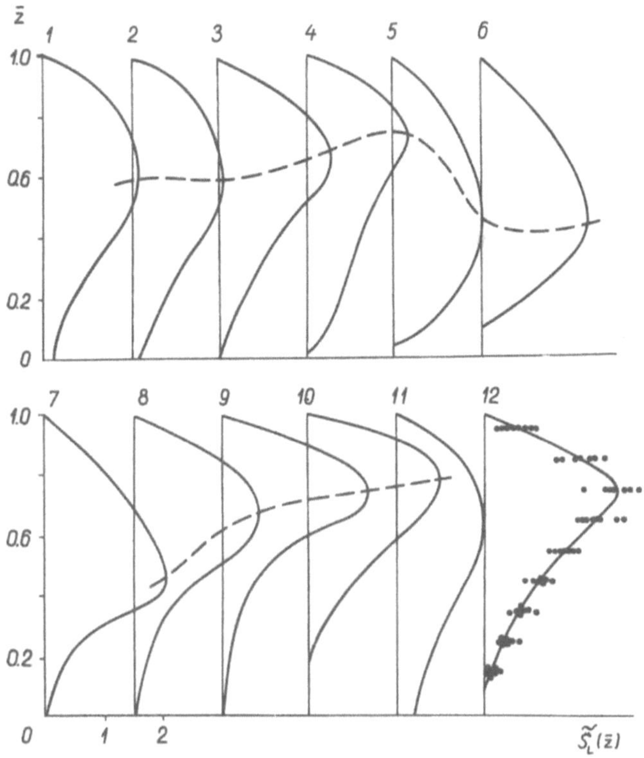

Fig. 10. Relative foliage area density of mean plant $\tilde{S}_L$ as a function of relative height $\bar{z}$ for maize.

Curve No.	Experiment No.	Date	Phenological stage
		Tadjikistan	
1	14	June 06, 1965	Emergence of 8–9th leaf
2	14	June 11	Emergence of 10th leaf
3	14	June 23	Emergence of 14th leaf
4	14	July 13	Tassel flowering
5	14	July 22	Ear flowering
6	14	Aug. 11	Milk ripeness
		Estonia	
7	19	June 28, 1966	Emergence of 5–6th leaf
8	19	July 03	Emergence of 8th leaf
9	19	July 24–Aug. 04	Emergence of 14th leaf
10	19	Aug. 26	Tassel flowering
11	24	July 12, 1967	Emergence of 5th leaf. Drought
12	24	Aug. 19–21	Tassel flowering

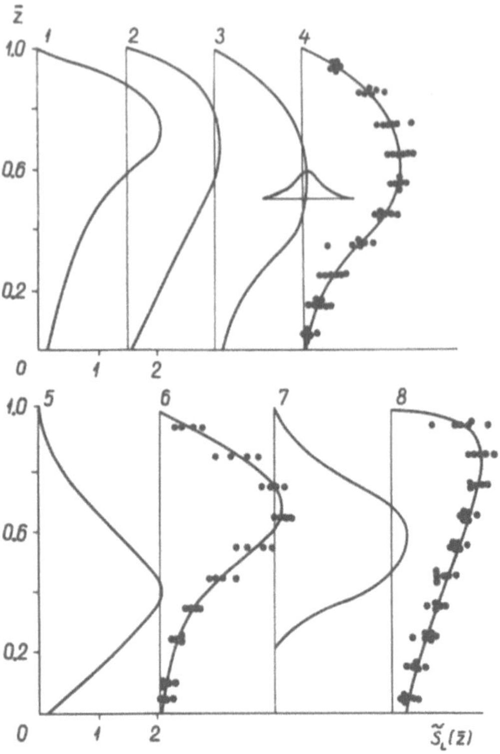

Fig. 11. Relative foliage area density of mean plant $\tilde{S}_L$ as a function of relative height $\bar{z}$.

Curve No.	Experiment No.	Crop	Date	Phenological stage
			Moldavia	
1	10	Maize	July 09–11, 1964	Emergence of 13th leaf
2	10	Maize	July 20–22	Tasseling
3	11	Maize	June 28–Aug. 14	Milk ripeness
4	11	Maize	Aug. 17–23	Wax ripeness
			Tadjikistan	
5	18	Cotton	Aug. 08, 1965	Flowering
			Estonia	
6	21	Hybrid swede	Aug. 08–09, 1966	Closure of rows
7	26	Hybrid swede	Aug. 23, 1967	Yellowing of lower leaves
8	22	Potatoes	Aug. 16, 1966	Flowering

the curve of $\tilde{S}_L$ to extend. As a result, a very dense layer of leaves is formed in a dense maize stand at a relative height of $\bar{z}=0.7$, in which overlapping of neighbouring plants is substantial. Below this layer the illumination is poor, consequently, photosynthesis is retarded and the leaves that are below this layer cannot possibly procedure phytomass throughout daytime.

A rice crop features a similar course of variation of foliage area vertical distribution during the growth period (Ito et al. 1973).

Curve 11 in Figure 10 characterizes the foliage area density function of a maize plant in Estonia during a drought, which results in a greater leaf inclination angle. The leaves droop, the maximum of S_L is shifted from the relative height of $\bar{z}=0.8$ to 0.6 and the curve resembles in appearance the $\tilde{S}_L$ curve in the period of maturity. With improved water supply conditions the previous situation is more or less regained (curve 12 in Figure 10).

The points at curves 12 in Figure 10 and at 4, 6 and 8 in Figure 11 describe experimental data scatter, each point corresponding to the mean value of the random function $\tilde{S}_L^*$ for intervals of heights $\Delta z = 10$ cm averaged over 5–20 plants. The curve drawn over curve 3 in Figure 11 represents schematically the normal distribution of points without averaging.

The dynamics of foliage area density function $\tilde{S}_L(\bar{z})$ for sunflower and sorghum during the growing season are shown in Figure 12. The relative height of maximum foliage area density is greater for sunflower than for maize and reaches 0.8–0.9. Relative height variations of a maximum density layer are not so pronounced in sunflower.

A peculiar shape of $\tilde{S}_L(\bar{z})$ is possessed by sorghum. According to the measurements made by Shamsiyev during the Tadjik expedition of 1965, in the early stages of plant development the curve of $\tilde{S}_L(\bar{z})$ (Figure 12, curve 5) resembles that of $\tilde{S}_L$ for sunflower at the ripening time (Figure 12, curve 4). But starting with the tasseling phase the vertical distribution of foliage area is bimodal (Figure 12, curves 6, 7, 8). The bottom maximum is at a relative height of $\bar{z}=0.4$ and slowly shifts down to $\bar{z}=0.2$ toward the flowering time, whereas the top maximum, which is more pronounced, shifts from 0.75 to 0.9. Among the species we studied sorghum was the only one to have the foliage area vertical distribution curve so vividly bimodal. But, generally, sorghum is characterized by a comparatively uniform vertical distribution of foliage area. The points at curves 3 and 6 of Figure 12 show the spread of data on some plants. Less consistent data on $\tilde{S}_L(\bar{z})$, illustrated in Figure 11, are available for cotton, hybrid swede and potatoes. Cotton, for instance, features the maximum density of foliage area at the flowering stage at a relative height of 0.4, while $\tilde{S}_L(\bar{z})$ for hybrid swede is similar to that of maize, despite the different appearance and habitus of these plants. Potatoes grown in a field among plants of hybrid swede have max $\tilde{S}_L$ at a relative height of $\bar{z}=0.85$, while on the whole the vertical distribution of foliage area is comparatively uniform.

To determine the relative plant foliage area density in grasses, we have used the data obtained by Alekseyenko (1967). It is seen from Figures 13 and 14 that the

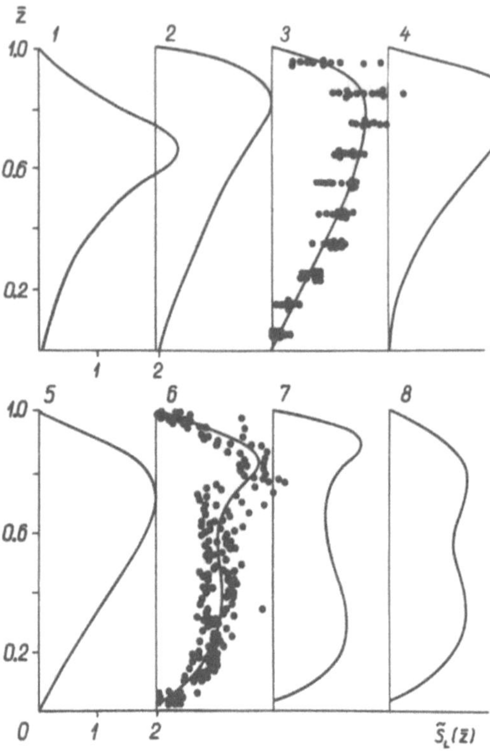

Fig. 12. Relative foliage area density of mean plant $\tilde{S}_L$ as a function of relative height $\bar{z}$ for sunflower and sorghum in Tadjikistan.

Curve No.	Experiment No.	Crop	Date	Phenological stage
1	16	Sunflower	June 09, 1965	Emergence of 16th leaf
2	16	Sunflower	June 27	Emergence of 28th leaf
3	16	Sunflower	July 12–13	Formation of inflorescence
4	16	Sunflower	Aug. 14	Ripening
5	15	Sorghum	1st decade, July, 1965	Emergence of 10th leaf
6	15	Sorghum	July 15–16	Tasseling
7	15	Sorghum	July 23	Tasseling
8	15	Sorghum	July 29–30	Flowering

vertical distribution of foliage area in meadow plants resembles similar curves for maize: max $\tilde{S}_L$ is associated with the middle layers at a height of $\bar{z} = 0.5 \div 0.7$, the 'elongation' of $\tilde{S}_L(\bar{z})$ is similar to that in dense maize stands.

Quite a different class of plants is constituted by grasses which have most foliage concentrated in the lowest layers at $\bar{z} = 0.05 \div 0.20$. The area density in the top layer of the plant foliage is negligibly low, correspondingly, the curves of $\tilde{S}_L(\bar{z})$

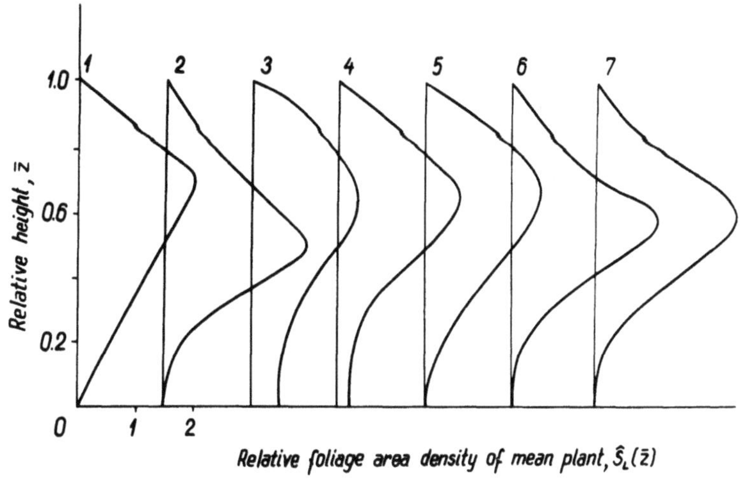

Fig. 13. Relative foliage area density of mean plant $\tilde{S}_L$ as a function of relative height $\bar{z}$ for meadow plants (Alekseyenko 1967):
1 – timothy, heading; 2 – same, flowering; 3 – meadow clover, budding; 4 – same, flowering; 5 – lucerne, budding; 6 – same, flowering; 7 – brome grass, flowering.

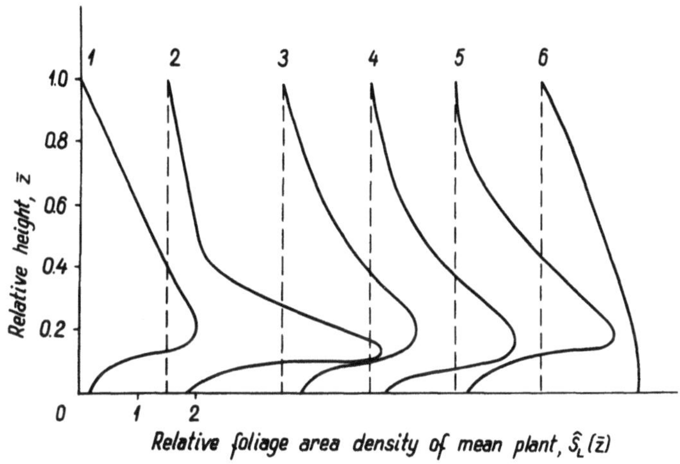

Fig. 14. Relative foliage area density of mean plant $\tilde{S}_L$ as a function of relative height $\bar{z}$ for meadow grasses (Alekseyenko 1967):
1 – fescue, heading; 2 – same, flowering; 3 – orchard grass, heading; 4 – same, flowering; 5 – Kentucky blue grass, flowering; 6 – bent grass, flowering.

are stretched in the lowest layers. Awnless brome grass is an exception with the maximum foliage area at a relative height of 0.6 (Figure 13, curve 7).

It is felt that grasses feature maximum foliage area density in the lower layers of the stand only in the period of vegetative growth. The period of reproductive growth is accompanied by max $\tilde{S}_L$ moving upwards due to yellowing of the leaves in the lower layers.

The function $\tilde{S}_L(\bar{z})$ while being well suited for making a comparative analysis of the foliage area vertical distribution in relative units, fails to present the real height and density of phytomass in a stand.

The basic quantity in describing the foliage area vertical distribution in a stand is either the foliage area density $u_L(z)$ or the downward cumulative leaf area index $L(z)$.

We shall now discuss area density of leaves, stems and other plant organs as a function of height for different stages of stand development.

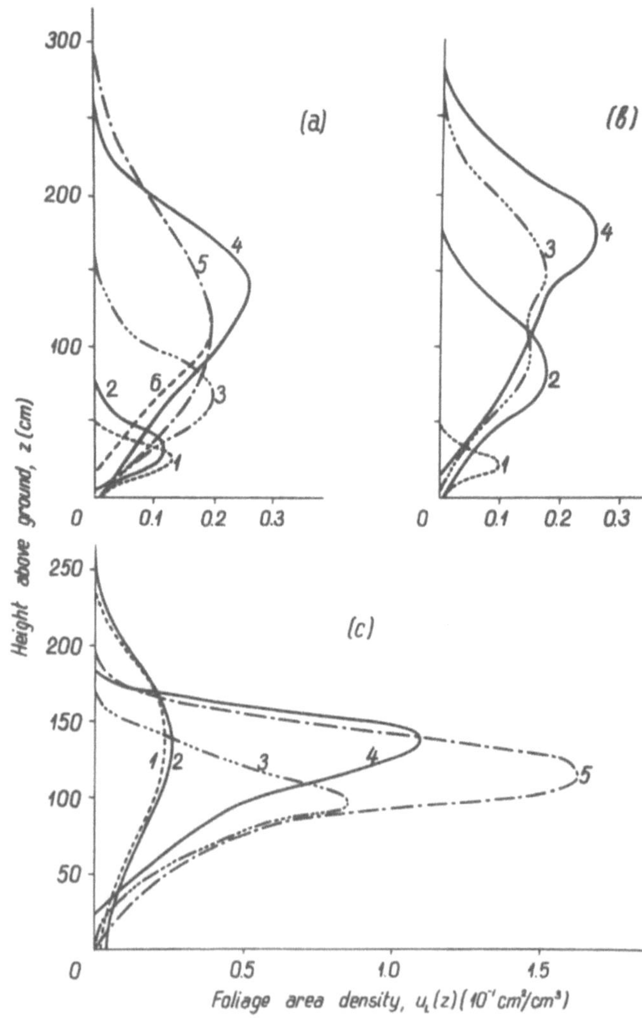

Fig. 15. Vertical distribution of foliage area density in maize and sunflower stands.
(a) – maize, experiment No. 14, Tadjikistan, 1965, see text; (b) – sunflower, experiment No. 16, Tadjikistan, 1965. 1 – June 9, emergence of 16th leaf; 2 – June 27, emergence of 28th leaf; 3 – July 12, formation of inflorescence; 4 – August 14, ripening; (c) – maize stands with various density at the stage of tasseling or tassel flowering, see Table 5.1.

Maize. Sparse stands of 4 to 6 plants/m^2 density grown for corn were investigated in Tadjikistan and Moldavia. As the general trends in the vertical distribution of foliage area were the same in both regions (see Figure 15B) we shall restrict ourselves to the case of Tadjikistan.

Figure 15A shows variation of $u_L(z)$ curves for living leaves during the growing season. Curve 1 corresponds to June 06 – the stage of emergence of the 7–8th leaf; curve 2, to June 11, emergence of the 10th leaf; 3, to June 23 emergence of the 14th leaf; 4, to July 13, tassel flowering; 5, to July 22, ear flowering, and curve 6, to Augst 06–11, milk ripeness. As the plants grow, the stand height, maximum density level z_m and max $u_L(z)$ increase, as well as the stand leaf area index L_0 equal to the area defined by the vertical axis and the curve of $u_L(z)$. The stand reaches its maximum foliage area $L_0 = 2.8$ during the tasseling and tassel flowering stages. The densiest layer of 0.025 cm^2/cm^3 is found at a height of 140–150 cm. Further on, as the plants enter the ear flowering stage, the foliage area in the top layers increases still a little more as does the uppermost level, while yellowing of the leaves in the bottom layers is beginning. Concurrently, the leaves start to droop increasing their inclination angle and max $u_L(z)$ is shifted downward (Figure 15A, curve 5). During the ripening stage, as the plants get yellow, the living leaf area is gradually diminishing starting with the bottom layers (curve 6).

More recent data by Japanese workers (Udagawa et al. 1968) agree in general with our data.

Dense stands of maize (15–45 plants/m^2) cultivated for ensilage were studied in Estonia (Figure 16). A drought in July retarded the stand development (curves 3 and 4), but favourable conditions in August (see also Ross & Vlasova 1966) gave it a new impetus. The vertical distribution of foliage area in a dense maize stand is less uniform as compared with that of a sparse stand. As the leaves in the bottom layers wither away and the plants come in full tassel, a dense layer is formed at a height of 150 cm where the bulk of the foliage area is concentrated.

As far as Estonia is concerned, the foliage area density in dense maize stands is 6 to 7 times higher than the maximum density in sparse stands (Table I.5.1), the height of the densiest stands in both cases being equal to 120–140 cm (Figure 15). The leaf area index differs correspondingly by a factor of 4, reaching 11 in the densiest stand.

Table I.5.1 *Characteristics of maize stands*

Curve No.	Experiment No.	Region	Date	N_P	L_0	$\frac{L_0}{N_P}$
1	10	Moldavia	July 20–22, 1964	5.4	3.2	0.59
2	14	Tadjikistan	July 13, 1965	4.5	2.8	0.62
3	3	Estonia	Sept. 01, 1962	16.5	5.4	0.33
4	4	Estonia	Aug. 20, 1963	30.5	7.8	0.26
5	20	Estonia	Aug. 01, 1966	44	11.2	0.39

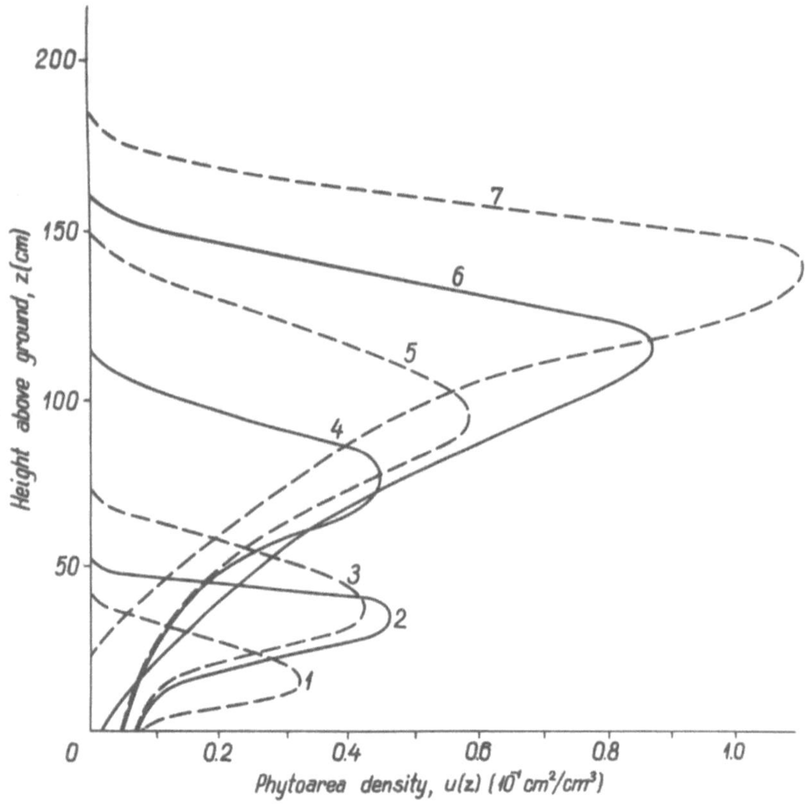

Fig. 16. Vertical distribution of phytoarea density in maize at various times during the growing season. Experiments Nos. 4 and 5, Estonia, 1963.

1 – emergence of 7th leaf; 2–8th leaf; 3–9th leaf; 4–10th leaf; 5–11th leaf; 6–12th leaf; 7–tasseling.

As seen from Figure 17 other organs of the maize plant do not contribute appreciably in terms of phytoarea. Expressed on a percentage basis the phytoarea of a whole plant is distributed as follows: the leaves will account for 87%, dead leaves for 1%, stalk for 8%, tassels for 3%, and ears for 1%. If instead of the axial section area we consider the entire lateral surface of the stalk, tassels and ears, modelled as cylinders, then the leaves will account for 70%, stalks for 20%, dead leaves for 1%, tassels for 6%, and ears for 3% of the total plant phytoarea. The leaves constituting the bulk of the phytoarea, all the other plant organs can be disregarded in solving many problems.

Sorghum. Although sorghum and maize are alike in respect of appearance and habitus they have a different vertical foliage area distribution in a stand. As previously mentioned in connection with the discussion of $\tilde{S}_L(\bar{z})$, the vertical foliage area distribution curve of an individual sorghum plant has two maxima (see Figure 12, curves 5–8). But under stand conditions owing to individual plant height and foliage area variations these maxima practically disappear (Figure 17A). For this reason $u_L(z)$ of a sorghum stand has a comparatively graduated

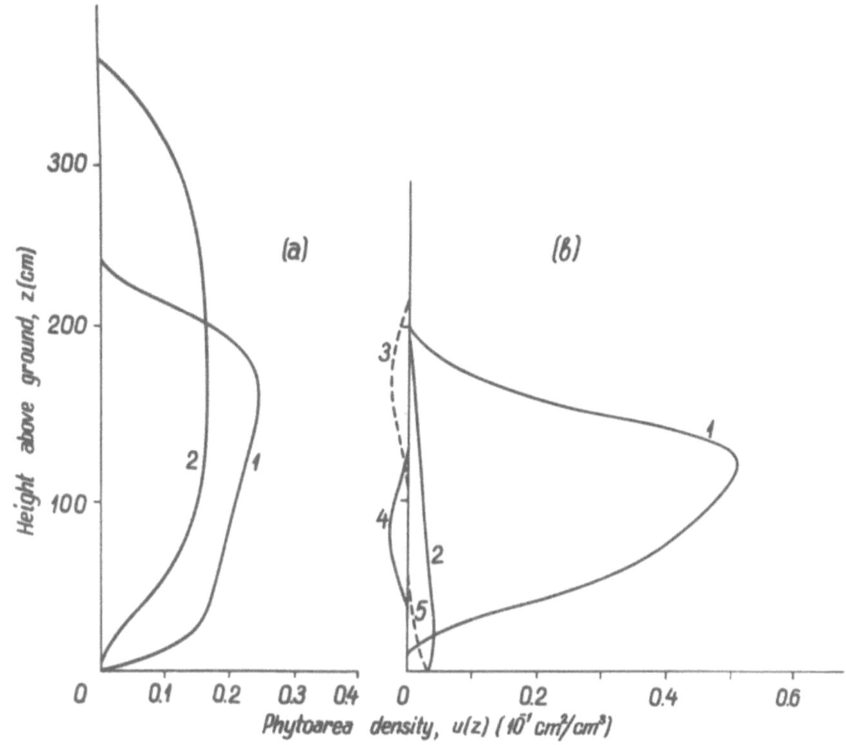

Fig. 17. Vertical distribution of phytoarea density in sorghum and maize stands.
(a) – sorghum leaves, experiment No. 15, Tadjikistan, 1965: 1 – July 12, 1965, tasseling; 2 – August 5, 1965, flowering; (b) – different maize organs at the stage of panicle flowering, experiment No. 5, Estonia, August 22, 1963: 1 – $u_L(z)$ of leaves; 2 – $u_{EC}(z)$ of stalk, 3 – $u_E(z)$ of tassels, 4 – $u_E(z)$ of ears, 5 – $u_L(z)$ of dead leaves.

maximum in the middle layers and an extremely uniform vertical foliage area distribution. From the viewpoint of the radiation regime the structure of a sorghum stand provides a more uniform vertical distribution of radiation as compared with that of a maize stand.

Sunflower. The vertical structures of maize and sunflower stands are similar despite the difference in the leaves and habitus of these plants (Figure 15B). In general, our own data have been confirmed by more recent data by Japanese authors presented in the work of Udagawa et al. (1970), where also the effect of the density on the form of $u_L(z)$ is discussed and where it is shown that $u_L(z)$ is practically independent of the stand density which in this experiment varies between 1 and 11 plants/m^2.

Cotton. (Hamidov 1972). At the budding stage, with plants 30 cm tall, $u_L(z)$ increases slowly to a height of 10–15 cm, whereupon it increases sharply to the uppermost layer where u_L reaches its maximum. At the flowering stage, with plants 50 cm tall, $u_L(z)$ increases slowly to a height of 35–40 cm, whereupon it slowly diminishes. In plants up to 50 cm tall the maximum of $u_L(z)$ moves upwards in proportion to the height but does not change as the plant grows taller.

Horse beans. The maximum density of leaf area in a horse bean stand is for most of the growing period associated with the upper layers, (Figure 18a). The effective phytoarea of the stalks, petioles and pods is insignificant as compared with that of the leaves and can therefore be disregarded.

Wheat (Ross & Nilson 1967a). The stalks and spikes play a significant role in the formation of phytoarea in cereals (Figure 18B).

Table I.5.2 illustrates the per cent distribution of total and effective phytoarea of the entire wheat stand separately for different plant organs.

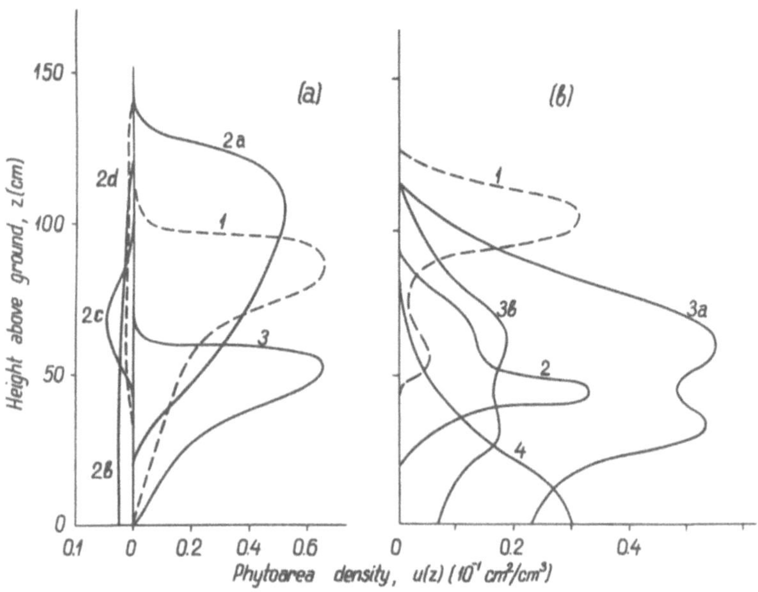

Fig. 18. Vertical distribution of phytoarea density $u(z)$ in horse bean and wheat stands.
(a) – horse bean: 1 – experiment No. 1, Estonia, August 17, 1962, flowering, $L_0 = 2.4$, $u_L(z)$ of leaves; 2 – experiment No. 1, September 10, 1962, flowering, $L_0 = 3.6$; 2a – $u_L(z)$ of leaves; 2b – $u_E(z)$ of stalks; 2c – $u_E(z)$ of pods; 2d – $u_E(z)$ of petioles; 3 – experiment No. 25, Estonia, July 19, 1967, flowering, $L_0 = 1.7$, $u_L(z)$ of leaves; (b) – wheat, experiment No. 7, Estonia, July 11, 1963, milk ripeness: 1 – $u_F(z)$ of ears; 2 – $u_L(z)$ of living leaves; 3a – $u_C(z)$ of stems; 3b – $u_{EC}(z)$ of stems; 4 – $u_L(z)$ of dead leaves.

Table I.5.2 *Per cent distribution of phytoarea among plant organs of wheat*

Phytoarea	Relative phytoarea of plants in stands	Proportion of total phytoarea (%) living leaves	dead leaves	stalks	spikes
Total	6.3	12	14	61	13
Effective	3.2	25	28	39	8

As will be apparent from Table I.5.2 the stalks contribute nearly 60% of the total and 40% of the effective phytoarea of the stand. Obviously, the stalks and spikes shall not be neglected in studying the radiation regime and phytosynthetic activity.

Vertical distribution of total assimilating surface over the growing season is presented in the work of Goisa et al. (1971). Comparison of these data with our own on barley indicates that variation of the vertical distribution of fresh phytomass is the same for wheat and barley.

The double maxima of $u_F(z)$ of spikes (curve 1) and of $u_S(z)$ of stalks (curve 3a) are due to the stand being composed of main axes and lateral shoots having a different height. The lower maxima, therefore, refer to lateral shoots.

According to Santhirasegaram and Black (1968) the general shape of the curve of $u_L(z)$ in a wheat stand is little affected by the rate of sowing, spacing and row direction.

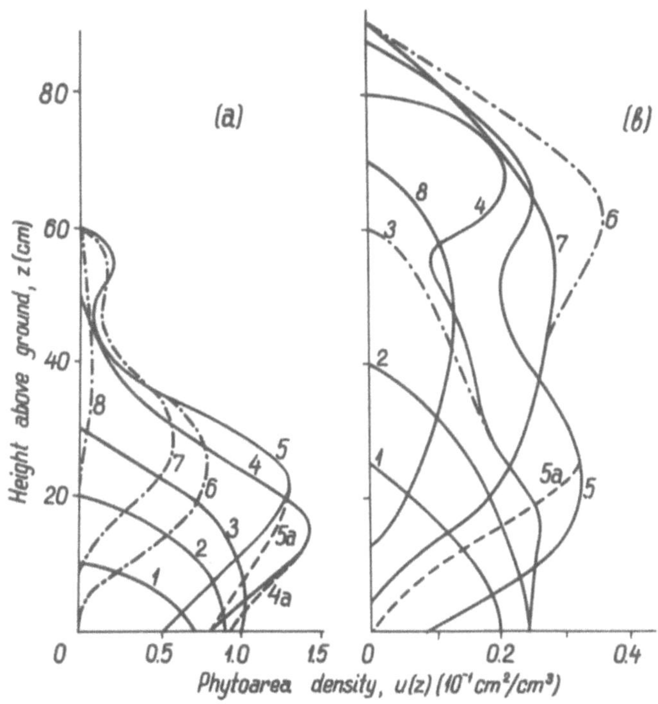

Fig. 19. Variation of vertical distribution of green phytoarea density in a barley stand. Experiment No. 12, Estonia, 1965.
(a) – $u_L(z)$ of leaves; 1 – June 5, emergence of 3rd leaf; 2 – June 10, emergence of 4th leaf; 3 – June 15, emergence of 5th leaf; 4 – June 25, emergence of 7th leaf; 5 – July 5, ear emergence; 6 – July 15, milk ripeness; 7 – July 20, milk ripeness; 8 – August 6, wax ripeness; 4a and 5a – total area density of living and dead leaves; (b) – $u_{EC}(z) + u_{EF}(z)$ total effective area density of stems and ears: 1 – June 20, 2 – June 25, 3 – June 30, ear emergence, 4 – July 5, 5 – July 15, 6 – July 26, milk ripeness, 7 – August 5, 8 – August 15, wax ripeness.

Barley. The variation of vertical distribution of phytoarea in a barley stand over the period of growing was studied by the author in cooperation with Mägi (Mägi & Ross 1969a, b) in Estonia.

Early after the emergence of seedlings, the maximum density of foliage area is found at the ground surface (Figure 19). As the stand reaches a height of about 20 cm, the layer of maximum density shifts to a relative height of 0.3–0.4, which is associated in time with the flowering stage. Further increase of the relative height, associated in time with the reproductive stage is dependent on the rate of senescence of the lower leaves. A second small mode in the uppermost layer is due to peculiarities inherent in barley as a species – a single plant has 1 or 2 lateral shoots which are 15 to 20 cm shorter than the main axis. In addition, the plant has its last, 'flag' leaf on the main axis located notably above the nearest leaf. Taken together these 'flag' leaves form a layer of their own in the stand. The data obtained by Duda and Kolek (1970) support this.

The curves for the total effective area of stalks and spikes have a more complicated shape. At the initial stage, preceding heading, the stalk area $u_{EC}(z)$ develops in the same way as the area of the leaves (curves 1–3 in Figure 19). Heading causes another maximum to appear at the level of spikes (curves 4 and 5) and curve $u_{EC}+u_{EF}$ at heading and flowering, resembles curve $\tilde{S}_L(\bar{z})$ for sorghum (see Figure 12, curve 6). But later, as the plants enter the reproduction stage, the stalk-dependent lower maximum at the relative height of 0.2–0.3 gradually diminishes owing to stand yellowing, and the bimodal curve for the green surface of the stalks and spikes becomes unimodal (curves 7 and 8).

The effect of thickening from 6 to 50 plants/m^2 was investigated in the work of Udagawa et al. (1970). The dynamics of $u_L(z)$ comply with our own data (see Figure 18). At the early stages of development the curves of $u_L(z)$ look similar for all densities. Starting with the heading stage dense stands show more rapid leaf area reduction in the lower layers than sparse stands.

It is felt that in general the outlined development is characteristic over the growing season not only of barley, but of all grasses.

Awnless brome grass (Ross & Nilson 1966b). The densiest layer of foliage area is clearcut and found at a relative height of 0.6 (Figure 20A). The stem area decreases almost linearly with increasing height and contributes noticeably to the total phytoarea. The area of flowers in the uppermost layers must also be taken into account.

Sweet clover (Ross & Nilson 1967b). The foliage area distribution at flowering is much the same as in the case of awnless brome grass (Figure 20), but for the upper layers where a very large proportion of the area is contributed by white flowers. The vertical distribution of the area of stalks and petioles is specific, with the maximum at a relative height of 0.6.

Meadow plants (Alekseyenko 1960). Of all stands described in the literature grass stands possess the highest foliage area density, where according in Alekseyenko (1967) the leaf area index L_0 reaches 17 in monocultures and 24 in grass mixtures. With grass stands being of comparatively small height (110–130 cm),

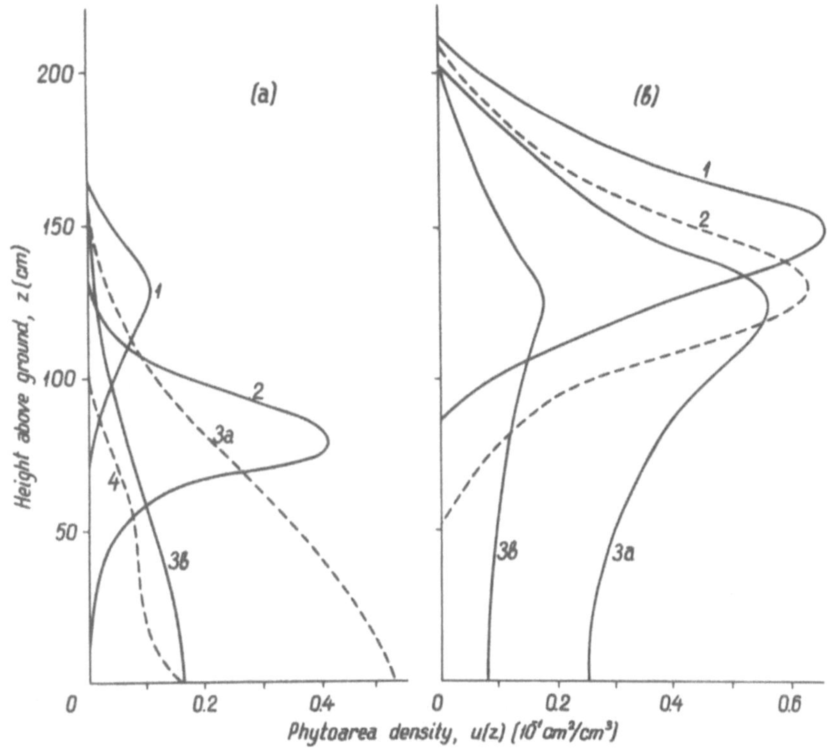

Fig. 20. Vertical distribution of phytoarea density in stands of awnless brome grass and arctic clover. (a) – awnless brome grass, experiment No. 8, Estonia, July 13, 1963, flowering: 1 – $u_F(z)$ of flowers, 2 – $u_L(z)$ of living leaves; 3a – $u_C(z)$ of stems and petioles, 3b – $u_{EC}(z)$ of stems and petioles, 4 – $u_L(z)$ of dead leaves, (b) – arctic clover, experiment No. 2, Estonia, August 20, 1962, flowering: 1 – $u_F(z)$ of flowers, 2 – $u_L(z)$ of living leaves, 3a – $u_C(z)$ of stems, 3b – $u_{EC}(z)$ of stems.

these figures characterize very high values of foliage area density. The value of $u_L(z)$ reaches 3.0–3.5 for monocultures and even 3.5–4.5 for grass mixtures. Thus, the leaf area index in the densiest 10-cm layer reaches 3–4, which is roughly ten times the density of typical agricultural stands. It is therefore of great interest to investigate in more detail the photosynthetic activity of a stand as thick as that.

Investigations in the dynamics of foliage area vertical distribution in fescue and timothy were undertaken in Sweden by Kornher (1971). In fescue the maximum value of $u_L(z)$ early in the growing season (with plants 30–50 cm tall) was observed at a height of 10 cm (similar to curve 1 of Figure 14), on the 60th day of the growing season, when plants were 60 cm tall, the maximum of $u_L(z)$ moved to a height of 25 cm, and on the 80th day, to a height of 38 cm (similar to curve 1 of Figure 13).

Stems account for a considerable proportion of area in the lower layers and a negligible proportion in the upper layers. Vertical distribution is more uniform in shaded stands. In the case of timothy the dynamics of $u_L(z)$ over the growing season are similar to those of fescue, although the maximum of $u_L(z)$ is less pronounced and is found at a lower level. So, the data by Alekseyenko and Kornher disagree somewhat on $u_L(z)$ of timothy. Stems constitute a small proportion of area in timothy.

Among tropical pasture plants (Heslehurst, Wilson 1971) $u_L(z)$ in Setaria sphacelate has a resemblance to curve 1 of Figure 12, and that of Desmodium intertum resembles curve 5 of Figure 12.

The work of Utekhin (1977) presents detailed data on the vertical distribution of leaf and stem areas in mixed plant communities in Central Russia and its dynamics during the growing season. For a meadow community the foliage area density $u_L(z)$ in standing grass 120 cm tall has a pronounced maximum at 15–20 cm height, with $L_0=3.7$, early in June, but by August $u_L(z)$ becomes more uniform, with $L_0=2.6$. In a watershed steppe plant community the foliage area density $u_L(z)$ in standing grass 70 cm tall also has a pronounced maximum at 15–20 cm height. The general character of $u_L(z)$ changes but little from May to August. Late in May $L_0=1.7$; and at the end of August $L_0=3.0$.

Much less information is available about the vertical distribution of foliage area in forests. According to Rauner (1972) $u_L(z)$ of 22–23 year old oak, lime and aspen trees first appaers at a height of 5–7 m reaching its maximum at 9–10 m and dropping off to zero at the foliage uppermost level, at a height of 12–13 m. The leaf area index L_0 of individual stands is within the limits of 4.6 to 7.1. A similar pattern of $u_L(z)$ is found in forests of larch, beech, and evergreen oak in Japan (Shinozaki & Kira 1977).

The cumulative leaf area index, $L(z)$ (see formula (I.2.32), though not as descriptive as $u(z)$, is more convenient for practical purposes, especially in the study of a stand radiation regime.

Figure 21 illustrates the variation of the curves of $L(z)$ during the growing season in a maize stand in Estonia. As defined earlier, the function $L(z)$ decreases monotonously from $L_0=L(0)$ at the ground surface to $L(z_{0L})=0$ at the foliage uppermost level.

The length of the horizontal line between the neighbouring curves of $L(z)$ in Figure 21 corresponds to the leaf area index increment in the above-lying layer, while the inclination of the $L(z)$ curve corresponds to the foliage area density at a given height and moment in the growing season, i.e. $\mathrm{d}L/\mathrm{d}z=-u_L(z)$.

A number of works on the stand structure and many works on the productivity use either the fresh phytomass density of plant's k-th organ as a function of height $u'_{mK}(z)$ or the absolute dry phytomass density of plant's k-th organ as a function of height $u_{mK}(z)$ instead of the phytoarea vertical distribution $u_K(z)$. Taking into account formulae (I.2.3) and (I.2.4) and the fact that the specific leaf dry weight σ_L varies little with time and space compared with u_L and u_{mL}, the general regularities can be assumed to be similar for $u_L(z)$, $u'_{mL}(z)$ and $u_{mL}(z)$. Quite the opposite is characteristic of stalks and reproductive organs. Here, the functions for transition from $u_K(z)$ to $u'_{mK}(z)$ (or to $u_{mK}(z)$) are strongly dependent on stalk diameter and development stage; they vary within broad limits, and a general regularity between $u_{mK}(z)$ and $u_K(z)$ may not exist. Consideration of Figure 22

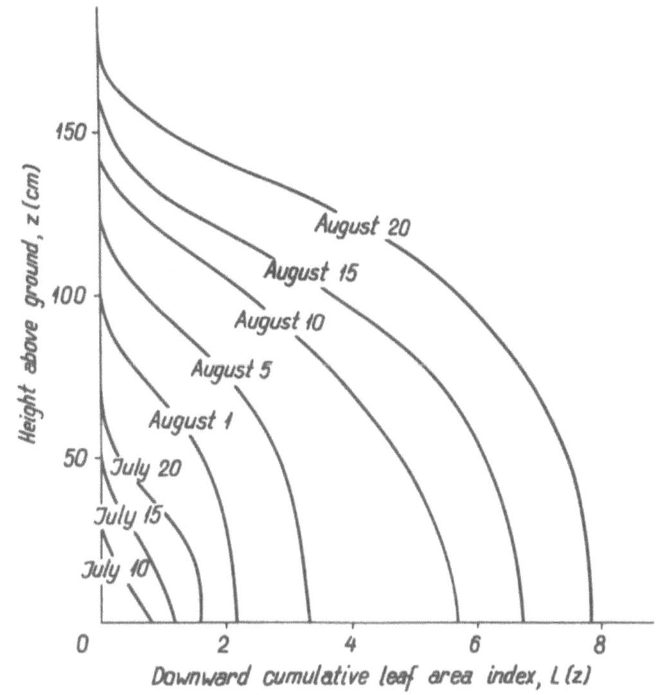

Fig. 21. Relationship between downward cumulative leaf area index $L(z)$ and height at various times during growing season. A maize crop, experiment No. 4, Estonia, 1963.

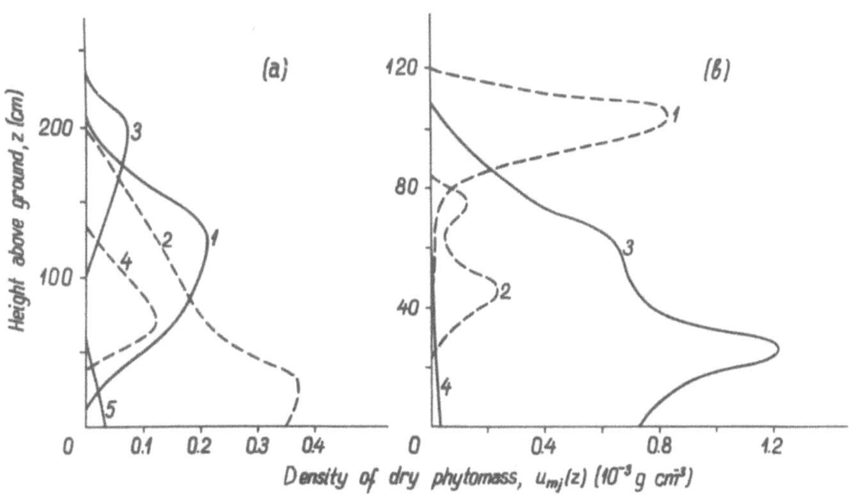

Fig. 22. Vertical distribution of dry phytomass density of different plant organs.
(a) – a maize stand, experiment No. 5, Estonia, August 22, 1963, tassel flowering: 1 – leaves, 2 – stalks, 3 – tassels, 4 – ears, 5 – dead leaves; (b) – a wheat stand, experiment No. 7, Estonia, July 11, 1963, milk ripeness: 1 – ears, 2 – living leaves, 3 – stems, 4 – dead leaves.

and its comparison with Figure 17B will prove this point. From the comparison of these Figures it is evident that while the area of non-synthesizing organs is negligible as compared with the leaf area, their phytomass, especially that of stalks, equals or exceeds the phytomass of leaves. That is why in solving problems in stand radiation regime, water supply and thermal conditions, which require the knowledge of stand phytoarea, one cannot rely on the phytomass density of non-synthesizing organs $u_{mK}(z)$ to adequately characterize the stand geometry.

Figure 22B illustrates the absolute dry phytomass density as a function of height $u_{mK}(z)$ distributed over the individual plant organs in a stand of wheat. Comparison of Figures 22B and 18B, the latter showing the vertical distribution of phytoarea for the same stand, indicates a notable difference between $u_K(z)$ and $u_{mK}(z)$. Variation of the total dry phytomass density in a winter wheat stand during the growing season (Goisa et al. 1971) is similar, in general, to that of a barley stand.

The above experimental data on the vertical distribution of phytoarea and phytomass in stands are indicative of a high variability and differences both between species and within a species during the growing season. However, towards the end of the growth certain specific features in the vertical distribution of $u_K(z)$ and $u_{mK}(z)$ for a given species become evident, so that the species can be said to have a peculiar vertical structure.

Attempts have been made in the literature to classify plants according to the vertical structure of their phytomass.

Alekseyenko (1968), for instance, proposes to subdivide both, the perennial tame meadow grass and natural meadow communities into two types according to the location of maximum thickness.

In her works Yefimova (1968, 1970) uses the total phytoarea density of plant canopy as a function of height $u(z)=\sum_k u_K(z)$ to characterize the vertical structure of agricultural crops and divides all crops into three classes with respect to this characteristic.

A more general, and a more detailed classifications of the vertical distribution for aerial vegetation have not yet been developed, the available data are fragmentary, and many species have not been studied. It is therefore essential that in the nearest future the acquisition of experimental data on many species and the data on intraspecific variability should be enhanced. Extended studies are also required of the dynamics of vertical structure variation of a given species over the growing season. It is desirable that all of these studies should be conducted using the same system of methods. It is our opinion that the most important characteristics required for the description and analysis of the stand vertical structure are the mean uppermost level of plant foliage z_{1L}, leaf area index L_0 and relative plant foliage area density as a function of relative height $\tilde{S}_L(\bar{z})$. A search should also be continued for ways of describing $\tilde{S}_L(\bar{z})$ by empirical functions not containing too many parameters.

In the general case, the phytometric characteristics, such as u_K, u_{mK}, etc. do not depend solely on the vertical coordinate z, but also on the horizontal coordinates. It is common practice to confine the study of horizontal variability of the stands to the consideration of either individual axisymmetrical plants or to row and square sown crops.

Experimental data on the horizontal variability and two-or three-dimensional distribution of phytomass and phytoarea are very scarce.

Some data on phytomass distribution in various rice and sweet potato cultivars furnished in the work of Tsunoda (1959) show a sharp decrease in the foliage area from the stalk to the plant's periphery for both species. Those species adapted to abundant mineral nutrition show lower plant heights and smaller basal area, while at the same time, greater volumetric leaf weight and total foliage phytomass. According to Tsunoda, plants adapted to different nutrition levels are different in habitus.

In a row crop of wheat (Santhirasegaram and Black 1968) 75% of the leaf area is confined in a belt 9 cm wide on both sides from the row centre, symmetrically to either side in the eastern and northern direction of rows; in the eastern row direction the leaf area of the southern side of the row exceeds that of the northern side of the row considerably.

A two-dimensional distribution of foliage area in a bean row crop and of individual organ area in a bunch of *Atriplex nummularia Lindl.* is presented in the work of Warren Wilson (1965b). Later those data were given a mathematical treatment by Philip (1965b).

The works of Heinicke (1963a, b) furnish two-dimensional sections of foliage area density $u_L(z, x)$ for various forms of the apple tree starting from 10-year old dwarf trees 2.5 m in height and ending with 25-year old standard trees 6 m in height.

The studies conducted by V. Ross at our laboratory were concerned with the horizontal distribution of phytomass and phytoarea in maize, sunflower and horse bean crops.

Table I.5.3 presents data on the two-dimensional distribution of foliage area density $u_L(z, x)$ and $u_L(z, y)$ in a square sown crop of maize, separately for the directions x and y. These data indicate a notable non-uniformity in a sparse maize crop (with $L_0 \approx 2$) at the milk ripeness stage in both directions of the rows. The foliage area density between rows is 2.5 to 3.0 times less than in the central part of the rows. The greatest density variations are found in the middle layers. Because of the lesser inter-row width (60 cm) the crop uniformity in the direction x was greater. A similar picture was observed in a sparse row crop at the milk ripeness stage in Tadjikistan. The contribution of stalks and ears to the effective area is small as compared with that of leaves.

The horizontal distribution of phytomass in a horse bean crop follows practically the same lines.

Table I.5.3 *Distribution of foliage area density $u_L(z, x)$ and $u_L(z, y)$ in horizontal direction in individual crop layers. Experiment No. 11, square sown crop of maize, 60 × 70 cm, milk ripeness stage, August 29, 1964, Moldavia*

Δz	Δx							$u_L(z)$
	0–10	10–20	20–30	30–40	40–50	50–60	60–70	
0–40	0.028	0.039	0.082	0.066	0.016	0.032	–	0.044
40–80	0.089	0.122	0.237	0.168	0.109	0.084	–	0.133
80–120	0.152	0.183	0.314	0.218	0.144	0.141	–	0.192
120–160	0.070	0.123	0.129	0.150	0.102	0.054	–	0.104
160–200	0.012	0.016	0.034	0.034	0.030	0.009	–	0.023
L_{0x}	1.41	1.93	3.14	2.54	1.61	1.28	–	1.98

Δz	Δy							$u_L(z)$
	0–10	10–20	20–30	30–40	40–50	50–60	60–70	
0–40	0.016	0.034	0.031	0.050	0.046	0.019	0.012	0.030
40–80	0.069	0.071	0.075	0.265	0.248	0.108	0.071	0.130
80–120	0.142	0.152	0.183	0.308	0.256	0.185	0.146	0.194
120–160	0.046	0.069	0.111	0.202	0.131	0.096	0.067	0.097
160–200	0	0	0.004	0.010	0.006	0.010	0.002	0.005
L_{0y}	1.08	1.28	1.60	3.40	2.76	1.68	1.20	1.85

Significantly different from the above data is the horizontal distribution of foliage area in a sunflower crop. As the plants enter the ripening stage, they start to lose their lowermost leaves, their stalks incline due to wind and the heavy weight of seed heads, while their tops with some pairs of leaves bend downward. The crop is no longer symmetrical about the row lines and the maxima of the leaf area index in different layers are displaced. A characteristic feature about sunflower is that there is little foliage on either side of the row centre, with the maximum of foliage area occuring between the rows (Figure 23, curve 2).

The highest non-uniformity in the direction y is in an open dense maize stand in Estonia at the early stage of development. The sunflower crop shows the highest uniformity.

The data on the horizontal distribution of phytoarea and phytomass discussed in the previous paragraphs concern either an individual plant or a pure stand arranged on a row or square cluster principle, and indicate that phytoelement density variations are great in an open dense stand. Usually, the central part of a plant is the densiest. Among the species discussed only the apple tree and sunflower are exceptions, the former having its maximum density in the lower part of the crown about 1 m from the centre, and the latter having its maximum density between the rows, which is due to its peculiar habitus and leaf shape.

However, the problem becomes much more complicated in the case of stands with irregular distribution of plants or with a plurality of species.

No systematic quantitative data on such most widely occuring types of natural vegetation are available at present.

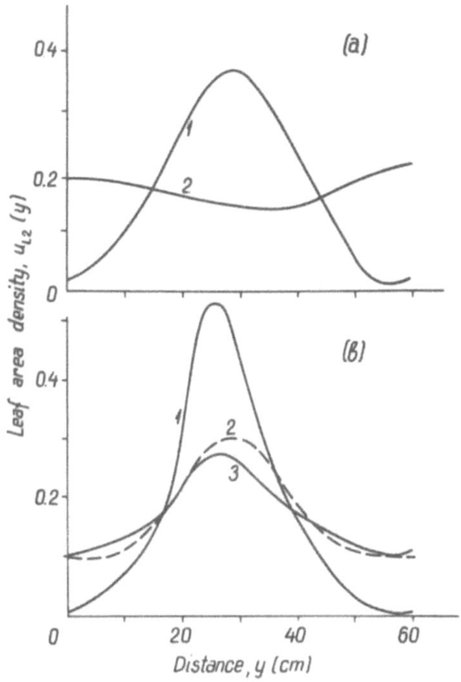

Fig. 23. Relative distribution of leaf area density $u_{2L}(y)$ in the direction y across rows in various plant stands.
(a) – horse bean and sunflower: 1 – horse bean, experiment No. 25, Estonia, July 19, 1967, flowering, 2 – sunflower, experiment No. 16, Tadjikistan, August 17, 1965, ripening; (b) – maize: 1 – experiment No. 24, Estonia, July 20, 1967, emergence of 7th leaf, 2 – experiment No. 14, Tadjikistan, August 7, 1965, milk ripeness, 3 – experiment No. 11, Moldavia, 1964, milk ripeness.

I.6 Foliage area orientation in stands

I.6.1 GENERAL

Literary data on foliage area orientation in stands bear primarily on the measurement of the mean inclination angle ϑ_L, or on the distribution function of leaf inclination $\hat{g}_L(\vartheta_L)$. Few works are available regarding azimuth distribution of foliage.

Of the pioneering attempts in this direction is the work by Nosatovsky (1947), concerned with azimuth dependence of leaf orientation, the range of variation of leaf inclination and illuminance of individual leaf parts in a wheat stand.

Experimental data on the distribution of foliage area according to three groups of inclination angle: 0–30, 30–60 and 60–90° for a number of crops are reported in the paper by Nichiporovich (1961), the first one of the kind that duly raises and discusses the question of stand structure as a peculiar dynamic optical system providing for a maximum absorption of solar radiation and its photosynthetic utilization. The stand optimal structure is given more consideration in further works by the same author (Nichiporovich, 1963, 1966).

The first works by the Australian school of phytometry (Warren Wilson 1959a, b, 1965a) report measurements of the variation of the mean angle ϑ_L with height for alfalfa, perennial ryegrass, white clover, common yarrow, and dandelion using the inclined point quadrat method and the appropriate mathematical approach.

Philip (1965a) perfected the method of mathematical treatment and using the data by Warren Wilson determined the distribution function of leaf inclination $\hat{g}_L(\vartheta_L)$ for alfalfa.

The integral distribution functions of inclination $G_W(\vartheta_L)$ for a number of crops (maize, rye, potato, sugar beet, rape, white clover, perennial ryegrass) have been provided by de Wit (1965) whose work, however, reports only the dates of measurements without mentioning either the phenophase or the time counted from the seedling emergence, which, in turn, makes comparison of these data with other results difficult. From de Wit's integral distribution functions we have calculated the functions $\hat{g}_L(\vartheta_L)$; the curves, however, appear only slightly smoothed, pointing thereby to the insufficient initial quantity of the leaves. Nevertheless, the data by de Wit added significantly to the investigation of leaf orientation in agricultural crops.

A number of experimental studies of orientation for several agricultural crops were performed in our laboratory within 1962–1967, with particular emphasis on space-time dynamics of leaf orientation in maize stands under diverse climatic conditions (Estonia, Moldavia, Tadjikistan). Measurements in Tadjikistan in 1965 included also sunflower, sorghum and cotton stands, in Estonia they included a horse bean stand in 1962, 1967, a barley stand during 1964–1966, a potato stand in 1967 and a stand of hybrid swede in 1966.

In recent years new data have been published by Ito (1969), Scott & Wells (1969), Ito & Udagawa (1971), Lemeur (1973b), Ito et al. (1973), Lang (1973), Shell & Lang (1975, 1976). A more detailed analysis of these papers will follow.

The results thus collected demonstrate that among different species in the plant kingdom there exists a large variety of leaf orientation, on the other hand, there is a space-time variation of orientation within a single species. The changes are due to the following factors:

1. Specific and cultivar differences of crops. As for example, leaves in potato, horse bean and cucumber stands approximate a horizontal position, while in cereal and grass stands they are more or less uniformly oriented. Thus, highly productive rice varieties (Tsunoda 1959) have leaves which are more vertical as compared with those of the cultivars of lower productivity. Specific differences in leaf orientation are the result of plant phylogenetic development, controlled modifications of orientation in the course of selection being closely associated with the stand optimal structure in its photosynthetic activity (Nichiporovich 1963, 1966).

2. Changes of the orientation in ontogeny. The phenomenon is still poorly understood quantitavely. However, measurements by de Wit (1965) in perennial ryegrass stand showed that in young grass vertical leaves are dominant. The inclination decreases with age so that a 50-day old crop has an orientation approximating horizontal; young tobacco leaves (Tanaka 1970) are largely horizontal, the new leaves, however, become progressively vertical, as the plant grows. Ontogenetic changes of orientation are also present in a maize stand, as can be seen further from the data of our laboratory.

3. Adaptive changes of orientation. As one would expect, leaf orientation raises productivity, facilitates reproduction and viability of a species. The vital processes of plant, and thereby, their leaf orientation are readjusted due to changes in the environment. It seems likely that the main environmental factor responsible for adaptive changes in leaf orientation is the stand radiation regime. As can be seen from our measurements in a dense row maize stand after its closure, the inclination angle is augmented and the habitus of individual plants appears elongated perpendicular to the rows. A stand of this kind normally adapts to a more rational utilization of radiation and its more uniform distri-

bution over different layers. The thickness and manner of sowing are, therefore, the factors that allow us to modify orientation within certain limits.

In addition to the above factors determining foliage area orientation during the growing season, short-term diurnal changes must also be considered. Of these are primarily:

1. Heliotropism of plants. According to the data by Polikarpov (1954), Ross & Ross (1969b) and Shell & Lang (1975, 1976), heliotropism is noted in sunflower leaves, particularly in the upper layers; data by Shiman (1967) point to this phenomenon in lettuce, while, conversely, maple leaves tend to become oriented parallel to the sun beams at noon.

2. Effects of the wind, whose action causes the leaves, the stems and the flowers to flutter round their middle position. The wind effects are reduced with depth and are less pronounced in dense stands.

3. Leaf withering in conditions of moisture deficiency. While wilting, the leaves occupy an almost vertical position, their orientation coming back to normal in cases of short disturbances of water supply.

In summary, leaf orientation in a stand is characterized by a definite space-time variability, which presents certain difficulties for the comparison of the data relating to individual species and varieties. The difficulties become all the greater when comparison is made of the data obtained by different authors at different times and localities. The calculations performed in our laboratory to evaluate the accuracy of the determination of the orientation functions g'_L and g''_L indicate that data scattering due to biological variation and to insufficient quantity of initial materials is considerable. Therefore, the conclusions that will be subsequently made, should be treated as merely tentative.

I.6.2 DISTRIBUTION FUNCTIONS OF LEAF INCLINATION AND AZIMUTH ORIENTATION

We have attempted to collect and bring into a pattern all the available information, both from the literature and our own, on leaf orientation in different stands. To achieve that goal, all the data were represented as the function $\hat{g}_L(z, \vartheta)$ with the normalizing condition (I.2.16a). For their approximate representation and comparison we calculated also part of foliage area $\hat{g}_{\mathrm{I}}$, $\hat{g}_{\mathrm{II}}$ and $\hat{g}_{\mathrm{III}}$ with inclination angle within the ranges 0–30, 30–60 and 60–90°, respectively. The data are summarized in Figures 25–31 and in Tables I.6.1, I.6.2.

De Wit (1965) suggested four classes of orientation according to leaf inclination:

(1) planophile orientation – leaf arrangement is largely horizontal;

(2) erectophile orientation – leaf arrangement is predominantly vertical;

(3) plagiophile orientation – inclined leaf arrangement with 45° inclination, predominantly;

Table I.6.1 *Distribution of foliage area inclination in maize stands*

Date	Region and experiment No.	Cultivar	Phenological stage	Layer	Density (plants/m^2)	Height (cm)	L_0	Inclination, deg. 0–30 $\hat{g}_{I}$	30–60 $\hat{g}_{II}$	60–90 $\hat{g}_{III}$	χ_L	Authors
Sept. 2–8, 1962	Estonia, No. 3	Bukovinsky 3	Tasseling	20–40	16.7	200	8.0	0.15	0.57	0.28	0.21	Ross & Nilson (1967)
				40–60				0.13	0.43	0.44	0.06	
				60–80				0.13	0.35	0.52	−0.02	
				80–100				0.12	0.30	0.58	−0.08	
				100–120				0.11	0.31	0.58	−0.08	
				120–140				0.11	0.30	0.50	−0.09	
				140–160				0.10	0.29	0.61	−0.11	
				160–180				0.11	0.33	0.56	−0.06	
				Entire crop				0.13	0.34	0.53	−0.03	
June 19, 1963	Estonia, No. 4	Bukovinsky 3	4th leaf	Entire crop	30.5	15	0.1	0.16	0.55	0.29	0.21	Ross & Nilson (1967)
July 4			5th leaf	Entire crop	30.5	40	0.3	0.17	0.56	0.27	0.23	Ross & Vlasova (1967)
July 12			7th leaf	Entire crop	30.5	40	0.5	0.18	0.52	0.30	0.20	
July 25			9th leaf			85	1.5	0.12	0.31	0.57	−0.07	
Aug. 2–3			10th leaf			125	2.8	0.11	0.30	0.59	−0.09	
Aug. 13			12th leaf			175	5.6	0.14	0.36	0.50	0.01	
July 24, 1963	Estonia, No. 5	Bukovinsky 3	9th leaf	Entire crop	31.7	25	1.4	0.10	0.37	0.53	−0.03	Ross & Vlasova (1967)

Table I.6.1 (*Continued*)

Date	Region and experiment No.	Cultivar	Phenological stage	Layer	Density (plants/m^2)	Height (cm)	L_0	Inclination, deg. 0–30 $\hat{g}_{I}$	30–60 $\hat{g}_{II}$	60–90 $\hat{g}_{III}$	χ_L	Authors
Aug. 6			11th leaf			40	3.8	0.12	0.30	0.58	−0.08	
Aug. 16			12th leaf			195	7.8	0.10	0.27	0.63	−0.13	
Aug. 22, 1963	Estonia, No. 6	Bukovinsky 3	Tasseling		27.4		5.5	0.14	0.36	0.50	0.01	Ross & Vlasova (1967)
July 21, 1964	Moldavia, Tiraspol	Bukovinsky 3	Tasseling	Upper, above 160	5.4		3.3	0.18	0.43	0.39	0.11	Ross & Ross (1969b)
				Middle, 80–160				0.26	0.41	0.35	0.15	
				Lower, 0–80				0.16	0.47	0.36	0.14	
				Entire crop				0.21	0.42	0.37	0.13	
Aug. 6, 1964	Moldavia, Chekolteny, No. 11	VIR-42	Milk ripeness	Upper, above 120	3.9	190	1.7	0.26	0.42	0.32	0.18	Ross & Ross (1969b)
				Middle, 60–120				0.20	0.45	0.35	0.15	
				Lower, 0–60				0.13	0.36	0.51	−0.01	
				Entire crop				0.20	0.43	0.37	0.14	
June 8–9, 1965	Tadjikistan, No. 14	VIR-338	9th leaf	Entire crop	4.5	50	0.3	0.16	0.50	0.34	0.16	Ross & Ross (1969b)
June 22, 1966	Estonia, No. 19	Bukovinsky 3	5th leaf	Entire crop	27.8	15	0.2	0.12	0.43	0.45	0.06	Ross & Ross (1969b)

Aug. 22, 1967	Estonia, No. 25	Bukovinsky 3	Tassel flowering	Entire crop	21.9	130		0.12	0.41	0.47	0.04	Ross & Ross (1969b)
	Moscow region						4.1	0.166	0.37	0.47	0.03	Nichipo-rovich (1961)
June 18	Holland		9th leaf (?)	Entire crop				0.32	0.44	0.24	0.26	de Wit (1965)
July 17	Holland		Tasseling (?)	Entire crop				0.30	0.47	0.23	0.27	de Wit (1965)
Oct. 1	Holland		Wax ripeness?	Entire crop				0.24	0.45	0.31	0.18	de Wit (1965)
Aug. 17, 1964	USA, Davis		Tassel flowering	Entire crop	4.9	320	5.1	0.59	0.33	0.08	0.46	Loomis & Williams (1969)

Table I.6.2 *Distribution of foliage area inclination in stands of different species*

Cultivar	Region	Date and time	Phenological stage	Density (plants/m^2)	L_0	Inclination deg. 0–30 $\hat{g}_{I}$	30–60 $\hat{g}_{II}$	60–90 $\hat{g}_{III}$	χ_L	Author
Strawberry	Moscow region		Beginning of fruiting		2.7	0.66	0.19	0.15	0.54	Nichiporovich (1961)
White clover	Moscow region		?			0.70	0.20	0.10	0.57	Nichiporovich (1961)
	Holland	Aug. 21	Flowering			0.54	0.41	0.05	0.45	de Wit (1965)
Sunflower	Tadjikistan	Morning July 28, 1965	Ripening	4.1	2.7	0.46	0.46	0.08	0.42	Ross & Ross (1969b)
		Evening June 24, 1965	Formation of efflorescence	4.1	1.2	0.27	0.53	0.20	0.30	Ross & Ross (1969b)
		Morning June 25, 1965	Formation of efflorescence	4.1	1.2	0.40	0.46	0.14	0.36	Ross & Ross (1969b)
	Japan	July 17		2.8	1.3	0.55	0.37	0.08	0.42	Horie & Udagawa (1970)
Horse beans	Estonia	Aug. 21, 1963	Flowering	12.8	2.5	0.42	0.47	0.11	0.39	Ross & Nilson (1966b)
	Estonia	Aug. 1, 1967	Flowering	24.5	1.5	0.30	0.50	0.20	0.30	Ross & Ross (1969b)
Potato	Moscow region		Aged 3 weeks		0.5	0.40	0.37	0.23	0.27	Nichiporovich (1961)
	Holland	May 28	?			0.57	0.38	0.05	0.45	de Wit (1965)
	Holland	June 17,	Flowering			0.41	0.52	0.07	0.43	de Wit (1965)
	Estonia	Aug. 2, 1967	Flowering	5.5	1.7	0.38	0.44	0.18	0.32	Ross & Ross (1969b)

Tobacco	Japan, Yonago	April 13	At planting	?	?	0.15	0.70	0.15	0.35	Tanaka (1969)
			20 days after planting			0.40	0.45	0.15	0.35	Tanaka (1969)
			40 days after planting			0	0.50	0.50	0.13	Tanaka (1969)
			60 days after planting			0.20	0.70	0.10	0.40	Tanaka (1969)
Cucumber	Moscow region		Flowering		3.2	0.52	0.37	0.11	0.39	Nichiporovich (1961)
Rape	Holland	April 20	?			0.36	0.50	0.14	0.36	de Wit (1965)
	Holland	April 27	?			0.28	0.55	0.17	0.33	de Wit (1965)
	Holland	May 12	Flowering			0.43	0.45	0.12	0.38	de Wit (1965)
Sugar beet	Holland	July 2 and 13	?			0.22	0.47	0.31	0.19	de Wit (1965)
	Holland	July 28	?			0.38	0.42	0.20	0.30	de Wit (1965)
	Holland	Oct. 1	?			0.30	0.52	0.18	0.32	de Wit (1965)
Hybrid swede	Estonia	June 30, 1966	Hypocotil thickening	2.8	0.6	0.26	0.54	0.20	0.30	Ross & Ross (1969b)
	Estonia	Aug. 9	Crop closure	2.8	4.4	0.20	0.43	0.37	0.13	Ross & Ross (1969b)
	Estonia	Aug. 22, 1967	Crop closure	6.0	?	0.29	0.43	0.28	0.22	Ross & Ross (1969b)
Lucerne	Australia	Dec. 7–12, 1964	50–60 cm high			0.18	0.54	0.28	0.22	Warren Wilson (1965a) Philip (1965a)
Sorghum	Tadjikistan	Aug. 5–6, 1965	Tasseling		4.4	0.13	0.55	0.32	0.18	Ross & Ross (1969b)
Timothy	Moscow region					0.10	0.43	0.47	0.06	Nichiporovich (1961)
Barley	Estonia	1964–1966	Shooting – milk ripeness			0.20	0.39	0.41	0.09	Mägi & Ross (1969a)
Wheat	Moscow region					0.13	0.35	0.52	−0.02	Nichiporovich (1961)
Rye	Holland	April 27				0.13	0.38	0.49	0.01	de Wit (1965)
		May 12	Shooting			0.11	0.28	0.61	−0.11	de Wit (1965)
		May 29	Earing			0.20	0.32	0.48	0.12	de Wit (1965)

Table I.6.2 (*Continued*)

Cultivar	Region	Date and time	Phenological stage	Density (plants/m^2)	L_0	Inclination deg.			χ_L	Author
						0–30 $\hat{g}_I$	30–60 $\hat{g}_{II}$	60–90 $\hat{g}_{III}$		
Perennial rye grass	Holland 1	May 6	Aged 30 days			0.06	0.17	0.77	−0.27	de Wit (1965)
	2	June 10	Aged 10 days			0.04	0.27	0.60	−0.19	de Wit (1965)
	3	June 10	Aged 30 days			0.16	0.49	0.35	0.15	de Wit (1965)
	4	June 10	Aged 50 days			0.44	0.45	0.11	0.45	de Wit (1965)
	5	Aug. 21	Aged 10 days			0.10	0.60	0.30	0.23	de Wit (1965)
	6	Aug. 21	Aged 20 days			0.19	0.55	0.26	0.24	de Wit (1965)
	7	Aug. 21	Aged 40 days			0.42	0.48	0.10	0.40	de Wit (1965)
	8	Aug. 21	Aged 60 days			0.49	0.39	0.12	0.38	de Wit (1965)
				Model curves						
Uniform orientation						0.13	0.37	0.50	0	
Horizontal leaves						1	0	0	1	
Vertical leaves						0	0	1	−1	
Planophile orientation						0.72	0.21	0.07	0.59	de Wit (1965)
Erectophile orientation						0.07	0.21	0.72	−0.22	de Wit (1965)
Plagiophile orientation						0.09	0.82	0.09	0.45	de Wit (1965)
Extremophile orientation						0.47	0.06	0.47	0.34	de Wit (1965)

(4) extremophile orientation – double peak curve of the distribution of inclination angles with maxima at 0 and 90°. For these orientation classes Lemeur (1973a) proposed the following empirical formulae

$$\hat{g}(\vartheta_L)=\frac{3\vartheta_L^2}{(\pi/2)^3}-\frac{6\vartheta_L}{(\pi/2)^2}+\frac{6}{\pi} \quad \text{for planophile canopy}$$

$$\hat{g}(\vartheta_L)=\frac{3\vartheta_L^2}{(\pi/2)^2} \quad \text{for erectophile canopy}$$

$$\hat{g}(\vartheta_L)=\frac{12\vartheta_L^2}{(\pi/2)^3}-\frac{12\vartheta_L}{(\pi/2)^2}+\frac{6}{\pi} \quad \text{for extremophile canopy}$$

The corresponding integral curves of distribution $G(\vartheta_L)$ were recalculated for differential curves $\hat{g}_L(\vartheta_L)$ (Figure 24). Comparison of these theoretical curves with the measured curves shows that the classification by de Wit is rather schematic, the fourth type, i.e. extremophile orientation, does not actually occur in nature.

Of all the theoretical distributions of leaf normals uniform distribution seems to be the simplest case, in which $g_L(z, r_L)\equiv 1$ and $\hat{g}_L(z, \vartheta_L)=\sin\vartheta_L$. Distribution of leaf normals is close to uniform in many real stands, especially grasses and herbs. It therefore, seems of interest, from the standpoint of the classification of the curves, to elucidate to what extent the leaf normals in real stands deviate from uniform distribution.

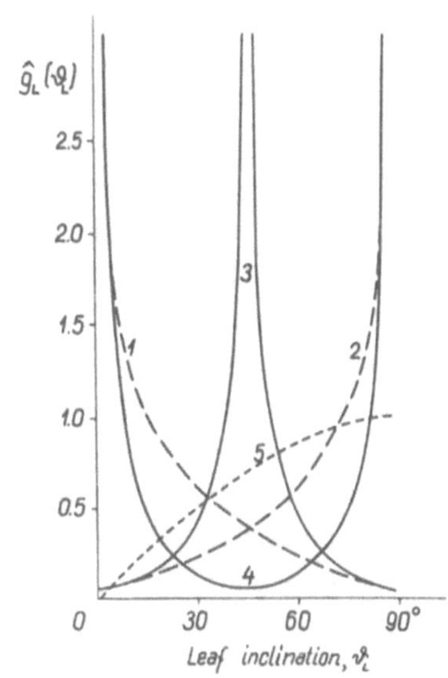

Fig. 24. Model distributions of foliage area inclination $\hat{g}_L(\vartheta_L)$ for different leaf orientations (from de Wit 1965):
1 – planophile, 2 – erectophile, 3 – plagiophile, 4 – extremophile, 5 – uniform.

Let us characterize this deviation using the new notion that we have introduced, i.e. *inclination index of foliage area* χ_L which is determined by the formula

$$\chi_L = \pm\tfrac{1}{2}\int_0^{\pi 2} |\sin \vartheta_L - \hat{g}_L(\vartheta_L)| \mathrm{d}\vartheta_L \tag{I.6.1}$$

or, approximately

$$\chi_L \approx \pm\tfrac{1}{2}[|\hat{g}_{\mathrm{I}P} - \hat{g}_{\mathrm{I}}| + |\hat{g}_{\mathrm{II}P} - \hat{g}_{\mathrm{II}}| + |\hat{g}_{\mathrm{III}P} - \hat{g}_{\mathrm{III}}|], \tag{I.6.1a}$$

where $\hat{g}_{\mathrm{I}P} = 0.13$, $\hat{g}_{\mathrm{II}P} = 0.37$, $\hat{g}_{\mathrm{III}P} = 0.50$ are parts of the leaf area with inclinations falling within angle classes 0–30, 30–60 and 60–90°, respectively, in the case of uniform distribution.

In terms of geometry 'inclination index of foliage area' roughly determines half the foliage area between the curves $\hat{g}_L(\vartheta_L)$ and $\hat{g}_{LP}(\vartheta_L) = \sin \vartheta_L$.

The sign '+' or '−' in in (I.6.1a) is dependent on that of the last difference $\hat{g}_{\mathrm{III}P} - \hat{g}_{\mathrm{III}}$ '+' showing deviation of inclination distribution $\hat{g}_L(\vartheta_L)$ from uniform towards horizontal leaves, '−' showing deviation towards vertical leaves. Greater deviations of actual distribution from the uniform distribution are found at greater values of $\chi_L . \chi_L = 0$ implies that the uniform distribution of leaf orientation is actually the case. The values of χ_L are presented in Tables I.6.1 and I.6.2.

With the data supplied by Figures 25–31 and Tables I.6.1, I.6.2 let us deal more closely with leaf orientation in different cultures.

Maize. Leaf orientation has been best studied in stands of maize. Besides the data collected by our laboratory during expeditions to Estonia, Moldavia and Tadjikistan, there are also the findings of de Wit (1965), Nichiporovich (1961), Loomis et al. (1968) and Udagawa et al. (1968) all of which are classified and presented in Table I.6.1 and Figures 25 to 30. Leaf orientation in dense maize stands grown for silage in Estonia will be dealt with first.

During 1962 and 1963 detailed measurements of leaf orientation in individual layers were carried out. Figure 25 shows clearly how $\hat{g}_L$ changes in different layers. The peak of $\hat{g}_L$ is shifted from $\vartheta_L = 50°$ in the lowermost layer to 90° in the middle layers and back from 90 to 75° in the top layers.

As can be seen from Figure 26, leaf area with inclination within 0–15° accounts for an average of 10% showing little variation in height. The lower and uppermost layers are dominated by inclined leaves with $\vartheta_L = 30-60°$, while in the middle layers vertical leaves with $\vartheta_L = 60-90°$ are predominant. Table I.6.1 (Sept. 2–8, 1962) shows that as the curves $\hat{g}_L$ change with height, so does the inclination index χ_L and, as a result, leaf orientation in all but the lowermust layer tend from uniform to vertical, although this tendency is not pronounced. It can, therefore, be concluded that the distribution of leaf area inclination in a dense stand of maize during the emergence of the 12th leaf and tasseling tends to approximate uniform orientation. A question arises whether this observed characteristic persists through the day or the growing season.

To find out whether diurnal variation of leaf orientation (heliotropism) occur

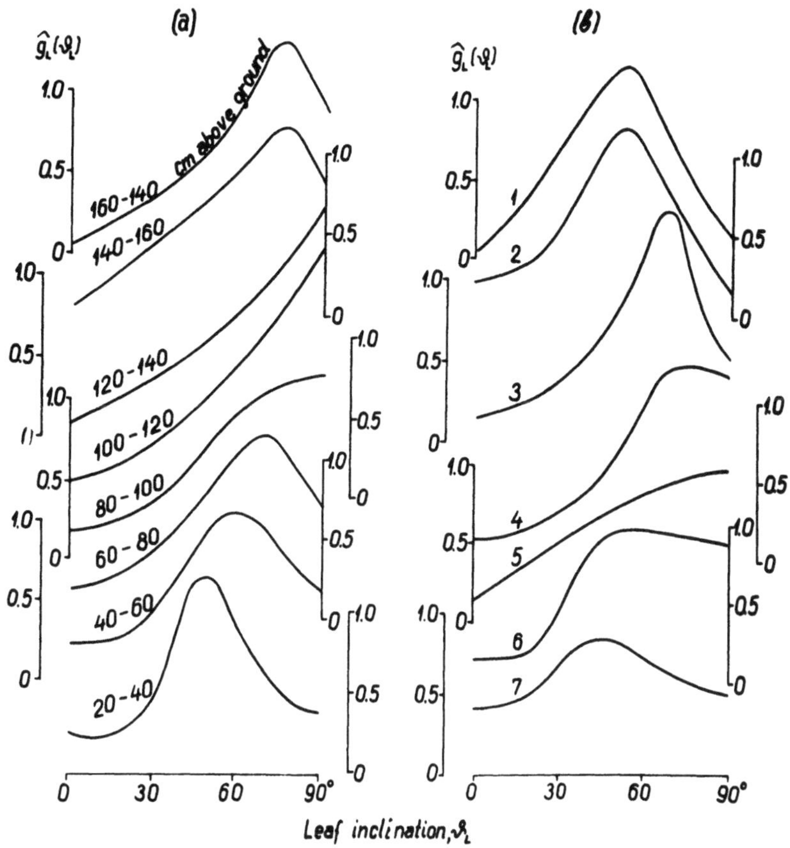

Fig. 25. Distribution function of foliage area inclination $\hat{g}_L(\vartheta_L)$ in a dense maize row crop. (a) – the canopy is divided into 20-cm thick horizontal layers. The stage of tasseling, experiment No. 3, Estonia, September 2–8, 1962. (b) – the upper 50–60-cm layer at various times during growing season. Experiment No. 4, Estonia, 1963. 1 – June 19, emergence of 4th leaf, 2 – July 4, emergence of 5th leaf, 3 – July 12, emergence of 7th leaf, 4 – July 25, emergence of 9th leaf, 5 – August 2–3, emergence of 10th leaf, 6 – August 13, emergence of 12th leaf, 7 – sparse square sown crop at milk ripeness. Moldavia, experiment No. 11, August 6, 1964.

in a maize stand, V. Ross undertook special measurements in Estonia, in 1966, which demonstrated that within the accuracy of measurements (the error being $\pm 20\%$) heliotropism is not observed in maize leaves.

This is not the case with leaf orientation during the growing season. A number of measurements carried out in Estonia in 1963 (Ross & Vlasova 1967; Ross & Nilson 1967a, b) allowed the dynamics of the variation of $\hat{g}_L(\vartheta_L)$ to be determined from the time of seedling emergence to the time of tasseling. The results obtained (Figure 27 and Table I.6.1) show few, if any, changes in leaf orientation in the early stage of development prior to the emergence of the 7th leaf, inclination of most leaves being $30-60°$, horizontal leaves (Figure 27, curve 1, $\Delta\vartheta_L = 0-15°$) accounting for $3-7\%$, vertical leaves (curve 6, Figure 27, $\Delta\vartheta_L = 75-90°$) – $8-10\%$. The stand, however, shows marked changes in leaf orientation starting with the

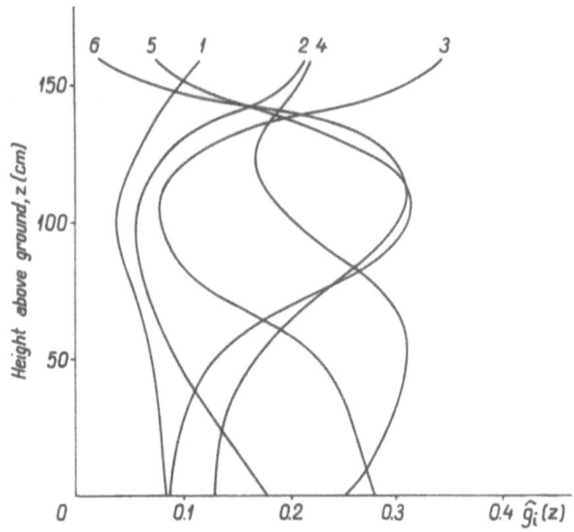

Fig. 26. Vertical distribution of parts of foliage area with different inclination ϑ_L. Maize crop, experiment No. 4, Estonia, August 13, 1963. The function $\hat{g}_i$ corresponds to the function averaged over the interval of inclination angle ϑ_L so that $i=1$ corresponds to interval 0–15°, $i=2$ to 15–30°, $i=3$ to 30–45°, $i=4$ to 45–60°, $i=5$ to 60–75°, $i=6$ to 75–90°.

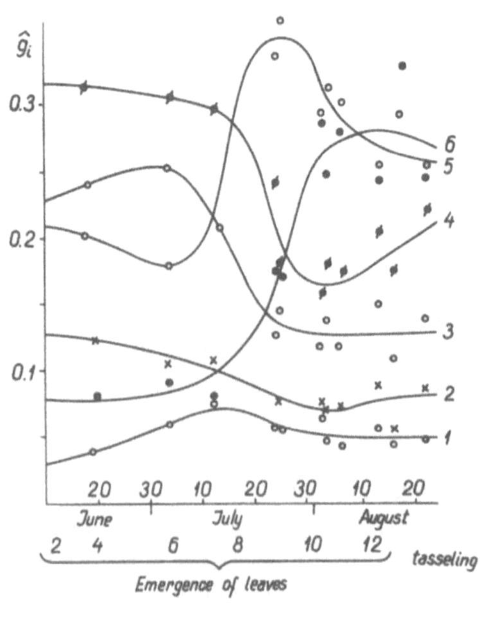

Fig. 27. Seasonal change of foliage area orientation in a maize crop. Experiment No. 4, Estonia, 1963. See the caption to Figure 26 for further explanation.

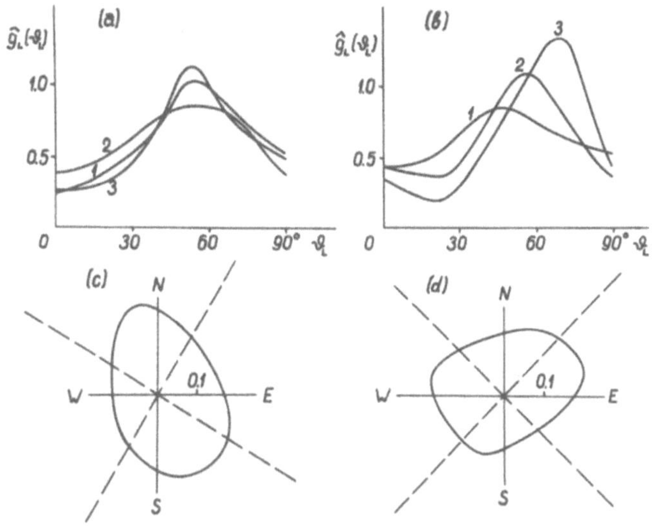

Fig. 28. Distribution functions of foliage area inclination (a), (b) and azimuth (c), (d) in a maize crop. (a) – tasseling, experiment No. 10, Moldavia, July 20–23, 1964: 1 – upper layer, 160–250 cm, 2 – middle layer, 80–160 cm, 3 – lower layer, 0–80 cm; (b) – milk ripeness, experiment No. 11, Moldavia, 1964: 1 – upper layer, 120–200 cm, 2 – middle layer, 60–120 cm, 3 – lower layer, 0–60 cm; (c) – experiment No. 10, Moldavia, July 20–23, 1964, experiment No. 11, Moldavia, August 6, 1964. Dash lines show row directions.

emergence of the 7th leaf. The proportion of vertical leaves increases sharply attaining 29%; leaves with inclination 60–75° (curve 5) increase in number from 20 to 35% during the period between the 7th and 9th leaves, then, towards the onset of tasseling their proportion is reduced to 26%. The proportions of leaves with inclinations 30–45° (curve 3) and 45–60° (curve 4) decrease also. Over the given period the number of horizontal leaves varies but slightly, while the proportion of leaves with inclination 15–30° (curve 2) gradually decreases before the emergence of the 10th leaf, after which time it slowly rises again. The changes start at $L_0 = 1.0 - 1.5$ and coincide with the beginning of stand closure.

Since attenuation of radiation within a stand occurs basically in the top layer, it seemed of interest to elucidate to what extent leaf orientation varies within that particular layer. To answer the question, we estimated the dynamics of the change of $\hat{g}_L(\vartheta_L)$ in the uppermost layer 50–60 cm thick (experiment No. 4, Estonia). The results presented in Figure 25B lead us to conclude that in the uppermost layer of a dense maize stand leaf orientation has a general pattern, following basically that of the entire stand. Curve 7 in Figure 25B corresponds to the milk ripeness stage of a sparse stand (experiment No. 11, Moldavia). One can expect $\hat{g}_L(\vartheta_L)$ to be the same for a dense stand.

Within the accuracy of measurements ($\pm$10–20%) we observed no disagreement between the distribution of leaf inclination $\hat{g}_L(\vartheta_L)$ for row and for square sown crops.

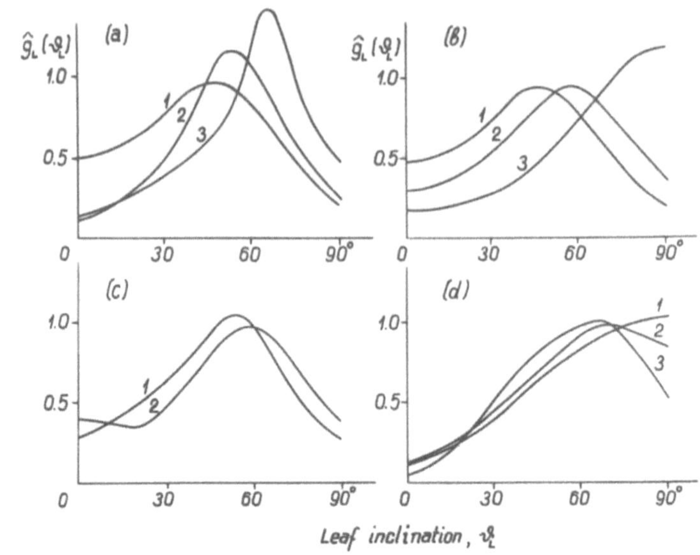

Fig. 29. Distribution function $\hat{g}_L(\vartheta_L)$ of foliage area inclination in a maize crop.
(a) – stage of 9th leaf emergence: 1 – Holland, June 18 (after de Wit, 1965), 2 – Tadjikistan, June 8–9, 1965, experiment No. 14, 3 – Estonia, July 25, 1963, experiment No. 4; (b) – tasseling: 1 – Holland, July 17 (after de Wit, 1965), 2 – Moldavia, July 21, 1964, experiment No. 10, 3 – Estonia, August 16, 1963, experiment No. 5; (c) – ripening: 1 – Moldavia, August 6, 1964, experiment No. 11, stage – milk ripeness, 2 – Holland, October 1, complete ripeness (after de Wit, 1965); (d) – tasseling: 1 – Estonia, September 2–8, 1962, experiment No. 3, 2 – Estonia, August 22, 1963, experiment No. 6, 3 – Estonia, August 22, 1967, experiment No. 24.

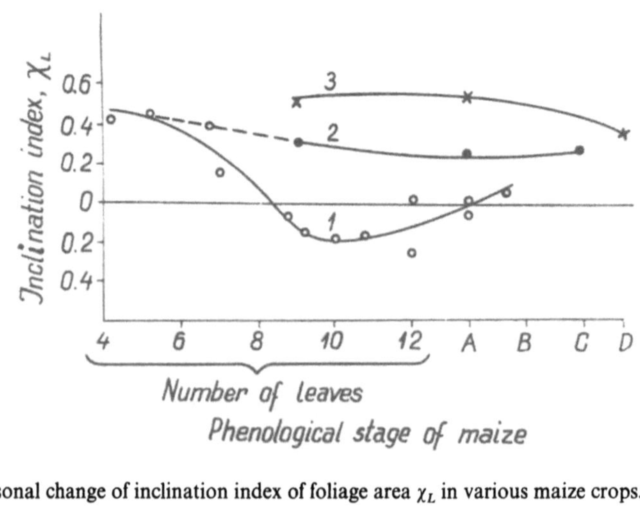

Fig. 30. Seasonal change of inclination index of foliage area χ_L in various maize crops.
1 – dense crop, 2 – sparse crop, 3 – data of de Wit (1965). A – tasseling; B – tassel flowering, C – milk ripeness; D – wax ripeness.

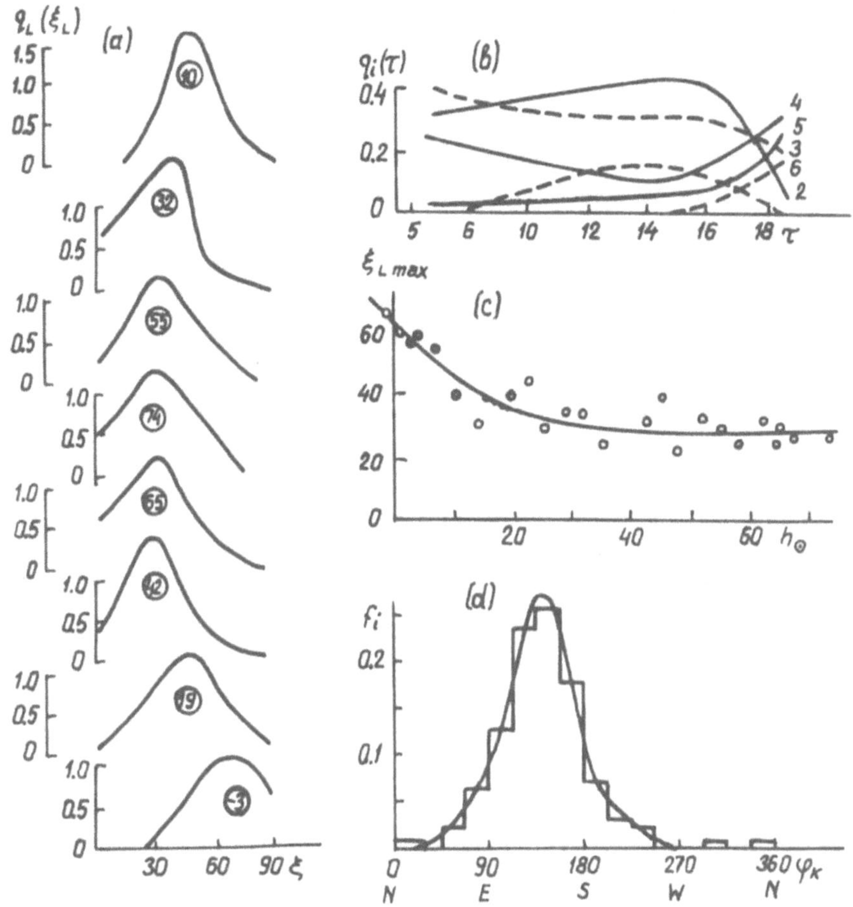

Fig. 31. Dynamics of $q_L(\xi_L)$ function in sunflower during the day (a), diurnal course of part of foliage area $q_i(\tau)$ oriented at different angles relative to sunrays (b), dependence of angle ξ_{Lmax} on solar elevation (c), preferred azimuth of sunflower seed heads (data of Morozov (1963), (d)).
1 – 0–15°, 2 – 15–30°, 3 – 30–45°, 4 – 45–60°, 5 – 60–75°, 6 – 75–90°, encircled numbers indicate solar elevation $h_\odot$ (degrees).

According to the authors themselves, data of Udagawa et al. (1968) concerning $\hat{g}_L(\vartheta_L)$ are in good agreement with ours. The situation is somewhat different with azimuthal dependence of orientation $(1/2\pi)g''_L(\varphi_L)$. Detailed studies were performed in 1963 (experiments Nos. 4 and 5, Estonia) of azimuthal dependence of leaf orientation in a dense row and square sown crops. The results obtained from these studies allow for the following conclusions.

1. Azimuthal dependence of orientation does not actually occur within the period from seedling emergence up to the time of stand closure (with the error in measurements being ± 10–20% and resulting primarily from the biological spread), i.e. $(1/2\pi)g''_L(\varphi_L) \approx \text{const}$.

2. In a row crop after closure there appears elongation of $(1/2\pi)g''_L(\varphi_L)$ in a direction perpendicular to the rows. The higher the leaf area density in a stand layer, the greater the elongation, i.e. the vertical distribution of foliage area determines also the extent of elongation of $(1/2\pi)g''_L(\vartheta_L)$. Since the maximum of $u_L(z)$ occurs in the middle layers, the elongation of $(1/2\pi)g''_L$ is to therefore maximum herein. This is presumably due to the stand 'tendency' to fill with leaves the empty space between the rows to better utilize the radiation. Moreover, there is a certain additional elongation of $(1/2\pi)g''_L$ in southern and south-eastern directions (Ross & Nilson 1967b), that can be observed in some cases, particularly in the stand top layers. The phenomenon again is obviously due to direct solar radiation, since duration of sunlight is less after midday than before it because of cloudiness.

3. In dense square sown crops the elongation is very slight in directions, perpendicular to the rows; however, in the southern and south-western directions it is manifested as much as in a row crop.

A different pattern of orientation dynamics is observed in thin maize stands.

Measurements of our laboratory carried out in Moldavia, Tadjikistan and Estonia, and also the data of de Wit (1965) demonstrate (Figure 29B, C) that during the growing season leaf inclination shows but a slight variation, the maximum of $\hat{g}_L(\vartheta_L)$ being within the range 45–60° and not shifting closely to $\vartheta_L = 90°$, as in the case of a dense stand. Also different is the dependence on height of the distribution of inclination angles.

The distribution functions of leaf inclination angles $\hat{g}_L$ in the top, middle and lower layers of a stand in Moldavia (Ross & Ross 1969b) are presented in Figure 28 (see also Table I.6.1). During tasseling the position of the maximum of $\hat{g}_L$, observed at $\vartheta_L = 50$–$60°$ is the same for all the layers. Individual layers differ only in the elongation of the curve $\hat{g}_L$ in the region of the maximum, the elongation being greatest in the lowermost layer, where also the amount of horizontal leaves is smallest. The course of $\hat{g}_L$ is smoothest in the middle layer, which, however, has the smallest amount of vertical leaves.

The pattern changes by the onset of milk ripeness. In the lower layer the maximum of $\hat{g}_L$ is shifted to angles as large as nearing 65°, and conversely, in the upper layer the maximum of $\hat{g}_L$ shifts to smaller angles, up to 45°, here also the course of being smoothest.

As can be seen from the comparison of curves 2 in Figures 29B and 29C, the distribution of inclination angles $\hat{g}_L$ for the entire stand (in Moldavia) is practically the same for both phases.

A regular pattern of variation of $\hat{g}_L(\vartheta_L)$ with the phases of maize growth can be noted in the upper layer of a sparse stand. Inclined leaf orientation with a corresponding $\hat{g}_L(\vartheta_L)$, whose maximum lies about $\vartheta_L = 55°$ is predominant during leaf emergence. Leaf inclination increases by the time of tasseling and tassel flowering. The maximum of $\hat{g}_L$ at $\vartheta_L = 55°$ shows little, if any, variation, increased, however, is the proportion of vertical and horizontal foliage areas. Towards maturity the leaves in the top layer become even more inclined. The maximum of

$\hat{g}_L$ shifts accordingly from $\vartheta_L = 55°$ to $\vartheta_L = 40-45°$ and there is an increase in the proportion of horizontal area, while the amount of vertical area remains unchanged.

Within the accuracy of measurements azimuthal dependence of orientation $(1/2\pi)g''_L(\varphi_L)$ in a sparse stand does not vary in height. As the stand grows $(1/2\pi)g''_L(\varphi_L)$ shows a slight elongation southward and in a direction perpendicular to the rows. On the whole, the azimuthal dependence of orientation in a sparse stand is considerably less pronounced than that in a dense stand.

Results somewhat different from ours are reported in the work by Japanese scientists Udagawa et al. (1968). These authors have also noted a certain south- and northward elongation $(1/2\pi)g''_L$, however, they have also observed an elongation $(1/2\pi)g''_L$ in the direction of the rows at the initial stage of the stand development ($L_0 = 0.3$–1.5), which is in disagreement with our data.

Variation in leaf orientation during the growing season also affects the inclination index of foliage area χ_L. Figure 30 shows a fundamental difference between sparse (curve 2) and dense (curve 1) sowings starting from the emergence of the 6–7th leaf. Curve 3 represents the data of de Wit (1965). There is a steady tendency from uniform towards horizontal orientation in a sparse stand, while in a dense stand the situation is more complex. From the time of sprouting to the emergence of the 8–9th leaf, orientation tends from uniform to horizontal; from the emergence of the 8–9th leaf up to tasseling it tends to vertical, then at later stages it turns back to horizontal. The reasons for the high values of χ_L in a stand studied by de Wit remain obscure, since adequate phenological characteristic is lacking.

It is also evident from Figure 30 that the inclination index of foliage area χ_L is at the same time a simple comparatively efficient characteristic in describing the geometry of a maize stand as regards phenophase, density and cultivar differences.

The results of stand structure studies in different countries are compared in Figures 29C, B, C. Figure 29C corresponds to the period of emergence of the 9th leaf. Unfortunately, de Wit does not state the phase; though, according to the date, his curve (June 18) roughly corresponds to the period in question. Figure 29B is a plot of function $\hat{g}_L$ during maize tasseling in Estonia, Holland and Moldavia. As can be judged from the Figure, the maize stand in Estonia appears to have the most marked vertical orientation, whereas the stand in Holland displays the most distinct horizontal position of leaves. The stands in Moldavia and Tadjikistan occupy an intermediate position. Comparison of Figures 29A and 29B indicates that within the period from the 9th leaf up to tasseling the amount of inclined leaves in Estonia declines, the amount of vertical leaves being increased accordingly. In Holland, however, the orientation within the same period does not change. The rate of $\hat{g}_L$ variation is most marked in the maize of Estonia. As follows from the records of the experiments, the density of the stand in Estonia is 5–6 times that of the stands in Moldavia and Tadjikistan. There is no indication as to the density of de Wit stand, though from Figure 30 it seems to be sparse.

A comprehensive study has been performed by the American scientists (Loomis & Williams 1969; Loomis et al. 1968) on the geometry of maize stand and its dynamics during the growing season involving seven kinds of density. The experiments were carried out in Davis (California) on the hybrid 'de Kalb 805', having a great height and a considerable foliage area. The stand density varied from 1.75 to 12.5 plants/m^2. The maximum values of L_0 varied from 3.5 to 8.5, respectively. The vertical distribution of foliage area was determined in layers 40 cm thick the leaf area inclination angle being, determined, though, only for three classes $\hat{g}_I$, $\hat{g}_{II}$ and $\hat{g}_{III}$. As compared with the above varieties, the American hybrid shows more horizontal leaves, with the inclination index χ_L varying between 0.3 and 0.4. It is therefore, more like crops with horizontal leaves, such as sunflower, horse beans and white clover. With the increase of the crop density L_0, the amount of vertical leaves grows, especially in the stand top layers; accordingly, the foliage area in the lower layers is being decreased.

It would not be easy to explain what the differences in the curves $\hat{g}_L$ in various countries are due to. Three factors affecting orientation can, however, be mentioned. These are: density of the stand, cultivars and weather conditions.

Three curves $\hat{g}_L(\vartheta_L)$ are presented in Figure 29D to estimate effects of weather conditions on leaf orientation. These refer to one and the same cultivar (Bukovinsky 3) during tasseling or tassel flowering for different years in Estonia. Curve 1 corresponds to the coldest and most rainy summer. There was a dry period at the end of June, 1963 (curve 2) which was then followed by a change, favourable for maize growth. The summer of 1967 (curve 3) was dry, warm and sunny. There was moisture deficiency in July which affected the stand adversely, so that towards late August it attained the height of no more than 130 cm. It seems likely that the diminished proportion of vertically oriented foliage area in 1963 and particularly, in 1967 is an obvious consequence of drought.

Unfortunately, cultivar differences cannot be established from foliage area orientation in maize stands since reliable quantitative data are lacking, although the discrepancy between our data and those of the American workers is presumably due to cultivar differences.

In summary, leaf area orientation for the majority of maize varieties is comparatively close to uniform, tending, on the whole, towards horizontal. During the growing season the distribution of leaf inclination $\hat{g}_L(\vartheta_L)$ shows specific dynamics, varying, though, for the sparse and dense stands. In a sparse stand, as the plant grows, its leaves naturally deviate from the stem, their upper portions tending downwards. As a result, the maximum of $\hat{g}_L(\vartheta_L)$ during leaf emergence decreases, shifting from $\vartheta_L = 50–60°$ to $\vartheta_L = 40–50°$. During maturation the horizontal and vertical areas increase simultaneously. In a dense stand the amount of vertical leaves increases as a result of closure, the maximum of $\hat{g}_L(\vartheta_L)$ shifting from $\vartheta_L = 50–60°$ at the appearance of the 7th leaf to $\vartheta_L = 90°$ during the emergence of the 10–11th leaf, then shifting back to smaller values of ϑ_L.. From the 8th leaf emergence up to tasseling leaf area deviates from uniform to vertical orientation.

A dense closed row stand shows also a marked elongation of the foliage area azimuth distribution $(1/2\pi)g_L''(\varphi_L)$ in a direction perpendicular to the rows, while in other cases the elongation is hardly noticeable, the leaf orientation azimuthal dependence, therefore, can be neglected.

Sorghum. In appearance and growth habit sorghum is rather similar to maize. Its orientation parameters measured in Tadjikistan (Ross & Ross 1969b) during tasseling are presented in Table I.6.2. Comparison of the leaf area inclination distribution curves for sorghum and maize shows that their maxima at $\vartheta_L = 55°$ coincide, although curve $\hat{g}_L$ for sorghum is more elongated, i.e. leaf area inclination in sorghum is more marked.

Sunflower. Of all the agricultural crops so far examined by us sunflower is one with a clearly pronounced heliotropism. There is a general opinion that heliotropism in sunflower is manifested in turning the flowering head following the motion of the Sun. More recent works (Pleshko 1948; Polykarpov 1954; Morozov 1963), however, disprove this opinion. Data of Morozov (1963) indicate that since the onset of flowering the head azimuth does not actually change diurnally. According to the same data, the heads show a drastic azimuth southeastward. On the whole, the data by Morozov are in line with those by Polykarpov (1954), who detected above all, that during the period following seedling emergence up to flowering turning with the Sun are the top end of the stem, the top leaves and the non-flowering head, their azimuth being 20–30° smaller than that of the Sun. When the Sun goes down below 20–25° the stem end assumes a vertical position while the leaves decline.

We studied heliotropism of sunflower leaves also in Tadjikistan (Ross & Ross 1969b). Under examination were young plants, whose heads had not yet come into flower. Orientation of the normals for all the leaves of the control plants was measured every two hours with an accuracy of 5° for the inclination, and 30° for the azimuth. Heliotropism is most conspicuous in the leaves of the top layer where the leaves follow the Sun azimuthally lagging behind by approximately 30°, which is in agreement with the data of Polikarpov. The inclination angle maximum remains within 35–45°. While during the morning and the evening hours the leaves are non-horizontally oriented, the bulk having orientation close to vertical, towards midday the proportion of horizontally oriented leaves is considerably, while vertical leaves do not occur.

When leaves are heliotropic, they are specially oriented in relation to the direct solar beams. If following the definition of the function $(1/2\pi)g_L$, we define a new function $q_L(\xi_L)$, where ξ_L is the angle between the leaf normal and direction of the Sun, we can then assume that due to heliotropism $q_L(\xi_L) \approx$ const. during the day. The inclination angle ξ_L is calculated from the data of the experiment as

$$\xi_L = \arccos \widehat{r_S r_L}.$$

Figure 31 shows that the function $q_L(\xi_L)$ has definite dynamics during the day with the maximum about $\xi_L = 30°$, except for the maximum in the morning and evening hours. There is a close relation between $\xi_{L\max}$ and the Sun elevation $h_\odot$

(Figure 31C). At $h_{\odot} > 25 \div 30°$, ξ_L remains within the range 25–35° in almost all the cases observed. It, however, shifts rapidly towards high values of ξ_L when the Sun goes down below 25°. Distribution of the leaf area parts oriented at different angles ξ_L during the day (t is the local time) can be seen from Figure 31B. The Figure also shows that the orientation course at midday is in some intervals not symmetrical.

From the above follows that maize leaves in the upper layer closely follow the Sun with a lag of about 30°, if $h_{\odot} > 25°$. As the Sun goes to and below 25°, the leaves gradually assume a more normal position. Heliotropism is less marked in leaves of the middle layer and totally absent in the lower layer.

Detailed studies of the sunflower leaf heliotropism have, in recent years, been performed by Australian workers (Lang 1973; Shell et al. 1974; Shell & Lang 1975, 1976; Lang & Shell 1976). The experiments were performed with plants grown as single specimens, in single rows and in a row stand before flowering when a plant bears 20 leaves. It was shown that the functions $g'_L(\vartheta_L)$ and $(1/2\pi)g''_L(\varphi)$ can fairly well be described using the Von Mises' distribution

$$g_L(\beta, \bar{\beta}, K_\beta) = \frac{\exp K_\beta(\beta - \bar{\beta})}{2I_0(K_\beta)},$$

where β stands for ϑ_L or for φ_L, $\bar{\beta}$ being the mean angle in relation to which the function is symmetrical, K_β, a parameter characterizing distribution dispersion and $I_0(K)$, the modified Bessel function of the first kind and order zero. The larger the value of the dispersion parameter K, the greater is the clustering about the mode.

The value of $\bar{\vartheta}_L$ varies between 29 and 42°, that of K_ϑ, between 0.8 and 1.2. The value of $\bar{\varphi}_L$ varies during the 24 hours due to heliotropism and direction of the rows. Parameter K_φ varies in the range 0.2–1.1. Laboratory experiments with a fixed source showed no heliotropism, which indicates that the sunflower leaf motion is governed by the Sun rather than by endogenic rythm. Leaf heliotropism in the above experiments had a pattern, basically similar to that in our observations.

According to data of Lemeur (1973c), the upper layers of a sunflower stand in Belgium have the greatest number of horizontal leaves, while in the lower layers g_L is maximum at $\vartheta_L = 30°$. The function $(1/2\pi)g''_L(\varphi_L)$ has three maxima: NNE, S and NNW whose existence is attributed by the author to spiral phyllotaxis, equal to 120°.

The dynamics of the changes of $\hat{g}_L(\vartheta_L)$ during the growing period and dependence of $\hat{g}_L$ on stand density for maize have been discussed in the work by Udagawa et al. (1969), and dependence $\hat{g}_L(z, \vartheta_L)$, in the work by Horie and Udagawa (1970). It has been found that the form of $\hat{g}_L(\vartheta_L)$ is independent of the density, but as the stand depth increases the number of vertical leaves increases too, so that at mid-growth the maximum of $\hat{g}_L(\vartheta'_L)$ in the upper layers at $\vartheta_L \approx 20°$ is shifted to $\vartheta_L \approx 60°$ in the lower layers. A similar shift of $\hat{g}_L$ occurs in ontogeny, being less pronounced, though. No mention of heliotropism is made in the works by the Japanese authors.

Horse beans. The studies were carried out in Estonia and covered the flowering period (Ross & Nilson 1967b; Ross & Ross 1969b). The leaves of horse beans show a clearly defined orientation that is close to horizontal (Table I.6.2), leaf inclination angles in the range $\vartheta_L = 15$–$45°$ dominating in all layers. The amount of horizontal leaves increases with depth, the amount of vertical leaves, small though it is, slowly diminishes with depth.

Horse bean plants show an insignificant preferred azimuth which is southerly and across the rows.

Data by Shell et al. (1974) suggest that horizontal posture of horse bean leaves is most pronounced in the morning and is less so towards evening. A northeasterly preferred azimuth is observed. Heliotropism of leaves was shown to occur, the leaf normals again, as in the case of sunflower, tending to lead the Sun's beam early in the morning, but in the afternoon the leaf normal lags.

An intricate phenomenon of leaf movement in response to that of a radiation source was observed in horse beans by Wien and Wallace (1973). If radiation is incident on horizontal leaves from above, they turn parallel to the flux, but if radiation is incident on horizontal leaves from the side, they turn perpendicular to it, pulvinules rather than the leaf blades, acting here as light receptors. This movement of the leaves provides for a better lighting of the lower leaves as well as for a more uniform distribution of radiation within the stand and more efficient utilization of the light in photosynthesis.

In young birch leaves (Araki 1972ab) the inclination angle of lit leaves grows with the increase of mean lighting. This effect is, however, not observed in shaded leaves.

Clover. Although data by de Wit (1965) pertain to white clover at the flowering stage, while those by Nichiporovich (1961) do not specify the phenological stage and also use the number of leaves rather than their area in order to determine $\hat{g}_L$, making a comparison difficult, the results, nevertheless, clearly indicate that clover is a culture with predominantly horizontal leaves.

Strawberry. Only arbitrary results by Nichiporovich (1961) are available which bear evidence to a pronounced horizontal posture of the leaves.

Potato. Data by Nichiporovich (1961) were obtained for the Moscow region, our data pertain to Estonia (the flowering stage), two curves of de Wit (1965), for May 28 and June 12, respectively (the flowering stage, too). The data of Nichiporovich are similar to ours, though different from those of de Wit, according to which the tendency for a horizontal posture in potato is more pronounced, especially at early stages. On the whole, potato is regarded as a definitely prostrate-leaved culture.

Cucumber. Data by Nichiporovich (1961) pertain to the Moscow region at the stage of flowering and point to the domination of horizontal leaves.

Data by Shell et al. (1974) obtained in Australia from experiments with plants having twenty leaves are in agreement with the data by Nichiporovich. They also report an insignificant northeasterly preferred azimuth in cucumber leaves.

Pepper. Data by Shell et al. (1974) obtained in Australia show that the function

$\hat{g}_L$ has a maximum at about $\vartheta_L = 47°$ and diminishes to either side. Leaves of young plants have an insignificant northeasterly preferred azimuth.

Rape. Data by de Wit (1965) and the course of the index indicate that the number of horizontal leaves increases as growth of this culture proceeds. At an early stage of its development rape belongs to plants with inclined leaves with the maximum of $\hat{g}_L$ at $\vartheta_L = 50°$, but near floral initiation it is classed among plants with more horizontal leaves having the maximum of $\hat{g}_L$ at $\vartheta_L = 30°$.

Jerusalem artichoke. Dominating are horizontal leaves (Lemeur 1973b), the curve $\hat{g}_L(\vartheta_L)$ having two maxima: at 22 and 55°, azimuthal distribution of leaves approaching uniform.

Soybean. Data by different authors concerning $\hat{g}_L(\vartheta_L)$ appear to be inconsistent. In Japanese cultivars (Ito & Udagawa 1971) the function maximum is between 30 and 45°, the percentage of horizontal leaves being larger than that of erect leaves. Inclination angle distribution changes little over the vegetative period and in different layers. American cultivars (Blad & Baker 1972) show variations in the inclination angle distribution over the vegetative period. Although soybean canopies are planophile through most of the growing season, the distribution function $\hat{g}_L(\vartheta_L) \approx$ const. in the early and late growth stages. Mature soybean leaves are more horizontal than new leaves. On the other hand, according to the data of Lemeur (1973b), Belgian cultivars have the maximum at 70° with leaves of vertical orientation prevailing. Different authors report azimuth dependence to be approaching uniform.

Hybrid swede. Measurements performed in Estonia (Ross & Ross 1969b) have shown that at the time of hypocotil swelling the major portion of foliage area has an inclination of 40°, but as growth proceeds, this is shifted to 50° by the time rows are closed. This is accompanied by an increase in the amount of erect leaves making the distribution less distinct. Hybrid swede is classed among plants with inclined leaf orientation.

Sugar Beet. Data by de Wit (1965) were collected at different development stages. The dominating inclination is about 60°, shifting to 25° by July 28 then going back to 55° towards October 1. The latter shift is presumably due to the tendency of the upper parts of leaves to droop later in the season under their own weight and thus assume a vertical position.

Lucerne. According to the data of Scott and Wells (1969) mean inclination in a stand of lucerne 60 cm at the time of floral initiation equals 33° and does not change significantly with height. The number of horizontal leaves exceeds that of vertical ones. The data relate to the distribution of the number of leaves rather than the leaf area. Heliotropism of leaves is reported to occur with leaf azimuth following the Sun movement with some lag. Somewhat different from the above data are the results of Warren Wilson (1965a), processed by Philip (1965a). Philip's technique for the evaluation of $\hat{g}_L$ differs from the conventional procedures.

Lupin. According to the data of Scott and Wells (1969) the mean inclination angle in a lupin stand 84 cm high is 17° in the upper layers, 37° in the middle layers and 30° in the lower layers. As in the case of lucerne, leaf heliotropism is reported to take place.

Tropical pasture plants (Heslehurst & Wilson 1971). In the uniform legume canopy (*Desmodium intortum*) ϑ_L is distributed between 25 and 75°, with a maximum about 45° and shows no variation down the profile. The distribution function $\hat{g}_L(\vartheta_L)$ of the grass Setaria sphacelata varies markedly with height. In the top grass layer (50–70 cm), ϑ_L is smoothly distributed between 5 and 80°, with a maximum at about 45°. At 40–50 cm layer, $\hat{g}_L$ is quite different from that of both higher and lower layer, there is a small proportion of leaves with ϑ_L between 0° and 40°, and a large proportion of erect leaves with ϑ_L greater than 60°. The two lower layers 0–20 cm and 20–40 cm have a higher portion of erect leaves than top layers, with ranging from 30° to 90°. Of the plants so far examined Setaria seems to be the only species whose leaves have mean inclination ϑ_L increasing with depth.

It must be noticed that Heslehurst and Wilson determined $\hat{g}_L(\vartheta_L)$ using the inclined point quadrat method. Their results are not therefore fully comparable with the data by other authors.

Barley. According to our data (Mägi & Ross 1969a) collected at different stages of development of crops with different mineral supplies no regular variations of the distribution function $\hat{g}_L(\vartheta_L)$ are observed either over the growing period or in response to different levels of mineral supply. A curve which is a mean of all cases and shows the variation ranges is bimodal with the main mode at $\vartheta_L \approx 70°$ and the other mode at $\vartheta_L = 0°$. As compared with uniform distribution a barley stand has more horizontal and less vertical leaves.

More recent, yet more detailed results of Abashina and Gorbachev (1977), obtained in the Kaluga and Novgorod regions of the USSR on cultivars Moskovsky-121 and Viner are consistent with our data in some general features. The above authors have established that with increasing stand density there is a greater proportion of steeply inclined leaves. Table I.6.3 illustrates variation of $\hat{g}_L(\vartheta_L)$ during the growing season.

Table I.6.3 *Variation of inclination angle distribution during growth period. Barley var. Moskovsky-121. Kaluga Region, USSR (Abashina & Gorbachev 1977)*

Time	Growth stage	Inclination angle, degree						
		0–15	15–30	30–45	45–60	60–75	75–90	χ
May 25	3rd leaf	0.047	0.107	0.107	0.300	0.120	0.320	0.18
June 5	5th leaf	0.198	0.135	0.068	0.185	0.122	0.293	0.22
June 15	7th leaf	0.080	0.104	0.064	0.173	0.184	0.390	−0.12
June 23	8th leaf	0.044	0.076	0.080	0.192	0.208	0.400	−0.15
July 3	Heading	0.130	0.068	0.087	0.106	0.186	0.422	−0.27
July 13		0.133	0.186	0.097	0.248	0.195	0.142	0.23
July 23		0.286	0.178	0.071	0.214	0.143	0.107	0.34

As can be seen from the table leaf orientation in the stand before emergence of the 7th leaf shows a slight tendency from uniform to horizontal, then upon emergence of the 7th leaf and up to heading there is an increase in the proportion of erect leaves, while after heading the proportion of horizontal leaves increases significantly.

Down the profile, the upper layers are dominated by erect leaves, the share of horizontal leaves increasing with depth.

Azimuth distribution is characterized by a northerly elongation, which the authors attribute to wind effects.

Our data differ markedly from those by Japanese workers (Udagawa et al. 1970), according to which barley leaves at an early stage of their growth are predominantly horizontal, but towards the end of the growth horizontal leaves almost entirely disappear making the acute maximum of $\hat{g}_L(\vartheta_L)$ shift to $\vartheta_L \approx 70°$. Such a change in barley leaves from a horizontal to a vertical position in ontogeny seems enigmatic. Unfortunately, it finds no explanation on the part of the authors. The same data suggest that stand density does not affect the form of $\hat{g}_L(\vartheta_L)$.

According to the data by Australian workers (Scott & Wells 1969) obtained before heading in a stand 58 cm high, the mean leaf inclination angle in the upper layers is 44° and $\hat{g}_L(\vartheta_L)$ diminishes to either side of this value more or less uniformly. In the lower layers $\bar{\vartheta}_L$ is 53° and the proportion of erect leaves increases. Azimuth distribution is uneven – most of the leaves are south-oriented. These data are not fully comparable with other data, since the functions g'_L and g''_L describe azimuth and inclination angle distribution of whole leaves, rather than of leaf area.

Rye. The three curves by de Wit (1965) show how orientation of the stand varies with time. Early in the season it is near uniform. At the time of shooting the share of inclined area decreases and correspondingly the share of erect leaves increases. The negative value of the inclination index χ_L (Table I.6.2) implies that there is a slight tendency from uniform towards vertical distribution of rye leaf orientation. Towards heading the amount of erect leaves is decreased, the maximum shifting to 65°, and the amount of horizontal leaf area increases resulting in a bimodal curve similar to that of barley. It is noteworthy that the barley stand in Estonia did not reveal a similar variation of orientation.

Works by Kimura (1974, 1975a, b) established that in the blue light in young wheat leaves the angle between leaf blade and leaf sheath increases up to 110° while in darkness this angle is 20°. The inclination of the leaf blade, in the case reported, began on the 7th day after germination, and increased rapidly and showed almost the maximum value 90 days after germination. Blue light of 1600 erg cm^{-2} sec^{-1} was fully effective and further increase of the light intensity had no effect. The photosensitive part of leaf in photonastic inclination response of wheat leaves is laminar joint itself.

Close to uniform is leaf orientation in wheat and timothy stands (Nichiporovich 1961).

Table I.6.4 Seasonal change of rice leaf area inclination index χ_L.

Variety	Date	July 17	Aug. 2	Aug. 25	Sept. 14	Sept. 28
Manryo		−0.141	−0.120	−0.188	−0.124	−0.058
IR-8		−0.108	−0.192	−0.399	−0.442	−0.378

Rice. Detailed studies of stand architecture were performed by Japanese scientists (Ito 1969; Ito et al. 1973). Rice was shown to be a culture with basically erect leaves. The change of the leaf area inclination index χ_L for two varieties Manryo and IR-8 during the growing season is presented in Table I.6.4.

The table shows that at heading the vertical orientation of the cultivar Monryo changes to uniform due to the increase of leaf inclination angle in the lower layers. A negative correlation was found to exist between χ_L and excess of spatial distribution of leaf area around the stem. This implies that the closer the leaves to the stem, the more erect they are. Distribution with respect to azimuth angle was found to be close to uniform in both cultivars.

Of the species investigated perennial ryegrass showed the widest range of leaf orientation variation.

As stated by de Wit (1965) stand age accounts for most of the changes in the orientation of ryegrass leaf area. Of all the cases examined in the paper young ryegrass in spring has the highest proportion of vertical leaves ($\chi_L = -0.27$). In summer, maximum leaf inclination in 10-day-old grass is at about 70°, but the proportion of vertical leaves is low as also is the amount of leaves with inclination angles between 0 and 40°.

As the stand ages the amount of horizontal leaves increases, the maximum shifting towards smaller angles, and the index χ_L increases. In early spring the vertical orientation of leaves of young ryegrass is less marked, distribution of inclination angles in 10-day-old grass practically coinciding with that of 30-day-old grass (May 5). At the age over 40 days the maximum remains at about 35°, however, the amount of near-vertical leaves diminishes, while the amount of horizontal leaves increases accordingly. In 50-60-day-old ryegrass the number of horizontal leaves is four times that of vertical ones. Leaf orientation in old ryegrass remains unchanged throughout summer and, typically, the vertical structure is replaced by horizontal as the stand ages.

Data concerning leaf orientation in tree stands are exceedingly scarce. Only Miller (1967) reports that the leaf mean inclination angle in young aspen 3 m high increases with height from 30 to 65° and in oak it increases from 30 to 80°. The same pattern has been observed by Grulois (1967) in oak and beech 16 m high. Dependence of the mean inclination angle ϑ_L on height z is expressed by the formula

$$\bar{\vartheta}_L = 17.75 + 16.64 \exp(-0.5406z) \quad \text{for oak}$$

$$\bar{\vartheta}_L = 3.94 + 30.26 \exp(-0.34647z) \quad \text{for beech}$$

The above data lead us to conclude that the form of $\hat{g}_L$ is variable within a wide range and representation of it in terms of simple empirical formulae cannot be easy and is hardly reasonable. As we shall see later, in the mathematical modelling of stand radiation regime and photosynthetic activity it is effective projection $G(z, r)$ rather than leaf normal distribution $(1/2\pi)g_L(z, r_L)$ that characterizes stand geometry. Therefore, $G(z, r)$ and not $(1/2\pi)g_L(z, r_L)$ should be given analytical representation.

Let us now pass to the consideration of the G-function correlated to $(1/2\pi)g_L$ by expression (I.2.18).

I.6.3 G-FUNCTION OF LEAF ORIENTATION

The G-function of leaf orientation is widely used in the description of stand radiation regime.

As can be seen from formula (I.2.18) G is a function of three variable quantities, viz., zenith angle ϑ, azimuth φ and height z (or depth $L(z)$). A study of stand radiation regime often requires calculation of the G-function for the direction of sunrays $r_S = (h_\odot, \varphi_\odot)$.

For the purpose of the present work the G-function was computed either from formula (I.2.18) in the Computing Centre of the Institute of Cybernetics of the ESSR Academy of Sciences, or approximately, from formula (I.2.23).

As has been shown under I.2.2, in the modelled situations $G = \frac{1}{2}$ for uniform orientation, $G = \cos\vartheta$ for horizontal and $G = (2/\pi)\sin\vartheta$ for vertical orientation. The above situations are presented in Figure 32A which also illustrates cases of predominant horizontal (curve 2a) and predominant vertical (curve 3a) orientations. Predominantly horizontal orientation is possible on condition that all leaves have their inclination angle ϑ_L within the range 0–15°, accordingly, the condition for predominantly vertical orientation is 75–90°.

Figure 32B illustrates the G-functions for the four types of modelled orientations according to de Wit (1965) which we have calculated from $\hat{g}_L(\vartheta_L)$ presented in Figure 24; Figure 32C provides the G-functions computed from the five classes of $\hat{g}_L(\vartheta_L)$ presented in our work (Ross & Nilson 1967b).

With reference to Figure 32D, it would be of interest to compare curve 1 for white clover with the curve for ryegrass, the former having the highest and the latter the lowest values of the inclination indices $\chi_L = 0.45$ and $\chi_L = -0.27$, respectively. The two curves represent extreme cases of the plant species so far examined, i.e. the species whose leaves are predominantly horizontal and those with leaves that are basically erect. The curves, therefore, allow for the evaluation of the maximum range of function $G(h_\odot)$ variation in relation to real plant types. At low solar elevations $G(h_\odot)$ varies within 0.30–0.55, at middle elevations the variation is within 0.4–0.6, while at high elevations of the sun $G(h_\odot)$ varies within 0.35–0.85. Curve 1 also illustrates the maximum range of naturally occurring variation of the leaf effective area with changing solar elevation. Where horizontal

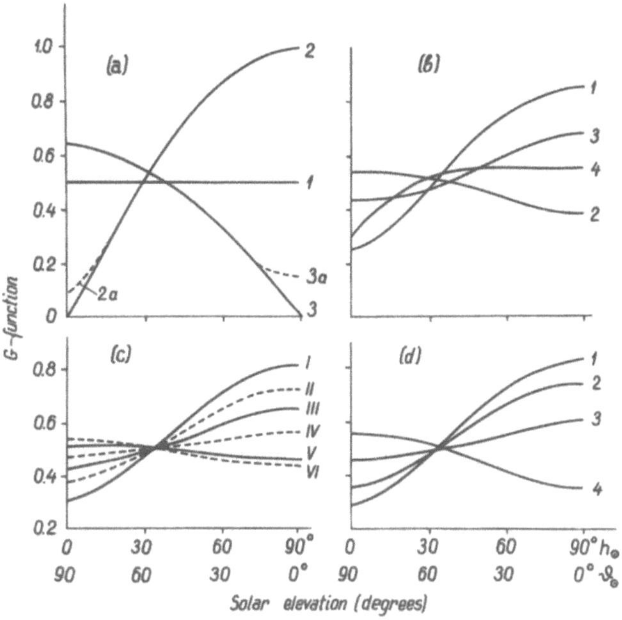

Fig. 32. Dependence of G-function on solar elevation $h_\odot$
(a) – theoretical orientations: 1 – uniform orientation, 2 – horizontal leaves, 2a – predominantly horizontal leaves, 3 – vertical leaves, 3a – predominantly vertical leaves; (b) – model orientation acc. to de Wit (1965): 1 – planophile, 2 – erectophile, 3 – plagiophile, 4 – extremophile; (c) – orientation classes acc. to Ross and Nilson (1966b); (d) – actual crops: 1 – white clover (acc. to de Wit (1965)), 2 – sunflower, experiment No. 16, Tadjikistan, July 28, 1965, flowering, 3 – maize, experiment No. 14, Tadjikistan, June 8, 1965, emergence of 10th leaf; 4 – perennial rye-grass 30 days old (according to de Wit (1965)).

orientation is marked, effective leaf projection varies from 0.3 at low elevations to 0.8 at high elevations, i.e. the effective leaf projection being, roughly, three times as high.

The dependence of $G(z, r_S)$ on height z is determined by the dependence of function $(1/2\pi)g_L(z, r_L)$ on height z. A comprehensive examination of the former was made by us for maize and horse bean stands (Ross & Nilson 1965), and for a maize stand (Udagawa et al. 1969). Dependence of the effective projection G on stand height is not too marked and can be neglected in a first approximation, the error being 20–30%. Horse beans, whose leaves are horizontally oriented, show a conspicuous dependence of G on solar elevation.

A clearly pronounced vertical profile is reported for maize in Japan (Udagawa et al. 1968). $G(z)$ diminishes around midday in the upper layers and then again increases in the lower layers, but this dependence gradually becomes less marked as the stand develops. It is noteworthy that the authors (Udagawa et al. 1968) obtained a linear dependence of $G(h_\odot)$ in 1966 and non-linear, in 1967.

The dependence of $G(z, h_\odot, \varphi_\odot)$ on azimuth $\varphi_\odot$ was studied in more detail for a stand of maize (Ross & Nilson 1965). As one would expect, $G(\varphi_\odot)$ is smoother than $(1/2\pi)g_L''(\varphi_L)$, neglect of the preferred azimuth of leaves affects the value of by 10 to 20%. Where the preferred azimuth is more pronounced, as is the case with sunflower, its effect on G is greater and it, therefore, cannot be neglected.

It should be pointed out that solar elevation $h_\odot$ and azimuth $\varphi_\odot$ are dependent quantities and their relationship is governed by astrogeophysical factors. For a given locality and a particular point of time there is always a definite couple of values for $h_\odot$ and $\varphi_\odot$, which vary in the course of the day. It is, therefore, of interest to study a diurnal course of G, determined by a diurnal course of $h_\odot$ and $\varphi_\odot$. This latter aspect was the subject of a detailed study in maize and horse bean stands (Ross & Nilson 1965) (Figure 33).

The top and bottom dashed lines in the Figure define the area where the effect of preferred azimuth is present, the middle dashed line shows the dependence of G on solar elevation with the effect of azimuth disregarded. It is seen from the figure that the effect of azimuth is most significant in the uppermost layer where diurnal variation of G is in the range 0.45–0.60. In other layers the daily course of G is within ± 0.05, the effective leaf area being greater after midday.

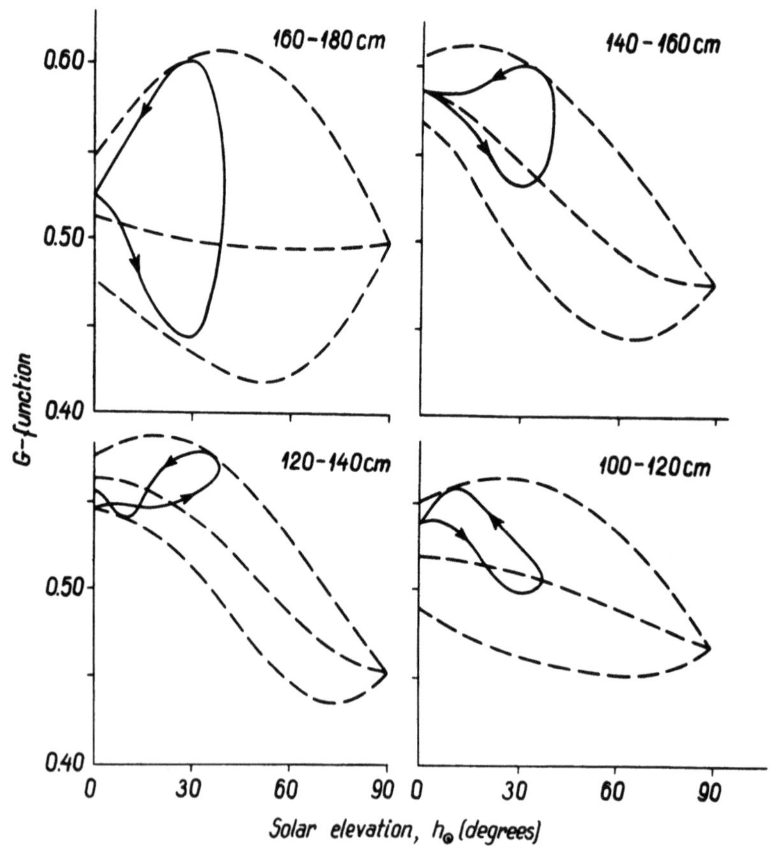

Fig. 33. Diurnal course and dependence of G-function on solar elevation in upper layers of maize crop. Experiment No. 3, Estonia, September 5, 1962, tasseling. See text.

In horse bean stands (Ross & Nilson 1965) the daily course of G is more pronounced. In Estonia the value for G rises from 0.3 to 0.7 with solar elevation, i.e. more than doubled.

In a rice stand (Ito 1969) the G-function varies during the growing season and is approximated by the following formulae:

July 10	$G(h_{\odot})=0.750-2.83\,10^{-2}h_{\odot}^{0.63}$
July 24	$G(h_{\odot})=0.600–1.32\,10^{-2}h_{\odot}^{0.61}$
August 12	$G(h_{\odot})=0.445+0.98\,10^{-2}h_{\odot}^{1.73}$

At about noon the G-function increases with depth from 0.2 in the upper layer to 0.6 in the lower layer.

In Japanese cultivars of soybean (Ito & Udagawa 1971) the G-function is markedly dependent on solar elevation and is approximated by the formula $G(h_{\odot})=a+b\sin^2 h_{\odot}$, a varying within $0.28-0.34$, b, within 0.38–0.53 as the stand ages. The vertical profiles of the G-function in soybeans are similar to those of the G-function in horse beans (Ross & Nilson 1965).

Thus, analysis of function $G(z, h_{\odot}, \varphi_{\odot})$ for agricultural crops shows that solar elevation has the most significant effect on G, especially in stands with horizontal leaves. The dependence of G on stand height and azimuth $\varphi_{\odot}$ can, in a first approximation, be neglected, the error involved being between 10 and 30%.

Representation of function $G(h_{\odot})$ as tables or graphs is inconvenient. We have, therefore, attempted (Ross & Ross 1969b) to obtain a simple empirical formula for $G(h_{\odot})$ which would make it possible to summarize the experimental data now available. Analysis of function $G(h_{\odot})$ has demonstrated that at $h_{\odot}>15°$ the experimental curves can be approximated by the formula

$$G(h_{\odot})=G_1+(G_2-G_1)\sin h_{\odot}, \tag{I.6.2}$$

where $G_1=G(0)$ and $G_2=G(\pi/2)$, i.e. G_1 and G_2 according to (I.2.23a) and (I.2.23b) are the effective leaf area projections on horizontal and vertical directions, respectively. In the case of uniform leaf area distribution $G=G_1=G_2=\frac{1}{2}$ while for horizontal leaves $G_1=0$, $G_2=1$ and $G=\sin h_{\odot}$. Formula (I.6.2) is inapplicable for erect leaves, since, following (I.2.18d), $G=(2/\pi)\cos h_{\odot}$, however, from (I.6.2) we obtain $G=(2/\pi)(1-\sin h_{\odot})$.

The quantities G_1 and G_2 are dependent on the form of $(1/2\pi)g_L(z, r_L)$ thus, they characterize leaf area orientation. Since for the arbitrary characteristic of leaf area orientation the inclination angle index χ_L was used instead of $(1/2\pi)g_L$ a question arises whether there is a correlation between G_1, G_2 and χ_L.

With the data on G and χ_L available for real stands we have compared the values of G_1, G_2 and G_2-G_1 with χ_L (Figure 34). The scatter of dots in the Figure is relatively small, despite the diversity of the leaf area orientation distribution function $(1/2\pi)g_L$. The crosses represent the values of G_1 and G_2 for the model orientations of de Wit. It appears that the planophile (1) and erectophile (2) types of orientation are in fairly good agreement with the experimental data, while the

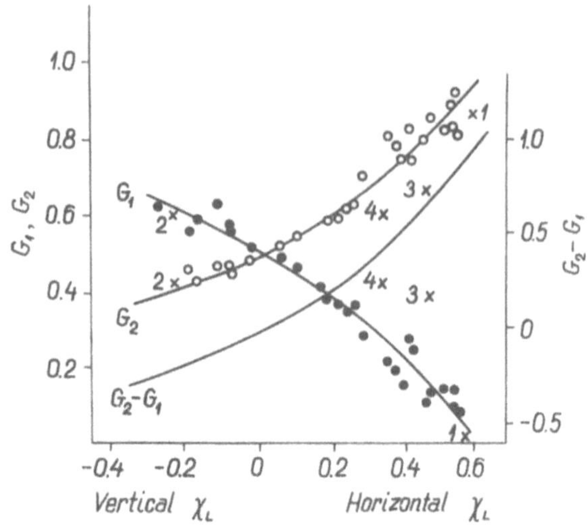

Fig. 34. Empirical relationships between foliage area orientation parameters G_1, G_2 and G_2-G_1 on foliage area inclination index χ_L. See text.

plagiophile (3) and extremophile (4) types are more artificial. The variation of G_2-G_1 suggests that with the leaf orientation becoming less horizontal and more uniform the dependence of G on $h_\odot$ diminishes, but it becomes stronger again and reverses its sign as uniform distribution changes into vertical.

The fairly closed relation between G_1, G_2 and χ_L allows us to suggest a simple and universal technique for a rough determination of $G(h_\odot)$ by using formula (I.6.2) according to which the orientation index χ_L is derived from phytometric measurements, G_1 and G_2, following the graph of Figure 34, while $G(h_\odot)$ is calculated from formula (I.6.2).

The data presented herein lead us to make some general conclusions about the nature of function $G(h_\odot)$:

1. A theoretical function $G(h_\odot)$ varies within the range from 0 to 1, but for real plants the range of G variations is notably narrower.

2. Function $G(h_\odot)$ is significantly smoother as compared with function $(1/2\pi)g_L$, the smoothing in formula (I.2.18) for $G(h_\odot)$ being accomplished by a multiplier $0 \leqslant |\cos \widehat{rr}_L| \leqslant 1$. The weak dependence of G on $h_\odot$ shown in real stands is partly due to the fact that the latter contain, with one weight or the other, all the leaf inclinations. This explains why the relatively large fluctuations and differences of leaf orientation in different stands do not affect radiation regime in a stand so strong as one would expect, the reason being the smoothing effect of the G-function.

3. With horizontal leaves prevailing in a stand, $G(h_\odot)$ is an increasing function of solar elevation. And, conversely, $G(h_\odot)$ is a decreasing function of $h_\odot$ if vertical leaves are predominant; however, the range of $G(h_\odot)$ variation is smaller in the latter case.

4. Within the solar elevation range $h_{\odot} = 30 \div 40°$ the dependence of the G-function on leaf orientation is weak and $G \approx \frac{1}{2}$.

This is a very important conclusion to be reckoned with in the assessment of the role of stand geometry in the radiation regime and photosynthetic activity. Solar elevations of 30 to 40° are frequent in moderate latitudes, hence the activity of a stand as a photosynthetic system under these radiation conditions is long enough. However, $G(h_{\odot})$ being little affected by leaf orientation, both the stand radiation regime and photosynthesis, as can be seen further, are not particularly dependent on leaf orientation. With solar radiation coming as above, one cannot expect a stand photosynthetic activity to improve drastically through a change in the stand geometry.

5. The effect of leaf orientation on $G(h_{\odot})$ is most pronounced at high solar elevations. Consequently, improvement of the stand photosynthetic activity through a modification of its geometry can bring about a maximum effect in the nearequatorial zone.

I.7 Plant stand architecture, photosynthesis and productivity

In terms of energy, an increase in crop yields is equivalent to an increase in the amount of PAR absorbed by crops and to its optimal utilization by plants in the process of photosynthesis. Since the stand architecture is closely related to its radiation regime, a question arises as to what role the stand architecture plays in photosynthesis and productivity and to what extent the stand photosynthesis and productivity can be improved by purposely changing its architecture.

In 1965—1970 de Wit (1965), Duncan et al. (1967), Loomis, Williams and Duncan (1967), Idso and Baker (1967), Ross and Bichele (1968, 1969) Idso & Baker 1967, and Kuroiwa (1968a, b, 1970) were independently developing mathematical models of stand photosynthesis with special emphasis on the study of the influence of architecture, basically, of leaf inclination angle and leaf area index L_0 on stand radiation regime and photosynthesis. Numerical experiments, performed on the models, showed that when L_0 is less than about 3 the stand productivity is closely related with percent cover and a horizontal foliage has advantage over inclined or vertical leaves, without regard to latitude and solar elevation. With further increase in L_0, stands with strongly erect leaves have a considerable advantage over other plant stands, especially in lower latitudes and high solar elevation. Calculations by Duncan (1971) demonstrated that by changing the inclination angle in different layers of a dense crop ($L_0=10$) photosynthesis can be increased twofold. However, in the daily total of photosynthesis, the course of the Sun during the day notably averages these differences. According to calculations by Ross and Bichele (1968, 1969) leaf angle variation shows little, if any, influence on daily photosynthesis in moderate latitudes, while in low latitudes the strongest effect of leaf orientation is associated with high values of L_0 and does not exceed 30%. Calculations of Tooming, summarized in his monograph (Tooming 1977) showed that in an 'ideal' plant stand with maximum productivity leaf inclination ϑ_L must vary from 0° in the upper layers to 90° in the lowermost layer. From the requirements formulated by Khosrowshahi (1970) for an 'ideal' cereal plant, follows that the plant stem must be comparatively short and thick-walled, it must have many small and predominantly vertical leaves, a big ear and long awns.

Results of computations based on mathematical models of stand photosynthesis have been supported by field experiments. Thus, for example, measurements by Pearce et al. (1967) demonstrated that at $L_0 > 2$ an increase in the vertical posture of leaves results in an increase of the rate of photosynthesis. By mechanically changing the leaf inclination angle Tanaka (1972) established that the total photosynthesis of a rice stand with predominantly horizontal leaves constitutes 80% of that of a stand with predominantly vertical leaves.

However, experimental data are controversial and cannot therefore be used to test the hypothesis that an increase of leaf angle leads to an increase of yield. Pendelton et al. (1968) reported yield increase of 15% when the upper leaves of maize in close spacing were tied to give them a more erect posture. Shaw and Weber (1967) found that in a soybean stand an increase in radiant energy penetration due to a modified architecture of the stand leads to an increase of yield. From experiments of Vidovič (1973) with maize hybrids follows that at high values of L_0 erect leaves lead to higher yields. Similar results have been obtained by Williams et al. (1968).

However, more recent experiments by Winter and Ohlrogge (1973) in maize involving a mechanical way of changing leaf angle showed that row crops ($L_0 = 3$–4) with erect leaves yielded somewhat less than usual, whereas dense crops ($L_0 > 5$) showed an increase in yield. The authors come to a conclusion that at optimal values of L_0 an increase in leaf steepness does not bring about an increase in yield. But where a higher plant density favours productivity, cultivars with erect leaves have an advantage. Field experiments of Angus, Jones and Wilson (1972), performed in Australia on two wheat cultivars with different leaf inclination showed that leaf erectness fails to bring about higher yields. A similar result was obtained by Heslehurst and Wilson (1971) for two pasture plants – *Setaria sphacelata* and *Desmodium intortum*. Trenbath and Angus (1975) in their detailed review of the effect of leaf inclination on productivity emphasize that among C_3-plants, particularly so in the case of rice, leaf erectness may prove to be helpful in raising productivity where there is a deficiency of assimilates retarding plant growth.

Presumably, the controversial data on the effect of leaf inclination on crop yields result from a misapprehension of the fact that the close correlation between the amount of PAR absorbed and plant growth exists at every stage of the growing season. It stands to reason, therefore, that a search for an association between stand yield and stand architecture over a short time at different stages of ontogeny may not always issue in satisfactory results. Ultimate solution of the problem calls for more prolonged and more comprehensive experiments covering the entire process of yield formation or for numerical experiments using more advanced mathematical models of the production process that take account also of the interrelation between photosynthesis, distribution of assimilates and growth.

PART TWO

Radiation regime in plant stand

II.1 Radiation field in a plant stand and the problem of its mathematical modelling

II.1.1 GENERAL

The radiation regime in a plant stand depents on the transfer of radiation in the biological environment formed by plant organs which possess certain geometrical and optical properties and are arranged in accordance with a certain pattern. The radiation is absorbed and scattered by plant organs (leaves, stems, flowers, etc.). Because of this interaction the incoming radiation is changed as it penetrates the plant stand. The plant stand is limited below by the ground surface, which absorbs one part of radiation and reflects the other part upwards. A horizontally uniform plant stand sufficiently large in area may, in a first approximation, be treated as an infinite plane-parallel turbid layer with the elements of vegetation treated as those of turbidity. A quantitative theory based on the radiative transfer equation and its solution for given boundary conditions and optical properties of turbidity elements has been evolved in astrophysics and atmospheric physics to describe the radiative transfer in such media. This theory has been used successfully in studies of solar radiation transmission through the Earth's atmosphere as well as of transmission of the radiation of stars through their upper strata.

Now the question is how this theory of radiative transfer in turbid media developed for the purposes of astrophysics and atmospheric physics could be applied in studying the radiation regime in plant stands. In this respect it should be pointed out that there exist considerable differences between a plant stand and the atmosphere, namely:

1. The absorbing and scattering elements of a plant stand are larger than molecules and aerosol particles in the air. This can limit the use of integro-differential calculus methods and the transfer theory in plant stands, and rather a large scale of spatial averaging is required if reliable results are to be obtained.

2. Plant stand architecture is manifold in form and varies largely. Plant parts are complicated in shape and their distribution in a plant stand is neither regular nor quite random. Therefore a plant stand can neither be treated as a geometrical nor as a random turbid medium amenable to mathematical modelling.

3. Unlike the uppermost level of the atmosphere, the upper level of a plant stand is exposed both to direct solar irradiance and to diffuse radiation from all points of the sky.

4. The optical depth of a plant stand (up to 8–10) is much greater than that of the atmosphere (0.2–0.6). Therefore a dense plant stand can be treated as layer of infinite depth. This simplifies the problem, and the optical properties of the ground surface below the plant stand need not be taken into account.

5. Green vegetation is characterized by a high absorptance in PAR (0.85) and a low absorptance in NIR (0.15). This is very important for the PAR regime. It is well known that absorption in the terrestrial atmosphere is not large and scattering is considerable, therefore great importance is attached to the shape of the scattering phase function and its mathematical modelling. A different situation exists in the case of plant stands: because of the large amount of absorption, single scattering may be assumed, which will substantially simplify calculation (of course, such a simplification is not admissible for NIR and short-wave radiation). In this regard the plant stand architecture, that is the shape and dimensions of plant parts, their arrangement within the plant and their orientation, the distribution of plants within the stand, etc., are of prime importance rather than the reflecting properties of plants. Therefore, the geometrical characteristics gain in importance in the radiative transfer theory, but only a statistical approach must be used to deal with the plant stand structure.

In view of the above features peculiar to plant stands the mathematical approach intended for studying the radiation regime of the atmosphere of starts and planets and based on the solution of the radiative transfer equation cannot be applied to the radiation regime in plant stands without a preliminary analysis.

Attempts to describe the radiation regime in plant stands mathematically have so far been made along two lines:

1. Modelling a plant stand as geometrical figures: cylinders, spheres, cones, etc. of given dimensions, spatial arrangement, reflectivity and transmissivity. The reflecting properties of the ground surface below the figures are also given.

Such models have been constructed with cylinders and spheres (Kastrov 1955; Szwarzbaum & Shaviv 1976), cones (Jahnke & Lawrence 1965) and ellipsoids (Charles-Edwards & Thornley 1973). A crop growing in rows is simulated as parallelopipeds in rows with a square (Allen 1974; Fukai & Davidson 1977; Goudriaan 1977) or ellipsoidal (Charles-Edwards & Thorpe 1976) cross section.

2. Modelling a plant stand as a turbid layer. The elements of vegetation are treated as small absorbing and scattering particles with given geometrical and optical properties. The plant stand is treated as a horizontally uniform plane-parallel layer, in which the radiation field depends only on the coordinate z and direction r and not on x or y. A turbid-layer model was first used by Schifrin (1953) to formulate the theory of albedo in plant stands. It was further developed in the works by Ross (1962, 1964), Ross and Nilson (1963, 1965, 1967c, 1968a, b), Nilson (1968a, b), Niilisk and Ross (1969), Niilisk et al. (1970), Isobe (1962a, b, c, 1969, 1971), Allen and Brown (1965), Chartier (1966), Cowan (1968), Verhagen and Wilson (1969), Idso and de Wit (1970), and others.

The first type of modelling appears to be more promising in the case of sparse plant stands (shrubs, sparse woods, orchards, maize and cotton crops in the initial period of growth, etc.), where multiple scattering may be neglected, and at high elevations of the Sun, when mutual shading of the individual figures need not be taken into account. A geometrical model is mainly applicable in the theory of albedo, but is of little use in describing the radiation regime in plant stands.

The denser and more uniform horizontally the plant stand and the smaller the phytoelements in comparison with the plant stand height, the more successful the use of the turbidlayer model. In view of the needs of agriculture and forestry, studies using a turbid-layer model have been prevalent.

II.1.2 BASIC CHARACTERISTICS OF THE RADIATION FIELD

Neglecting the polarizing properties and statistical character of the radiation field, its most detailed characteristic is *the monochromatic intensity* $i_\lambda(\tau, P, r)$.

In order to define the monochromatic intensity of radiation let us consider Figure 35. Suppose there is at the point $P(x, y, z)$ a surface element $d\sigma_L$ whose orientation is shown by its normal r_L. Beams of radiation fall on this surface from all possible directions. Consider the beams confined to the solid angle $d\Omega$ around the direction r and constituting a volume confined in the truncated cone contiguous to the base $d\Omega$. Let dE_λ express the amount of radiant energy in the spectral band λ to $\lambda + d\lambda$ penetrating through the surface $d\sigma_L$ in the time increment $d\tau$ within the segregated volume. The intensity of radiation for the wavelength λ at the point P in the direction r is defined by the relation

$$dE_\lambda = i_\lambda(\tau, P, r)|\cos \widehat{r_L r}| d\sigma_L d\Omega d\tau d\lambda. \quad \text{(II.1.1)}$$

Thus the monochromatic intensity of radiation $i_\lambda(\tau, P, r)$ is the amount of radiant energy in a unit wavelength band within a unit solid angle around the direction r, which passes through a unit surface perpendicular to the beam direction during a unit time.

The radiation field is uniform if i_λ is independent of the point P position, and horizontally uniform if i_λ does not depend on the coordinates x and y but depends on the coordinate z. The latter case is often considered in actinometry and in

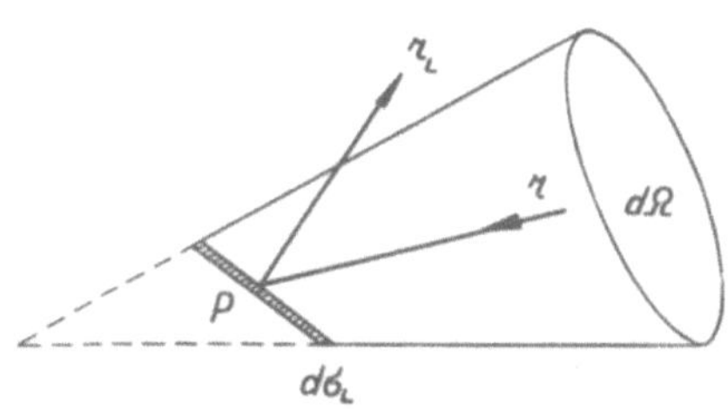

Fig. 35. Drawing to illustrate the definition of radiation intensity. See text.

studying the radiation regime in plant stands. A more general case is that of a radiation field, horizontally uniform in the statistical sense, where the mathematical expectance of intensities obtained by averaging over the horizontal area does not depend on x or y, but the variability of intensities within the same area is given by statistical characteristics. This case will be discussed under II.1.6 and II.1.9.

The radiation field is isotropic if i_λ is independent of the direction r.

Sometimes it is convenient to divide the radiation field $i_\lambda(P, r)$ into two: the downward part $i_{\lambda 1}(P, r)$ and the upward one $i_{\lambda 2}(P, r)$, so that

$$\left.\begin{aligned} i_{\lambda 1}(P, r) &= i_\lambda(P, -r) \\ i_{\lambda 2}(P, r) &= i_\lambda(P, r) \end{aligned}\right\} \tag{II.1.2}$$

and

$$\left.\begin{aligned} -r &= (\theta, \pi+\varphi) \\ \theta &= \pi - \vartheta \end{aligned}\right\}. \tag{II.1.3}$$

When studying the radiation regime in plant stands we often have to determine the amount of radiation coming from all possible points to the upper or lower side of a leaf oriented in a specified direction r_L. This problem is reduced to determining the irradiance $J_\lambda(\tau, P, r_L)$ on a plane surface. *The irradiance* $J_\lambda(\tau, P, r_L)$ at the point P in the direction r_L at the instant τ is related to the intensity by the equation

$$J_{\lambda 1}(\tau, P, r_L) = \int_{\Omega_1^L} i_\lambda(\tau, P, r_L r)|\cos \widehat{r_L r}|\, d\Omega, \tag{II.1.4}$$

in which $\pi/2 \leqslant \widehat{r_L r} \leqslant \pi$.

The irradiance on the lower side of the same surface is expressed by

$$J_{\lambda 2}(\tau, P, r_L) = \int_{\Omega_2^L} i_\lambda(\tau, P, r_L r)|\cos \widehat{r_L r}|\, d\Omega, \tag{II.1.4a}$$

where $0 \leqslant \widehat{r_L r} \leqslant \pi/2$.

Taking into account (II.1.2) and Figure 36 we obtain instead of (II.1.4) and (II.1.4a)

$$J_{\lambda 1}(P, r_L) = \int_{\Omega_{11}^L} i_{\lambda 1}(P, r)|\cos \widehat{r_L r}| \mathrm{d}\Omega + $$

$$+ \int_{\Omega_{21}^L} i_{\lambda 2}(P, r)|\cos \widehat{r_L r}| \mathrm{d}\Omega, \tag{II.1.4b}$$

$$J_{\lambda 2}(P, r_L) = \int_{\Omega_{12}^L} i_{\lambda 1}(P, r)\cos \widehat{r_L r}\, \mathrm{d}\Omega + $$

$$+ \int_{\Omega_{22}^L} i_{\lambda 2}(P, r)\cos \widehat{r_L r}\, \mathrm{d}\Omega, \tag{II.1.4c}$$

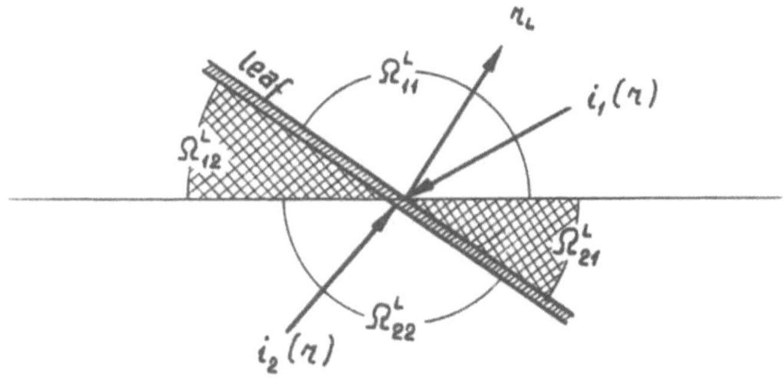

Fig. 36. A leaf in radiation field.

where in accordance with Figure 36, Ω_{11}^L and Ω_{12}^L are solid angles in the upper hemisphere equal to $2(\pi-\vartheta_L)$ and $2\vartheta_L$, respectively, and Ω_{21}^L and Ω_{22}^L are solid angles in the lower hemisphere equal to $2\vartheta_L$ and $2(\pi-\vartheta_L)$, respectively.

In the particular case where the field is uniform and isotropic but different in the upper and lower hemispheres, we obtain from (II.1.4) and (II.1.4a)

$$J_{\lambda 1}(\vartheta_L)=\frac{\pi}{2}(i_{\lambda 1}+i_{\lambda 2})+\frac{\pi}{2}(i_{\lambda 1}-i_{\lambda 2})\cos\vartheta_L, \tag{II.1.4d}$$

$$J_{\lambda 2}(\vartheta_L)=\frac{\pi}{2}(i_{\lambda 1}+i_{\lambda 2})-\frac{\pi}{2}(i_{\lambda 1}-i_{\lambda 2})\cos\vartheta_L \tag{II.1.4e}$$

and

$$J_{\lambda 1}+J_{\lambda 2}=\pi(i_{\lambda 1}+i_{\lambda 2}), \tag{II.1.5}$$

since

$$\left.\begin{aligned}\int_{\Omega_{11}^L}|\cos\widehat{r_L r}|\,\mathrm{d}\Omega=\int_{\Omega_{22}^L}|\cos\widehat{r_L r}|\,\mathrm{d}\Omega=\frac{\pi}{2}(1+\cos\vartheta_L),\\ \int_{\Omega_{12}^L}|\cos\widehat{r_L r}|\,\mathrm{d}\Omega=\int_{\Omega_{21}^L}|\cos\widehat{r_L r}|\,\mathrm{d}\Omega=\frac{\pi}{2}(1-\cos\vartheta_L).\end{aligned}\right\} \tag{II.1.6}$$

From the standpoint of the field theory it is immaterial whether the radiation field is expressed in terms of intensities $i_\lambda(P, r)$ or irradiances $J_\lambda(P, r_L)$.

If the field is horizontally uniform, *the downward and upward flux densities* (i.e. irradiance on the upper and lower side of the horizontal surface, respectively) are usually considered enough for the purpose of a more approximate radiation regime analysis:

$$\left.\begin{aligned}F_{\lambda 1}(z)=J_{\lambda 1}(z, z)=\int_{\Omega_1} i_{\lambda 1}(z, r)\cos\theta\,\mathrm{d}\Omega,\\ F_{\lambda 2}(z)=J_{\lambda 2}(z, z)=\int_{\Omega_2} i_{\lambda 2}(z, r)\cos\vartheta\,\mathrm{d}\Omega.\end{aligned}\right\} \tag{II.1.7}$$

Hereafter the subscript λ will be omitted in discussing the radiation characteristics.

In the case of an isotropic radiation field, the vertical flux densities will be

$$F_1 = \pi i_1 \qquad F_2 = \pi i_2 \tag{II.1.7a}$$

In traversing a path ds in a turbid medium one part dE_κ of the initial amount dE of radiant energy is absorbed, another part dE_ω is scattered in other directions, and still another part dE_P is transmitted without interacting with the medium. In accordance with the radiant energy conservation law

$$\mathrm{d}E = \mathrm{d}E_\kappa + \mathrm{d}E_\omega + \mathrm{d}E_P \tag{II.1.8}$$

the absorbed and scattered parts constitute the attenuated part dE_K of the radiant energy, i.e.

$$\mathrm{d}E_K = \mathrm{d}E_\omega + \mathrm{d}E_\kappa. \tag{II.1.9}$$

In the transfer theory it is generally understood that the absorbed and scattered parts of energy are proportionate to the path ds and the initial amount of energy dE. In this case the proportionality coefficients are characteristic of the absorptivity and scattering power of the turbid medium.

Thus for absorption:

$$\mathrm{d}E_\kappa = \kappa^* \mathrm{d}E \mathrm{d}s, \tag{II.1.10}$$

for scattering:

$$\mathrm{d}E_\omega = \omega^* \mathrm{d}E \mathrm{d}s \tag{II.1.11}$$

and for attenuation

$$\mathrm{d}E_K = k^* \mathrm{d}E \mathrm{d}s, \tag{II.1.12}$$

where κ^*, ω^* and k^* are *volume absorption*, *scattering*, and *attenuation coefficients*, respectively.

It follows from (II.1.8)–(II.1.12) that

$$k^* = \kappa^* + \omega^* \tag{II.1.13}$$

and

$$p^* = 1 - k^* = 1 - \kappa^* - \omega^*, \tag{II.1.14}$$

where p^* is *the volume transmission coefficient.*

One of the radiation characteristics of a turbid medium is *the volume emission coefficient* η^* defined by

$$\mathrm{d}E_j = \eta^* \mathrm{d}V \mathrm{d}\Omega, \tag{II.1.15}$$

where dE_j is the radiant energy emitted by the volume dV to the solid angle dΩ in a unit time.

In the transfer theory the coefficients κ^*, ω^*, k^*, p^* and η^* are usually considered as dependent on the position of the point P and independent of the direction r.

The angular distribution of the scattered radiation is described by means of the scattering phase function $(1/4\pi)\gamma(P, r', r)$ defined by

$$dE_\gamma = \frac{\gamma(P, r', r)}{4\pi} d\Omega dE_\omega = \frac{\omega^*}{4\pi} \gamma(P, r', r) d\Omega ds dE', \tag{II.1.16}$$

where $dE' = i'(P, r')d\sigma d\Omega' d\tau$ is the energy of the beam coming from the direction r', and r is the scattered beam direction.

The scattering phase function $(1/4\pi)\gamma$ defines that part of the total scattered radiation dE_ω which is scattered in the direction r to a unit solid angle. By integrating dE_γ over all scattering directions r we must obtain the total scattered radiation, i.e.

$$\int_{4\pi} dE_\gamma = dE_\omega = \omega^* ds dE',$$

from which we obtain

$$\frac{1}{4\pi} \int_{4\pi} \gamma(P, r', r) d\Omega \equiv 1, \tag{II.1.17}$$

which is a normalizing relation for the scattering phase function.

The above radiation characteristics of the medium are local ones which refer to an elementary path ds, an elementary area $d\sigma$, or an elementary volume dV at the point P. In addition to the above-indicated characteristics the more integral characteristics, viz. the attenuation and transmission functions will be used also.

The attenuation function K_i for radiation intensities is defined by

$$K_i(P, P', r) = \frac{i(P, r) - i(P', r)}{i(P, r)} \tag{II.1.18}$$

and *the transmission function*, by

$$P_i(P, P', r) = 1 - K_i(P, P', r) = \frac{i(P', r)}{i(P, r)}. \tag{II.1.19}$$

Analogous characteristics for radiation flux densities are determined similarly.

When studying the radiation regime in plant stands, however, the above-indicated local characteristics must be made more concrete by taking into account the geometry and radiation properties of plant organs in the stand and their distribution pattern in an elementary plant stand volume.

Reflection from the ground surface below a stand is specified in most detail by *the bidirectional reflection function* $\gamma_B(r', r)$ defined by

$$dE_\gamma = \gamma_B(r', r)|\cos \widehat{r'r}| d\Omega dE'. \tag{II.1.20}$$

The reflection function is normalized either to the *brightness coefficient* $b_B(r)$ by the relation

$$b_B(r)=\frac{\int_{\Omega_1'}\gamma_B(r',r)i(r')|\cos\vartheta'|d\Omega'}{\int_{\Omega_1'}i(r')|\cos\vartheta'|d\Omega'} \tag{II.1.21}$$

or to *the albedo* A_B by

$$A_B=\int_{\Omega_1}b_B(r)\cos\vartheta d\Omega=$$

$$=\frac{\int_{\Omega_1}\cos\vartheta d\Omega\int_{\Omega_1'}\gamma_B(r',r)|\cos\vartheta'|i(r')d\Omega'}{\int_{\Omega_1'}i(r')|\cos\vartheta'|d\Omega'} \tag{II.1.22}$$

II.1.3 OPTICAL CHARACTERISTICS OF PHYTOELEMENTS

One of the most detailed radiation characteristics of a phytoelement in the short-wave radiation region is the spectral scattering phase function $(1/\pi)\gamma_L$. In order to define it, consider Figure 37 in which a leaf elemental area $d\sigma_L$ is oriented in the direction r_L. Let monochromatic radiation fall on the leaf from the direction r', radiation intensity being $i(r')$.

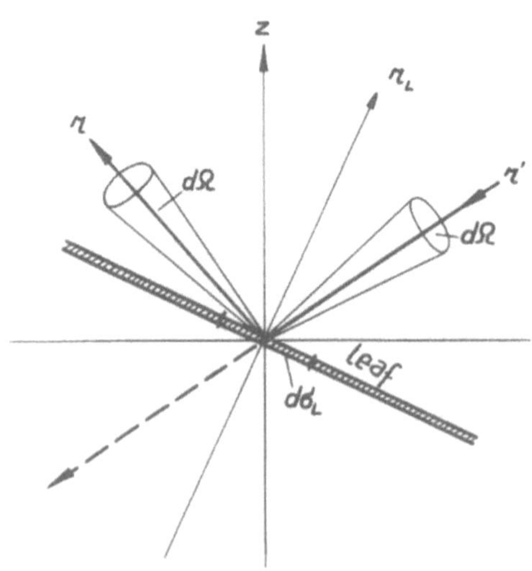

Fig. 37. Radiation scattered by leaf.

Then according to (II.1.1), $dE'_L = i(r')|\cos \widehat{r_L r}| d\sigma_L d\Omega' d\tau$ is the amount of radiation energy contained in the solid angle $d\Omega'$ and flowing through the leaf elemental area $d\sigma_L$ within the time $d\tau$. One part dE_κ of this energy is absorbed by the leaf element, the rest dE_ω is scattered in all possible directions. Let $dE_\kappa = \kappa_L dE'_L$ and $dE_\omega = \omega_L dE'_L$ where κ_L and ω_L are the leaf's absorption and scattering coefficients, respectively, for the incident monochromatic radiation intensity $i(r')$. According to the definition $\kappa_L + \omega_L = 1$.

Let us consider another direction r and, around it, a solid angle $d\Omega$ within which some part dE_γ of the radiant energy is scattered upon interacting with the leaf.

The leaf scattering phase function $(1/\pi)\gamma_L$ is defined by the relation

$$dE_\gamma = \frac{1}{\pi} \gamma_L(r_L, r', r) |\cos \widehat{r_L r}| d\Omega dE'_L. \tag{II.1.23}$$

According to (II.1.16) $(1/\pi)\gamma_L$ defines the part of incident radiation which is scattered from the leaf unit area perpendicular to the direction r_L to the unit solid angle around the direction r. The leaf scattering phase function $(1/\pi)\gamma_L$ is a nondimensional quantity.

In the general case, depending on the internal structure and optical properties of leaves, the leaf scattering function $(1/\pi)\gamma_L$ depends both on the coordinates of the incident beam and on those of the scattered one.

Now we will set a coordinate system so that the z axis coincides with the leaf normal r_L, and the x axis lies along the leaf's mid-rib from the leaf stalk to the leaf end point.

The direction r in the new coordinate system is also specified by the angle of inclination α and the azimuth angle β, so that with respect to the former coordinates

$$\alpha = \widehat{r_L r} = \arccos [\cos \vartheta_L \cos \vartheta + \sin \vartheta_L \sin \vartheta \cos \varphi_2], \tag{II.1.24}$$

$$\beta = \operatorname{arccot} \frac{\cot(\varphi_L - \varphi_1)[\cos \varphi_2 \cos \vartheta_L - \cot \vartheta \sin \vartheta_L] + \cos \vartheta_L \sin \varphi_2}{\cos \vartheta_L [\cot \vartheta \sin \vartheta_L - \cos \vartheta_L \cos \varphi_2] - \sin \varphi_2 \cot(\varphi_L - \varphi_1)}, \tag{II.1.25}$$

where $\varphi_2 = \varphi_L - \varphi$ and φ_1 is the azimuth angle of the leaf midrib relative to the former coordinate system.

By analogy with r the direction r' is specified by coordinates α' and β'.

Let us discuss the different cases of scattering using Figure 37.

If $0 \leqslant \alpha' \leqslant \pi/2$, then radiation is incident on the lower side leaf surface and we have the scattering phase function of the lower side leaf surface. Analogously, the condition $\pi/2 \leqslant \alpha' \leqslant \pi$ defines the scattering phase function for the upper side leaf surface. Since in the general case, the upper and lower sides of a leaf may scatter differently, the scattering phase function is determined separately for the upper (subscript 1) and the lower (subscript 2) side:

Table II.1.1 *Conditions of reflection from, and transmission through, a leaf*

	Limits of α'	Limits of α
Reflection from upper surface	$\frac{\pi}{2}, \pi$	$0, \frac{\pi}{2}$
Transmission through upper surface	$\frac{\pi}{2}, \pi$	$\frac{\pi}{2}, \pi$
Reflection from lower surface	$0, \frac{\pi}{2}$	$\frac{\pi}{2}, \pi$
Transmission through lower surface	$0, \frac{\pi}{2}$	$0, \frac{\pi}{2}$

$$\frac{1}{\pi}\gamma_L(\alpha', \beta', \alpha, \beta) = \begin{cases} \frac{1}{\pi}\gamma_{L1}, \frac{\pi}{2} \leqslant \alpha' \leqslant \pi; \\ \frac{1}{\pi}\gamma_{L2}, 0 \leqslant \alpha \leqslant \frac{\pi}{2}. \end{cases} \tag{II.1.26}$$

Scattering can be divided into reflection and transmission depending on the scattered beam direction. The different types of reflection and transmission are given in Table II.1.1.

In the general case the leaf scattering phase function is difficult to measure and too unwieldy for practical application. Therefore more integral characteristics of the leaf are used, viz. the coefficients of scattering ω_L, reflection R_L, transmission T_L and absorption κ_L. They are related to the scattering phase function by the following equations:

(a) *leaf's reflection coefficient*

$$R_{LK}(\alpha', \beta') = \frac{1}{\pi}\int_{\Omega_R} \gamma_{LK}(\alpha', \beta', \alpha, \beta)|\cos\alpha| d\Omega, \tag{II.1.27}$$

(b) *leaf's transmission coefficient*

$$T_{LK}(\alpha', \beta') = \frac{1}{\pi}\int_{\Omega_T} \gamma_{LK}(\alpha', \beta', \alpha, \beta)|\cos\alpha| d\Omega, \tag{II.1.28}$$

(c) *leaf's scattering coefficient*

$$\omega_{LK} = T_{LK} + R_{LK}, \tag{II.1.29}$$

(d) *leaf's absorption coefficient*

$$\kappa_{LK} = 1 - \omega_{LK}, \tag{II.1.30}$$

where the solid angles Ω_R and Ω_T equal to 2π are determined from Table II.1.1 and $k = 1$ and 2, the subscripts 1 and 2 denoting the upper and lower leaf side surfaces, respectively.

For very thick leaves and other plant parts $T_L = 0$, and scattering is reduced to reflection.

In the case of mat leaves $(1/\pi)\gamma_{L1}$ and $(1/\pi)\gamma_{L2}$ are constant, and scattering is independent of the incident and scattered beam direction, therefore R_{LK} and T_{LK} will not change.

When a leaf is exposed to diffuse radiation, the reflection and scattering coefficients will be expressed, respectively, by

$$R_{LJ_K} = \frac{\frac{1}{\pi}\int_{\Omega_K^L} i(r')|\cos\alpha'|d\Omega \int_{\Omega_R} \gamma_L|\cos\alpha|d\Omega}{\int_{\Omega_K^L} i(r')|\cos\alpha'|d\Omega'} \tag{II.1.27a}$$

and

$$T_{LJ_K} = \frac{\frac{1}{\pi}\int_{\Omega_K^L} i(r')|\cos\alpha'|d\Omega \int_{\Omega_T} \gamma_L|\cos\alpha|d\Omega}{\int_{\Omega_K^L} i(r')|\cos\alpha|d\Omega'} \tag{II.1.28a}$$

From (II.1.27a) and (II.1.28a) it follows that in the case of diffuse radiation the integral optical characteristics of a leaf R_{LJ_K}, T_{LJ_K}, ω_{LJ_K} and κ_{LJ_K} depend both on the leaf scattering function and on the radiation field structure.

In the long-wave radiation region phytoelements act as radiation sources in accordance with the Stefan-Boltzmann law

$$E_{LK} = \delta_{LK}\sigma(273 + t_{LK})^4, \tag{II.1.31}$$

where $k = 1,2$, and E_{LK} is the radiant energy emitted by a unit area of the leaf surface to the hemisphere above the leaf in a unit time, t_{LK} is the leaf surface temperature, δ_{LK} is the emissivity of the leaf k-th side, and σ is the Stefan-Boltzmann constant.

The long-wave radiation emitted by phytoelements is supposed to be isotropic, thus obeying the Lambert law. Therefore, taking into account (II.1.15) we obtain the following expressions for *the emission coefficient of phytoelements:*

$$\eta_{LK}(r_L) = \frac{1}{\pi} E_{LK}(r_L) \tag{II.1.32}$$

and

$$di_{LK} = \eta_{LK}|\cos \widehat{r_L r}|d\sigma d\Omega, \tag{II.1.33}$$

where di_{LK} is the amount of radiation energy emitted by a leaf elemental area $d\sigma_L$ perpendicular to the direction r_L to the solid angle $d\Omega$ around the direction r.

Radiation emission to both sides is

$$\eta_L = \eta_{L1} + \eta_{L2} = \frac{E_{L1} + E_{L2}}{\pi}. \tag{II.1.32a}$$

In expression (II.1.31) $E_{LK} = E_{LK}(r_L)$, i.e. leaf temperature can depend on leaf orientation.

Consider a turbid medium composed of tiny thin plates arranged in a certain fashion in space. We shall not specify the size and thickness of the plates nor their location in space. The quantity of elemental plates will be characterized not by their mass but by the total upper side surface area of all plates within a unit volume around the point P.

Let $u_L(P)$ denote the total upper side surface area of all plates within a unit volume around the point P. Suppose that the plates are oriented in all possible directions. We will characterize their orientation by means of the distribution function of normals to the upper surface of the plates, all normals being assumed to fall in the upper hemisphere.

As was defined under 1.2.2 $(1/2\pi)g_L(P, r_L)$ is the distribution function of the plate upper side area orientation, with the normalizing condition (I.2.16). The plates absorb, reflect and transmit radiation. Their scattering properties are defined by the scattering phase function $(1/\pi)\gamma_{Lk}(P, \alpha', \beta', \alpha, \beta)$ discussed in the preceding subsection.

Suppose that the plates also emit long-wave radiation which is isotropic but may be dependent on plate orientation. As shown in the preceding subsection, the emitting power of such plates is specified by the emission coefficient $\eta_{Lk}(P, r_L)$. Thus the geometrical structure of the plate medium is characterized by the functions $u_L(P)$ and $(1/2\pi)g_L(P, r_L)$, and the radiation properties of plates by the scattering phase function $(1/\pi)\gamma_{Lk}(P, \alpha', \beta', \alpha, \beta)$ and the emission coefficient $\eta_{Lk}(P, r_L)$.

Taking into account the above structural and optical characteristics we will derive some local optical characteristics of the plate medium.

Volume attenuation coefficient k_M^* due to shading. Consider Figure 38. Let a monochromatic elemental beam of intensity $i(P, r)$ fall on the surface $d\sigma$ of an

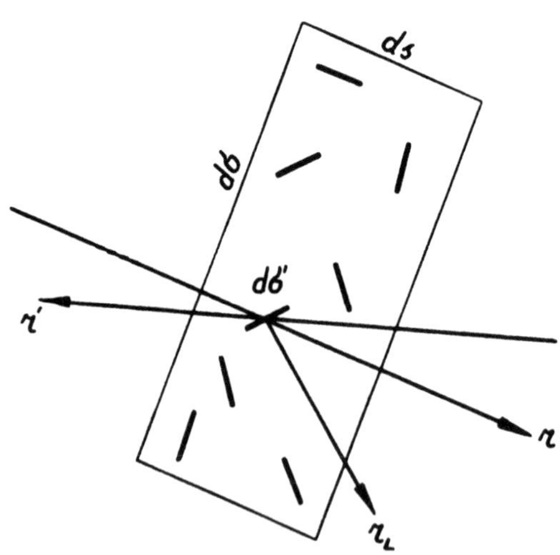

Fig. 38. Drawing to illustrate the definition of 'elemental' volume of plant stand.

elementary volume $dV = d\sigma ds$. In traversing the path ds it is so attenuated that now its intensity is $i(P, r) - di(P, r)$. The intensity attenuation di is in this case determined by the shading caused by the plates $d\sigma_L$ inside the volume dV. Consider a plate oriented in the direction r_L inside the volume dV. As seen from Figure 38, the shading effect of this plate is determined by the projection of its area on the direction r, i.e. its shading area is equal to $d\sigma_L|\cos \widehat{r_L r}|$. Suppose that the plates are so small and thin and are so arranged in dV that their mutual shading and lateral area can be neglected within the volume dV. Then the shading equal to the attenuation is determined by the projection of the area of all plates located in the volume dV on the direction r. The total area of plates within dV is $u_L dV$. The area of those plates whose normals fall within the solid angle $d\Omega_L$ around the direction r_L is equal to $(1/\pi) g_L u_L dV d\Omega_L$. The projection of this area on the direction r is $(1/2\pi) u_L g_L |\cos \widehat{r_L r}| dV d\Omega_L$. The projection of the area of all the plates on the same direction is found by summing the projected areas over all solid angles $d\Omega_L$, i.e. taking into account (I.2.18) we obtain

$$u_L(P) dV \frac{1}{2\pi} \int_{\Omega_1} g_L(P, r_L) |\cos \widehat{r_L r}| d\Omega_L = u(P) G(P, r) d\sigma ds.$$

Consequently we obtain the following expression for the attenuation of radiation intensity, di, over the path ds, due to the shading effect of plates:

$$di(P, r) = -u_L(P) G(P, r) i(P, r) ds. \tag{II.1.34}$$

Comparing expressions (II.1.12) and (II.1.34) we can conclude that the volume attenuation coefficient k^* for the plate medium is

$$k^*_M = u_L(P) G(P, r), \tag{II.1.35}$$

does not depend on the optical characteristics of the plates and is only determined by the structural characteristics of the medium. A significant feature of the attenuation coefficient k^*_M for a plate medium is its dependence not only on the point P but also on the direction r.

In the radiative transfer theory the function $G(P, r)$ discussed in greater detail in subsection 1.2.2 is the attenuation coefficient of the plate medium, related to a unit plate area, i.e.

$$k_M(P, r) = \frac{k^*_M(P, r)}{u_L(P)} = G(P, r). \tag{II.1.35a}$$

Expression (II.1.34) also defines the optical depth of the medium

$$\tau_M(s, r) = \int_{P_1}^{P_2} u_L(P) G(P, r) ds, \tag{II.1.35b}$$

which in the general case also depends on the direction r.

If the medium comprises several varieties of plates with the characteristics $u_k(P)$ and $G_K(P, r)$, then

$$\tau_M(s, r) = \int_{P_1}^{P_2} \sum_k u_k(P) G_k(P, r) \mathrm{d}s. \tag{II.1.35c}$$

Scattering phase function Γ_L^*. Refer to Figure 38 again. Let a monochromatic beam of diffuse radiation of intensity $i'(P, r')$ fall from the direction r' onto the plate of area $\mathrm{d}\sigma_L$ normal to the direction r_L. Suppose that the scattering properties of the plates are specified by the scattering phase function $(1/\pi)\gamma_L$. Determine the amount of radiation energy $\mathrm{d}E''_\gamma$ scattered by the elemental area $\mathrm{d}\sigma_L$ to the solid angle $\mathrm{d}\Omega$ in the direction r. According to (II.1.23) and (II.1.1),

$$\mathrm{d}E''_\gamma = \frac{1}{\pi} \gamma_L i' |\cos \widehat{r' r_L}| \cdot |\cos \widehat{r r_L}| \mathrm{d}\sigma_L \mathrm{d}\Omega \mathrm{d}\Omega'.$$

By integrating $\mathrm{d}\sigma_L = (1/2\pi) u_L g_L \mathrm{d}\Omega \mathrm{d}\sigma \mathrm{d}s$ over the orientations of all plates within the volume $\mathrm{d}V$ we obtain that part of radiant energy which is scattered from the beam $\mathrm{d}E' = i \mathrm{d}\sigma \mathrm{d}\Omega'$ incident on $\mathrm{d}\sigma$ in the direction r in the path length $\mathrm{d}s$ to the solid angle $\mathrm{d}\Omega$, i.e.

$$\mathrm{d}E_\gamma = \int_{\Omega_1} \mathrm{d}E''_\gamma = u_L \mathrm{d}E' \mathrm{d}\Omega \mathrm{d}s \int_{\Omega_1} \frac{\gamma_L}{\pi} \frac{g_L}{2\pi} \times$$
$$\times |\cos \widehat{r_L r'}| \cdot |\cos \widehat{r r_L}| \mathrm{d}\Omega_L \tag{II.1.36}$$

Comparing (II.1.16) and (II.1.36) we obtain

$$\omega^*(P) \frac{\gamma(P, r, r')}{4\pi} = u_L(P) \frac{\Gamma_L(P, r', r)}{\pi} = \frac{\Gamma_L^*(P, r', r)}{\pi}, \tag{II.1.37}$$

where

$$\frac{1}{\pi} \Gamma_L^*(P, r', r) = u_L(P) \int_{\Omega_1} \frac{g_L(P, r_L)}{2\pi} \times$$
$$\times \frac{\gamma_L(P, \alpha', \beta', \alpha, \beta)}{\pi} \cos \alpha' \cos \alpha \mathrm{d}\Omega_L \tag{II.1.38}$$

is the volume scattering phase function of the plate medium, which defines that part of the radiant energy coming in the direction r' which is scattered by a unit volume of the plate medium in the direction r within the unit solid angle. The quantity

$$\frac{1}{\pi} \Gamma_L(P, r', r) = \int_{\Omega_1} \frac{g_L}{2\pi} \frac{\gamma_L}{\pi} |\cos \widehat{r_L r'}| \cdot |\cos \widehat{r_L r}| \mathrm{d}\Omega_L \tag{II.1.38a}$$

is the area scattering phase function of the plate medium, i.e. the part of the radiation energy coming in the direction r' which is scattered by a unit area of the plate medium in the direction r within a unit solid angle.

From (II.1.38) it is clear that $(1/\pi)\Gamma_L$ contains both the radiation and structural characteristics of the plates.

Scattering and absorption coefficients. We obtain the normalizing relation for $(1/\pi)\Gamma_L^*(P, r', r)$ by multiplying (II.1.37) by $\mathrm{d}\Omega$ and integrating over all directions inside the solid angle 4π. Then, taking into account (II.1.17) we find

$$\omega_M^*(P, r') = \frac{1}{\pi} \int_{4\pi} \Gamma_L^*(P, r', r) \mathrm{d}\Omega \tag{II.1.39}$$

and correspondingly

$$\omega_M(P, r') = \frac{\omega_M^*(P, r')}{u_L(P)} = \frac{1}{\pi} \int_{4\pi} \Gamma_L(P, r', r) \mathrm{d}\Omega, \tag{II.1.40}$$

where $\omega_M(P, r')$ is the area scattering coefficient of the plate medium.

From (II.1.39) it is seen that the scattering phase function is normalized to the scattering coefficient of the plate medium.

Changing the sequence of integration over $\mathrm{d}\Omega$ and $\mathrm{d}\Omega_L$ in (II.1.40) and taking into account (II.1.27) and (II.1.28) we obtain

$$\begin{aligned} \omega_M(P, r') = & \int_{\Omega_{11}^L} \frac{g_L(P, r_L)}{2\pi} [R_{L1}(P, r', r_L) + \\ & + T_{L1}(P, r', r_L)] |\cos \widehat{r_L r'}| \mathrm{d}\Omega_L + \\ & + \int_{\Omega_{12}^L} \frac{g_L(P, r_L)}{2\pi} [R_{L2}(P, r', r_L) + \\ & + T_{L2}(P, r', r_L)] |\cos \widehat{r_L r'}| \mathrm{d}\Omega_L. \end{aligned} \tag{II.1.41}$$

If in a special case R_{Lk} and T_{Lk} are constant, i.e. we have matt plates,

$$\omega_M(P, r') = \omega_{L1} G_{11}(P, r') + \omega_{L2} G_{12}(P, r'), \tag{II.1.42}$$

where the functions

$$G_{11}(P, r') = \int_{\Omega_{11}^L} \frac{g_L(P, r_L)}{2\pi} |\cos \widehat{r_L r'}| \mathrm{d}\Omega_L \tag{II.1.43}$$

and

$$G_{12}(P, r') = \int_{\Omega_{12}^L} \frac{g_L(P, r_L)}{2\pi} |\cos \widehat{r_L r'}| \mathrm{d}\Omega_L \tag{II.1.44}$$

are only dependent on the geometrical characteristics and $G = G_{11} + G_{12}$.

If $\omega_L = \omega_{L1} = \omega_{L2}$, i.e. scattering is the same on both sides of the plates, then

$$\omega_M(P, r') = \omega_L G(P, r'). \tag{II.1.45}$$

An important conclusion can be drawn from (II.1.42): the scattering and absorption coefficients of a plate medium in the case of mat leaves still depend on the incident beam direction, this dependence being determined by the orientation of plates relative to the incident beam.

Taking into account (II.1.13), (II.1.35) and (II.1.39) we obtain accordingly for the volume absorption coefficient of a plate medium:

$$\kappa_M^*(P, r') = u_L(P) \left[G(P, r') - \frac{1}{\pi} \int_{4\pi} \Gamma_L(P, r', r) \mathrm{d}\Omega \right] \tag{II.1.46}$$

and correspondingly

$$\kappa_M(P, r') = G(P, r') - \frac{1}{\pi} \int_{4\pi} \Gamma_L(P, r', r) \mathrm{d}\Omega. \tag{II.1.47}$$

If $\omega_L = \omega_{L1} = \omega_{L2}$, then taking into account (II.1.13), (II.1.30) and (II.1.45) we obtain from (II.1.46) and (II.1.47) respectively:

$$\kappa_M^*(P, r') = \kappa_L u_L(P) G(P, r') \tag{II.1.46a}$$

and

$$\kappa_M(P, r') = \frac{\kappa_M^*}{u_L} = \kappa_L G(P, r'). \tag{II.1.47a}$$

It will be observed from (II.1.45) and (II.1.47a) that if the plates are mat and their optical properties are the same for both the upper and lower side surfaces, the absorption and scattering coefficients of the plate medium can be expressed through the absorption and scattering coefficients of separate plates, respectively. However, coefficients of the medium depend on the incident beam direction and are determined by the orientation of plates.

Emission coefficient η_M^*. Emission from the plates of $\mathrm{d}\sigma_L$ is defined by formula (II.1.33). Emission $\mathrm{d}E$ of the elemental volume $\mathrm{d}V$ is obtained by integrating (II.1.33) over all the plates in $\mathrm{d}V$:

$$\mathrm{d}E = \mathrm{d}\Omega \mathrm{d}V \frac{u_L}{2\pi} \int_{\Omega_1} g_L \eta_L |\cos \widehat{r_L r}| \mathrm{d}\Omega_L. \tag{II.1.48}$$

Comparing (II.1.48) and (II.1.15) we obtain the following expression for the volume emission coefficient of the plate medium:

$$\eta_M^*(P, r) = u_L(P) \eta_M(P, r), \tag{II.1.49}$$

where

$$\eta_M(P, r) = \frac{1}{2\pi} \int_{\Omega_1} g_L(P, r_L) \eta_L(P, r_L) |\cos \widehat{r_L r}| \mathrm{d}\Omega_L \tag{II.1.50}$$

is the area emission coefficient of the plate medium.

It is seen from (II.1.50) that the area emission coefficient of a plate medium is determined both by its radiation properties and its structure.

It is a distinguishing feature of all radiation characistics of a plate medium that they not only depend on the point P but also on the direction r. Consequently, a plate medium is optically anisotropic, which of course complicates the mathematical modelling of its radiation field.

Another feature of a plate medium is that the amount of scattering elements is best expressed by their area rather than by their mass.

II.1.5 THE RADIATION TRANSFER EQUATION FOR AN OPTICALLY ANISOTROPIC PLATE MEDIUM

The equation of the radiation transfer in a turbid medium in the case of a steady radiation field is (Marchuk 1961)

$$\nabla i\cdot r = -k^*i + \eta^* + \frac{\omega^*}{4\pi}\int_{4\pi} i(P, r')\gamma(P, r', r)\mathrm{d}\Omega', \qquad \text{(II.1.51)}$$

where ∇ is the Hamiltonian operator. For short-wave radiation (also for PAR and NIR), $\eta^*=0$; and for long-wave radiation, $\omega^*=0$ and $k^*=\kappa^*$.

In order to describe the radiation field in a plate medium the transfer equation (II.1.51) must be modified by replacing the turbid-medium characteristics k^*, η^* and $(\omega^*/4\pi)\gamma$ with the corresponding characteristics k_M^*, η_M^* and $(1/\pi)\Gamma_L^*$ appropriate to the plate medium. This has been done for shortwave radiation (Ross 1964) and for long-wave radiation (Budagovsky, Ross & Tooming 1968).

In the general form the modified equation of radiation transfer in a plate medium is

$$\nabla i(P, r)\cdot r = -u_L(P)G(P, r)i(P, r) + u_L(P)\eta_L(P, r) + $$
$$+\frac{1}{\pi}u_L(P)\int_{4\pi}\Gamma_L(P, r', r)i(P, r')\mathrm{d}\Omega'. \qquad \text{(II.1.52)}$$

Equation (II.1.52) differs from (II.1.51) in that the optical characteristics of a plate medium G, η_L and Γ_L, are more complex in form and depend not only on the point P but also on the direction r.

Consider the special case where the plate medium is planeparallel and horizontally homogeneous, i.e. the functions contained in equation (II.1.52) do not depend on the x and y coordinates. Then taking into account (I.2.39),

$$\nabla i\cdot r/u_L = \frac{\partial i}{\partial z}\cos\vartheta/u_L = -\frac{\partial i}{\partial L}\cos\vartheta$$

and instead of (II.1.52) we obtain

$$-\frac{\partial i(L, r)}{\partial L}\cos\vartheta = -G(L, r)i(L, r) + \eta_L(L, r) + $$
$$+\frac{1}{\pi}\int_{4\pi}\Gamma_L(L, r', r)i(L, r')\mathrm{d}\Omega'. \qquad \text{(II.1.53)}$$

The boundary conditions for (II.1.53) are

$$\left.\begin{aligned} & i(0, r) = q_0(r), \qquad \frac{\pi}{2}\leqslant\vartheta\leqslant\pi, \\ & i(L_0, r) = b_B(r)\int_{\Omega_2} i(L_0, r')|\cos\vartheta'|\mathrm{d}\Omega', \\ & \frac{\pi}{2}\leqslant\vartheta'\leqslant\pi \qquad 0\leqslant\vartheta\leqslant\frac{\pi}{2}, \end{aligned}\right\} \qquad \text{(II.1.54)}$$

where $q_0(r)$ is the intensity of the total solar radiation above the stand, and $b_B(r)$ is the brightness coefficient of the ground surface below the stand given by (II.1.21).

Dividing the short-wave radiation field into two components in accordance with (II.1.2) and taking into account (II.1.3) we obtain a set of two equations instead of (II.1.53):

$$\left.\begin{aligned}
&\cos\theta\frac{\partial i_1(L,r)}{\partial L}=-G(L,r)i_1(L,r)+\\
&+\frac{1}{\pi}\int_{\Omega_1}\Gamma(L,r',r)i_1(L,r')\mathrm{d}\Omega'+\\
&+\frac{1}{\pi}\int_{\Omega_2}\Gamma(L,r',r)i_2(L,r')\mathrm{d}\Omega',\\
&\cos\vartheta\frac{\partial i_2(L,r)}{\partial L}=G(L,r)i_2(L,r)-\\
&-\frac{1}{\pi}\int_{\Omega_1}\Gamma(L,r',r)i_1(L,r')\mathrm{d}\Omega'-\\
&-\frac{1}{\pi}\int_{\Omega_2}\Gamma(L,r',r)i_2(L,r')\mathrm{d}\Omega'
\end{aligned}\right\} \qquad (II.1.55)$$

with the boundary conditions

$$\begin{aligned}
&i_1(0,r)=q_0(r),\\
&i_2(L_0,r)=b_B(r)\int_{\Omega_1}i_1(L_0,r')\cos\theta'\mathrm{d}\Omega'.
\end{aligned} \qquad (II.1.56)$$

The transfer equation for long-wave radiation is given in Chapter II.10.

II.1.6 PLANT STAND AS A PLATE MEDIUM

In deriving the radiative transfer equation for a plate medium we proceeded from a few contradictory assumptions. On the one hand the assumed elementary volume is so small that no mutual shading exists in any direction within it. On the other hand the number of plates in $\mathrm{d}V$ was assumed to be so great that the functions $u_L(P)$ and $(1/2\pi)g_L(P, r_L)$ are realized to the necessary degree of accuracy. The radiation intensities for which the transfer equation is set down are essentially mean values for $\mathrm{d}V$. It was also assumed that the functions contained in the transfer equation (II.1.52) and dependent on five variables x, y, z, ϑ and φ are continuous.

Let us discuss how these basic assumptions are realized for vegetation and how a plant stand can be modelled as a plate-type turbid medium.

Suppose we have a sufficiently dense and horizontally comparatively uniform plant stand. Segregate a volume ΔV within the stand so that this volume contains several leaves without any considerable mutual shading in any direction. Denote the area of leaves in this volume as ΔS_L. Divide the leaf surface within ΔV into small elementary areas and define the law of distribution of their normals by $g_L(\Delta V, r_L)$. Displacing the volume ΔV in any direction within the stand we obtain the values of $u_{Li} = \Delta S_{Li}/\Delta V$ and the function $g_{Li}(\Delta V, r_L)$.

It is shown by experiment that the quantities u_{Li} and g_{Li} vary irregularly and discontinuously at any displacement of ΔV. In view of this it can be concluded that if the requirement of no mutual shading within the elementary volume ΔV in a plant stand is maintained, the functions u_L and g_L are not continous functions of x, y and z and should be treated as statistically random variables. Consequently, the structural characteristics and the radiation regime in plant stands are statistical in character and require an appropriate mathematical approach. But if this statistical character is neglected and average values are considered as a first approximation, there is the problem of how this averaging is to be done and what scales of average are to be used.

Treating a plant stand as an infinite plane-parallel horizontal layer and taking into account the fact that most radiation fluxes come downwards to the stand, we find that the largest variations in the average characteristics of the radiation field occur in directions close to the vertical. In this case it would be wise to perform an averaging over a horizontal plane and consider the averaged values as functions of z. Then the statistical character of these values is revealed as ΔV is displaced along x and y. The elementary volume ΔV is a thin vertical cylinder with a height Δz and base area σ_0, i.e. $\Delta V = \sigma_0 \Delta z$. The value of Δz is determined by the size of leaves and apparently is of the same order of magnitude. The scale of the horizontal average $\sqrt{\sigma_0} \gg \Delta z$, and the base area dimensions are determined both by the plant habitus and the distribution of plants in the stand.

It is now clear that the requirement of there being no mutual shading of leaves in any direction is no longer satisfied in this elementary volume. This requirement is satisfied well enough in directions close to the vertical, but shading will increase as directions approach the horizontal. That is why expressions (II.1.35), (II.1.49) and (II.1.38) for the coefficients of attenuation k_L^*, emission η_L^*, and the scattering function $(1/\pi)\Gamma_L^*$, respectively, are inaccurate and overestimate values in the case of directions close to the horizontal.

The structural and optical characteristics of the plate medium u_L, $(1/2\pi)g_L$, $(1/\pi)\gamma_{Lk}$ and η_{Lk} are determined for each layer Δz by averaging over the horizontal area σ_0. Consequently, the values of the radiation field intensity determined from the transfer equation are also averaged over the area σ_0.

Unfortunately, practically no evaluations of the limits of application of the radiative transfer equation for the description of the radiation field in plant stands are available. Nor are there sufficient data to show how the scale of average for σ_0 is related to the structural characteristics of plants and their distribution within the stand. If average characteristics of the medium do not vary in a horizontal plane, the plant stand can be treated as statistically homogeneous in horizontal directions. Only comparison between calculated and experimental data can provide an indirect solution to the problem.

In the case of sparse plant stands (sparse crops, a sparse forest, a plantation, etc.) it would be wise to reject the plane-parallel horizontally-homogeneous layer concept and to model a plant stand as geometrical figures filled with the plate medium. Then the average values of structural characteristics u_L and $(1/2\pi)g_L$ will be assigned within the figures, and the radiation field calculated both for them and for the stand.

In a more simplified treatment the mutual shading of figures may be neglected, particularly for high solar elevations. Some studies carried out on these lines are discussed under II.4.4.

Though in most plant stand types the scattering area is primary provided by leaves, in many cases other plant parts, such as the flowers and heads in sunflower, the stems and ears in cereals, etc. should also be taken into account. In such cases the plant stand should be treated as a multicomponent medium, with each component characterized by its structural and radiation functions u_K, $(1/2\pi)g_K$, $(1/\pi)\gamma_{KK}$ and η_{KK}. The corresponding functions in the transfer equation (II.1.52) are to be replaced with sums: $u_L G_L \rightarrow \sum_K u_K G_K$, $u_L \eta_L \rightarrow \sum_K u_K \eta_K$, and $u_L \Gamma_L \rightarrow \sum_k u_k \Gamma_k$. A similar course should be taken if the plant stand includes several species with distinctly different characteristics (mixed crops, mixed forests, etc.).

The plate-medium model is obviously not applicable to a coniferous forest. As a first approximation, the orientation function $(1/2\pi)g_L$ should be discarded and the classical transfer equation (II.1.51) used instead.

II.1.7 THE RADIATION TRANSFER EQUATION FOR A PLANT STAND

Under steady-state irradiation conditions the short-wave radiation field in a plant stand is determined by the following factors.

1. The spectral quality and spatial distribution of the direct solar radiation incident on the plant stand at its upper level and those of the diffuse sky radiation, which are characterized by the irradiance of direct solar radiation above the stand, $S_0(r_S)$, and the sky brightness $d_0(r)$ in the direction r. Zenith angles $\theta_\odot$ and θ are measured in the range from zero to $\pi/2$, and azimuth angles $\varphi_\odot$ and φ from zero to 2π.

The monochromatic irradiance of total radiation is defined by the expression

$$q_0(r) = S_0(r_S)\delta(r - r_S) + d_0(r), \qquad \text{(II.1.57)}$$

where $\delta(r - r_S)$ is Dirac's delta-function, provided that

$$\int_{\Omega_1} f(r)\delta(r - r')\mathrm{d}\Omega = f(r')$$

and the total solar monochromatic irradiance above the stand is expressed by

$$Q_0 = \int_{\Omega_1} q_0(r)\cos\theta \mathrm{d}\Omega = S_0'(r_S) + D_0, \qquad \text{(II.1.58)}$$

where

$$S_0'(r_S) = S_0(r_S)\sin h_\odot \tag{II.1.59}$$

and

$$D_0 = \int_{\Omega_1} d_0(r)\cos\theta \, d\Omega \tag{II.1.60}$$

is the monochromatic irradiance of diffuse sky radiation above the stand.

2. Spectral reflection properties of the ground surface below the stand characterized by the spectral brightness coefficient $b_B(r)$ given by (II.1.21).

3. Spectral scattering properties of plant organs (predominantly leaves) specified by the spectral scattering phase functions $(1/\pi)\gamma_{k_1}$ and $(1/\pi)\gamma_{k_2}$.

4. Plant stand architecture which is generally interpreted as the aggregate of characteristics defining the shape and dimensions of leaves, stems and other plant parts, their spatial orientation and location, the geometry of the habitus of individual plants, their distribution within the stand, total number of scattering elements, spatial distribution of phytoarea, etc.

Note that most difficulties in the mathematical modelling of the radiation regime in plant stands are caused by the qualitative description of the plant stand architecture.

From a mathematical standpoint factors 1 and 2 are the boundary conditions for the radiation transfer equation, and factors 3 and 4 determine the properties of the scattering medium. In view of the radiation field complexity it is convenient to divide a radiation field into four components and write a separate equation for each of them.

1. The direct solar radiation $S(z, r_S)$ which penetrates as parallel beams from the upper level of the stand to a height z without interacting with the stand. This component is discrete in character at a point within the stand. Neglecting the penumbra effect it is equal to S_0 in the direction r_S and on condition that there are no plant parts between the sun and the given point at the height z. It is zero in all other cases.

The attenuation of $S(z, r_S)$ in a plant stand is measured by the decrease in the number and size of the horizontal areas irradiated by direct irradiation (gaps) with depth. Assuming $\Gamma_L = \eta_L = 0$ in the transfer equation (II.1.53), we obtain a transfer equation for the irradiance of direct solar radiation $S(z, r_S)$ in the form

$$\frac{\partial S(z, r_S)}{\partial z} = \sec\theta_\odot \, \partial \cdot S(z, r_S) \sum_k u_k(z) G_k(z, r_S). \tag{II.1.61}$$

The boundary condition for S: at the plant stand upper level, $z = z_0$

$$S(z_0, r_S) = S_0(r_S) \tag{II.1.61a}$$

If the plant stand comprises leaves, stems and reproductive organs, then according to (II.1.35c) its optical depth is

$$\tau=\int_{z}^{z_0}(u_L G_L+u_C G_C+u_F G_F)\mathrm{d}z'.$$

Assuming that $G_L(z, r)$, $G_C(z, r)$ and $G_F(z, r)$ are independent of height, τ may be expressed by means of the relative area of plant organs $L(z)$, $L_C(z)$ and $L_F(z)$ in accordance with (I.2.32). In this case

$$\tau=G_L L+G_C L_C+G_F L_F=\bar{G}\bar{L}, \tag{II.1.62}$$

where

$$\left.\begin{aligned}\bar{L}&=L+L_C+L_F\\ \bar{G}&=\beta_L G_L+\beta_C G_C+\beta_F G_F\end{aligned}\right\} \tag{II.1.62a}$$

and

$$\beta_k=\frac{L_k}{\bar{L}} \tag{II.1.62b}$$

The penetration function of direct solar radiation, i.e. the relative irradiance of direct solar radiation inside the stand a_S for S is defined, in accordance with (II.1.19), by the expression

$$a_S(L, r_S)=\frac{S(L, r_S)}{S_0(r_S)}. \tag{II.1.63}$$

The penetration function a_S should be interpreted as the probability that at an arbitrarily chosen point at the depth L the sky is visible in the direction r_S. From (II.1.63) it follows that $a_S=1$ at the plant stand upper level.

2. The mean sky brightness $d(z, r)$ which reaches the height z as diffuse radiation without interacting with the stand. All that has been said about the penetration of direct radiation likewise holds for diffuse radiation, except that diffuse sky radiation reaches a plant stand from all directions of the upper hemisphere. Therefore by analogy with (II.1.61) we can write a transfer equation for d in the form

$$\frac{\partial \mathrm{d}(z, r)}{\partial z}=\sec\theta\cdot d(z, r)\sum_k u_k(z)G_k(z, r) \tag{II.1.64}$$

with the boundary condition

$$d(z_0, r)=d_0(r). \tag{II.1.65}$$

The penetration function of the diffuse solar radiation intensities is

$$a_d(L, r)=\frac{d(L, r)}{d_0(r)}. \tag{II.1.66}$$

The irradiance of diffuse solar radiation inside the stand averaged over the horizontal surface at the depth L, $D(L)$, is defined by the expression

$$D(L)=\int_{\Omega_1} d(L, r)\cos\theta\,\mathrm{d}\Omega \tag{II.1.67}$$

and the penetration function of diffuse solar radiation, i.e. the relative irradiance of diffuse solar radiation is defined by the expression

$$a_D(L) = \frac{\int_{\Omega_1} d(L, r) \cos\theta \, d\Omega}{\int_{\Omega_1} d_0(r) \cos\theta \, d\Omega}. \tag{II.1.68}$$

As seen from the above formulae, S and d are practically independent of the spectral scattering properties of plant stand elements and are determined only by the geometrical structure of the stand and the angular structure of the incident radiation. Therefore a_S and a_D too are independent of the radiation wavelengths and are determined by geometrical characteristics.

From the condition $\Gamma_L = 0$ assumed in deriving the transfer equations (II.1.61) and (II.1.64) it follows that for the S and D radiation $\omega_k = 0$, i.e. the plant organs are absolutely black.

3. The intensity of the complementary direct solar radiation due to the scattering of direct radiation on the phytomass elements and reflection from the ground surface.

The complementary radiation field i_S is defined by a transfer equation similar to (II.1.52). However, in many cases multiple scattering may be neglected and first-order scattering assumed only. Therefore in the term under the integral sign it is convenient to separate out that intensity $i_S(r')$ which represents the initial beam of direct solar radiation reaching the level z without attenuation, i.e. the beam $S(z, r_S)$. Then for the complementary field i_S we obtain a transfer equation in the form

$$\begin{aligned} \cos\vartheta \frac{\partial i_S(z, r)}{\partial z} &= -\sum_k u_k(z) G_k(z, r) i_S(z, r) + \\ &+ \sum_k u_k(z) \int_{4\pi} \frac{\Gamma_k(z, r', r)}{\pi} i_S(z, r') d\Omega' + \\ &+ \sum_k u_k(z) \frac{\Gamma_k(z, r_S, r)}{\pi} S(z, r_S). \end{aligned} \tag{II.1.69}$$

The boundary conditions for i_S are

$$\left.\begin{aligned} & i_S(z_0, r) = 0 \\ & i_S(0, r) = \int_{\Omega_1} \gamma_B(r', r)[i_S(0, r') + \\ & + S(0, r_S)\delta(r' - r_S)] \cos\theta' d\Omega' \end{aligned}\right\} \tag{II.1.70}$$

provided that

$$0 \leqslant \vartheta \leqslant \frac{\pi}{2}, \qquad 0 \leqslant \theta' \leqslant \frac{\pi}{2}, \qquad 0 \leqslant \theta \leqslant \frac{\pi}{2}.$$

If the plant stand consists of leaves only, then taking into account (II.1.2) and (II.1.55) we obtain the following set of equations for i_{S1} and i_{S2} from (II.1.69):

$$\left.\begin{aligned}\cos\theta\frac{\partial i_{S1}}{\partial L}&=-Gi_{S1}+\int_{\Omega_1}\frac{\Gamma_L}{\pi}i_{S1}\mathrm{d}\Omega'+\\&+\int_{\Omega_2}\frac{\Gamma_L}{\pi}i_{S2}\mathrm{d}\Omega'+\frac{\Gamma_L}{\pi}S,\\\cos\vartheta\frac{\partial i_{S2}}{\partial L}&=Gi_{S2}-\int_{\Omega_1}\frac{\Gamma_L}{\pi}i_{S2}\mathrm{d}\Omega'-\\&-\int_{\Omega_2}\frac{\Gamma_L}{\pi}i_{S2}\mathrm{d}\Omega'-\frac{\Gamma_L}{\pi}S\end{aligned}\right\}\tag{II.1.69a}$$

with the boundary conditions

$$\begin{aligned}&i_{S1}(0,r)=0\\&i_{S2}(L_0,r)=\gamma_B(r_S,r)S'(L_0,r_S)+\\&+\int_{\Omega_1}\gamma_B(r',r)i_{S1}(L_0,r)\cos\theta'\mathrm{d}\Omega'\end{aligned}\tag{II.1.70a}$$

In the case of a mat ground surface, i.e. $\gamma_B=A_B/\pi$

$$i_{S2}(L_0,r)=\frac{A_B}{\pi}F_{S1}(L_0),$$

where

$$F_{S1}(L)=S'(L,r_S)+\int_{\Omega_1}i_{S1}(L,r')\cos\theta'\mathrm{d}\Omega'.\tag{II.1.71}$$

The downward and upward irradiances of complementary direct solar radiation i_S, according to (II.1.7), will be

$$\left.\begin{aligned}H_{S1}(L)&=\int_{\Omega_1}i_{S1}(L,r)\cos\theta\mathrm{d}\Omega\\H_{S2}(L)&=\int_{\Omega_2}i_{S2}(L,r)\cos\vartheta\mathrm{d}\Omega\end{aligned}\right\}\tag{II.1.72}$$

and the penetration functions (relative irradiances),

$$\begin{aligned}a^H_{S1}(L)&=\frac{H_{S1}(L)}{S'_0}\\a^H_{S2}(L)&=\frac{H_{S2}(L)}{S'_0}.\end{aligned}\tag{II.1.73}$$

4. The intensity of the complementary diffuse solar radiation $i_D(z,r)$ due to the scattering of diffuse sky radiation d on the phytomass elements and its reflection

from the ground surface. By analogy with i_S the transfer equation for i_D has the form

$$\cos\vartheta\frac{\partial i_D(z,r)}{\partial z} = -\sum_k u_k(z)G_k(z,r)i_D(z,r)+ + \sum_k u_k(z)\int_{4\pi}\frac{\Gamma_k(z,r',r)}{\pi}i_D(z,r')\mathrm{d}\Omega' + + \sum_k u_k(z)\int_{\Omega_1}\frac{\Gamma_k(z,r',r)}{\pi}d(z,r')\mathrm{d}\Omega' \tag{II.1.74}$$

with the boundary conditions

$$\left.\begin{aligned} & i_D(z_0,r)=0 \\ & i_D(0,r')=\int_{\Omega_1}\gamma_B(r',r)[i_D(0,r)+d(0,r)]\cos\theta\mathrm{d}\Omega \\ & 0\leqslant\vartheta\leqslant\frac{\pi}{2},\qquad 0\leqslant\theta\leqslant\frac{\pi}{2} \end{aligned}\right\} \tag{II.1.75}$$

If the plant stand consists of leaves only, then from (II.1.74) we obtain the following set for i_{D1} and i_{D2}:

$$\left.\begin{aligned} & \cos\theta\frac{\partial i_{D1}}{\partial L} = -Gi_{D1}+\int_{\Omega_1}\frac{\Gamma_L}{\pi}i_{D1}\mathrm{d}\Omega'+ \\ & +\int_{\Omega_2}\frac{\Gamma_L}{\pi}i_{D2}\mathrm{d}\Omega'+\int_{\Omega_1}\frac{\Gamma_L}{\pi}d\cdot\mathrm{d}\Omega', \\ & \cos\vartheta\frac{\partial i_{D2}}{\partial L} = Gi_{D2}-\int_{\Omega_1}\frac{\Gamma_L}{\pi}i_{D1}\mathrm{d}\Omega'- \\ & -\int_{\Omega_2}\frac{\Gamma_L}{\pi}i_{D2}\mathrm{d}\Omega'-\int_{\Omega_1}\frac{\Gamma_L}{\pi}d\cdot\mathrm{d}\Omega' \end{aligned}\right\} \tag{II.1.74a}$$

with the boundary conditions

$$\left.\begin{aligned} & i_{D1}(0,r)=0, \\ & i_{D2}(L_0,r)=\int_{\Omega_1}\gamma_B(r',r)d(L_0,r)\cos\theta'\mathrm{d}\Omega'+ \\ & +\int_{\Omega_1}\gamma_B(r',r)i_{D2}(L_0,r')\cos\theta\mathrm{d}\Omega'. \end{aligned}\right\} \tag{II.1.75a}$$

Expressions analogous to (II.1.72) and (II.1.73) will be obtained for the irradiances $H_{D1}(L)$ and $H_{D2}(L)$ and the penetration functions $a^H_{D1}(L)$ and $a^H_{D2}(L)$.

Transfer equations (II.1.69) and (II.1.74) are more complicated in comparison with the equations commonly used in astrophysics and atmospheric physics.

First, they include the sums $\sum_k u_k G_k$ and $\sum_k u_k \Gamma_k$ which take into account the plant stand architecture, the scattering properties and the amount of the phytomass both of separate plants parts (leaves, stems, flowers, etc.) and of different species of plants in the stand. Second, the generalized scattering phase function $(1/\pi)\Gamma_K(z, r', r)$ is not governed by the angle $\widehat{r'r}$ between the incident and reflected rays, as is usually the case, but the directions r' and r involved in $(1/\pi)\Gamma_K$ as independent arguments. In the general case, therefore, $(1/\pi)\Gamma_K(z, r', r)$ cannot be developed as a series in Legendre polynomials. Third, G_K functions depend on the direction r.

The second and the third circumstance imply that the turbid medium (plant stand) is anisotropic and that in solving the transfer equation no transition is possible to a direction independent optical depth of the type $\tau = \int \rho k \mathrm{d}z'$, where ρ is the turbid medium density and $k(z')$ is the mass coefficient of attenuation.

Thus the intensity of downward total solar radiation averaged over a horizontal surface inside the plant canopy $i_{Q1}(z, r)$ has four components and is expressed by their sum:

$$i_{Q1}(z, r) = S(z, r_S)\delta(r - r_S) + d(z, r) + i_{S1}(z, r) + i_{D1}(z, r). \qquad \text{(II.1.76)}$$

The intensity of upward total solar radiation averaged over a horizontal surface inside the plant canopy $i_{Q2}(z, r)$ has two components:

$$i_{Q2}(z, r) = i_{S2}(z, r) + i_{D2}(z, r) \qquad \text{(II.1.77)}$$

Expressions (II.1.76) and (II.1.77) finally define the radiation field inside the plant canopy.

Of course, we cannot distinguish all components of radiation by direct measurement, except for S, though experiments can be conceived for their indirect determination.

Hereafter it will be suitable to use irradiances in all possible directions instead of radiation intensities for solving biogeophysical problems.

Transition from intensities $i(z, r)$ to irradiances $J(z, r_L)$ is performed in accordance with the formulae (II.1.2) and (II.1.3). Taking into account (II.1.76) we obtain the following expression for the average irradiance on the upper side of a phytoelement:

$$\begin{aligned} J_{Q1}(z, r_L) &= \int_{\Omega_1^L} i(z, r)|\cos \widehat{r_L r}|\mathrm{d}\Omega = \\ &= S(z, r_S)|\cos \widehat{r_L r}| + \int_{\Omega_{11}^L} d(z, r)|\cos \widehat{r_L r}|\mathrm{d}\Omega + \\ &+ \int_{\Omega_1^L} i_S(z, r)|\cos \widehat{r_L r}|\mathrm{d}\Omega + \int_{\Omega_1^L} i_D(z, r)|\cos \widehat{r_L r}|\mathrm{d}\Omega, \end{aligned} \qquad \text{(II.1.78)}$$

where

$$\frac{\pi}{2} \leqslant \widehat{r_L r_S} \leqslant \pi, \qquad \frac{\pi}{2} \leqslant \widehat{r_L r} \leqslant \pi.$$

The average irradiance $J_{Q2}(z, r_L)$ on the lower side of á phytoelement, in accordance with (II.1.3) and (II.1.77), will be expressed by

$$J_{Q2}(z, r_L) = \int_{\Omega_{12}^L} d(z, r) \cos \widehat{r_L r} d\Omega + \int_{\Omega_2^L} i_S(z, r) \cos \widehat{r_L r} d\Omega +$$

$$+ \int_{\Omega_2^L} i_D(z, r) \cos \widehat{r_L r} d\Omega, \qquad \text{(II.1.79)}$$

where

$$0 \leqslant \widehat{r_L r} \leqslant \frac{\pi}{2}$$

If $0 \leqslant \widehat{r_L r_S} \leqslant \pi/2$, then the term $S(z, r_S)|\cos r_L r_S|$ in the expression (II.1.78) changes to expression (II.1.79). The irradiance on a horizontal surface alone is usually sufficient for approximate calculations.

From (II.1.78) and (II.1.79), with (II.1.7), (II.1.72) and (II.1.73) taken into account, we obtain the following expressions for the total irradiance inside the plant canopy averaged over a horizontal surface:

(a) for clear-day conditions

$$Q(L) = F_1(L) = S'(L) + D(L) + H_{S1}(L) + H_{D1}(L) =$$

$$= S'_0[a_S(L) + a_{S1}^H(L)] + D_0[a_D(L) + a_{D1}^H(L)], \qquad \text{(II.1.80)}$$

$$F_2(L) = H_{S2}(L) + H_{D2}(L) = a_{S2}^H(L) S'_0 + a_{D2}^H(L) D_0 \qquad \text{(II.1.81)}$$

(b) for overcast-day conditions

$$Q^C(L) = D(L) + H_{D1}(L) = [a_D(L) + a_{D1}^H(L)] D_0, \qquad \text{(II.1.82)}$$

$$F_2(L) = H_{D2}(L) = a_{D2}^H(L) D_0. \qquad \text{(II.1.81a)}$$

Dividing clear-day radiation inside a plant canopy into two parts, that in gaps and that in shade, we obtain this expression for the gap region:

$$Q^0(L) = S'_0 + D(L) + H_{S1}(L) + H_{D1}(L) = S'_0 + Q^T(L), \qquad \text{(II.1.83)}$$

where

$$Q^T(L) = D(L) + H_{S1}(L) + H_{D1}(L) =$$

$$= a_{S1}^H S'_0 + [a_D + a_{D1}^H] D_0 \qquad \text{(II.1.84)}$$

is the total irradiance in shade, with

$$Q(L) = S'(L) + Q^T(L) \qquad \text{(II.1.85)}$$

The determination of i_1 and i_2 or J_{Q1} and J_{Q2} in the general case by solving the corresponding transfer equations offers considerable mathematical difficulties. Even when the calculations are accomplished successfully, the information obtained is too detailed for practical purposes. That is why it is of great practical interest to find ways of simplifying the problem. This will be discussed in the following chapters.

II.1.8 LEAF AND PLANT STAND ABSORPTION FUNCTIONS

Let a unit leaf surface oriented in the direction r_L be located at the height z in a radiation field specified by intensities $i_{Q1}(z, r)$ and $i_{Q2}(z, r)$ (formulae (II.1.76) and (II.1.77)).

Determine the amount of radiation absorbed by the leaf separately for the upper and the lower leaf side. The absorption function of the upper side leaf surface $\Pi_1(z, r_L)$ will be expressed by

$$\Pi_1(z, r_L) = \int_{\Omega_{11}^L} \kappa_{L1}(z, \alpha, \beta) i_{Q1}(z, r)|\cos\alpha| d\Omega + \\ + \int_{\Omega_{21}^L} \kappa_{L1}(z, \alpha, \beta) i_{Q2}(z, r)|\cos\alpha| d\Omega, \quad \text{(II.1.86)}$$

and that of the lower side leaf surface $\Pi_2(z, r_L)$ by

$$\Pi_2(z, r_L) = \int_{\Omega_{12}^L} \kappa_{L2}(z, \alpha, \beta) i_{Q1}(z, r)|\cos\alpha| d\Omega + \\ + \int_{\Omega_{22}^L} \kappa_{L2}(z, \alpha, \beta) i_{Q2}(z, r)|\cos\alpha| d\Omega, \quad \text{(II.1.87)}$$

where $\alpha = \widehat{r_L r}$ and the solid angles Ω_{iK}^L are defined as in Figure 36.

Taking into account (II.1.76) and (II.1.77) the leaf absorption function will be expressed as

$$\Pi_L(z, r_L) = \Pi_1 + \Pi_2 = \Pi_{LS} + \Pi_{LD} + \Pi_{LIS} + \Pi_{LID}. \quad \text{(II.1.88)}$$

If the absorption coefficients κ_{L1} and κ_{L2} are independent of the incident beam direction r and are equal for both leaf sides,

$$\Pi_{LS}(z, r_L, r_S) = \kappa_L(z) S(z, r_S)|\cos\alpha_S|, \quad \text{(II.1.89)}$$

$$\Pi_{LD}(z, r_L) = \kappa_L(z) J_D(z, r_L) \quad \text{(II.1.90)}$$

and

$$\Pi_{LIS}(z, r_L) = \kappa_L(z) I_{S1+S2}(z, r_L), \quad \text{(II.1.91)}$$

where

$$J_D=\int_{\Omega_{11}^L} d(z, r)|\cos\alpha|\mathrm{d}\Omega+\int_{\Omega_{12}^L} d(z, r)\cos\alpha\mathrm{d}\Omega \tag{II.1.92}$$

and

$$J_{Sk}=\int_{\Omega_1} i_S(z, r)|\cos\widehat{r_L r}|\mathrm{d}\Omega=$$

$$=\int_{\Omega_{1k}^L} i_{S1}(z, r)|\cos\widehat{r_L r}|\mathrm{d}\Omega+\int_{\Omega_{2k}^L} i_{S2}(z, r)|\cos\widehat{r_L r}|\mathrm{d}\Omega, \tag{II.1.93}$$

where Ω_{jk} is defined as in Figure 36.

The functions J_D and J_{S1+S2} are the irradiances on both leaf surfaces, due to $d(z, r)$ and $i_S(z, r)$, respectively.

In order to determine the absorption function of a horizontal layer of a unit thickness, we will add up the transfer equations (II.1.61), (II.1.64), (II.1.69) and (II.1.74), multiply both sides by $\mathrm{d}\Omega$ and integrate over all directions within 4π.

Taking into account (II.1.67), (II.1.72) and (II.1.27)–(II.1.30) and changing the order of integration, we obtain in case of $\kappa_k=$const. for direct solar radiation

$$\Pi_S(z, r_S)=-\frac{\mathrm{d}S'(z, r_S)}{\mathrm{d}z}=\sum_k u_k(z)G_k(z, r_S)S(z, r_S), \tag{II.1.94}$$

for diffuse sky radiation

$$\Pi_D(z)=-\frac{\mathrm{d}D(z)}{\mathrm{d}z}=\sum_k u_k(z)\int_{\Omega_1}\frac{g_L(z, r_L)}{2\pi}J_D(z, r_L)\mathrm{d}\Omega_L=$$

$$=\sum_k u_k(z)\int_{\Omega_1} d(z, r)G_k(z, r)\mathrm{d}\Omega \tag{II.1.95}$$

and for complementary radiation i_S

$$\Pi_{IS}(z)=-\frac{\mathrm{d}J_S}{\mathrm{d}z}=$$

$$=\sum_k u_k(z)\kappa_k(z)\int_{\Omega_1}\frac{g_k(z, r_L)}{2\pi}[J_{S1}(z, r_L)+J_{S2}(z, r_L)]\mathrm{d}\Omega_L=$$

$$=\sum_k u_k(z)\kappa_k(z)\int_{4\pi} i_S(z, r)G_k(z, r)\mathrm{d}\Omega. \tag{II.1.96}$$

If the absorption coefficients of phytoelements are equal for both sides and are independent of the incident beam direction r, the effect of the plant stand geometry (specified by $u_k(z)$ and $G_k(z, r)$) and that of the optical properties of phytoelements (specified by the absorption coefficients κ_k) is revealed as separate multiplying factors in the absorption function $\Pi_{IS}(z)$ given by (II.1.96). Where κ_k is inconstant, this separation of influential factors is impracticable.

Formulae (II.1.94)–(II.1.96) allow calculating the radiation absorbed in a horizontal layer of a unit thickness if the radiation field, plant stand geometry, and the absorption coefficients of phytoelements are known.

An expression analogous to (II.1.96) will also be obtained for the absorption function $\Pi_{ID}(z)$ of the complementary radiation i_D.

II.1.9 STATISTICAL CHARACTER OF THE RADIATION FIELD IN PLANT STANDS

The radiation field within plant stands has both temporal and spatial variability.

The time frequency spectrum of fluctuations in radiation field intensity may be divided into five intervals:

1. Rapid fluctuations with a period of about a second due to leaf flutter in the wind.
2. Fluctuations with a period of about 10 min due to broken clouds.
3. Fluctuations with a period of about an hour due to the movement of sunflecks and shaded areas inside the stand caused by the sun traversing the sky.
4. Diurnal variation of radiation in the stand.
5. Annual variation of radiation in the stand.

The first and third of these intervals are specific of the radiation field in stands. The second, fourth and fifth intervals are generally typical of the ground surface radiation regime, but within a plant stand the statistical characteristics of the radiation field also involve the effects of the stand.

Experimental data on the temporal statistical variability of radiation in plant stands are scarce. Data on the integral function of radiation inside plant stands measured with a pyranometer are found only in the work by Kornher and Rodskjer (1967). Dadykin et al. (1964) point out that in a forest deeply shaded in the absence of wind, even a slight breeze causes the appearance of sunflecks. The lifetime of such sunflecks amounts to 25–30% of the observing interval.

The spatial statistical variability of radiation in plant stands is determined by the following factors:

(a) properties of the plant species in the stand;
(b) plant stand age structure within a single species;
(c) statistical character of plant distribution in the stand;
(d) statistical character of habitus formation and the arrangement of plant organs inside the plant volume;
(e) variation in the shape and size of individual plant organs.

A quantitative description of each of these factors is difficult because of the complicated structure of a plant stand. A qualitative description of their combined effect on the radiation regime in the plant stand is still more complicated. That is why the task of studying the statistical character of the radiation regime in plant stands has not yet been properly formulated.

Some aspects of the spatial statistical characteristics are discussed in works by Roussel (1962), Alekseyev (1963a, 1975), Laisk (1965b, 1968), Vygodskaya (1967), Zelniker (1969), Isobe (1969), Acock et al. (1970).

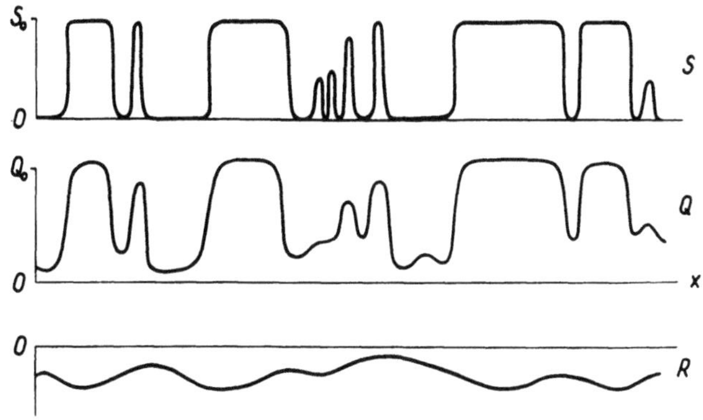

Fig. 39. Horizontal variability of radiative fluxes within plant stand. S – direct solar radiation, Q – total radiation, R – reflected radiation.

In view of the complexity of this task, an approximation to elementary characteristics appears to be suitable, with only the spatial variability of radiation fluxes taken into consideration.

Figure 39 shows records of direct, total and reflected solar radiation in a stand, obtained with radiation detectors which were moved horizontally. Direct solar radiation was measured with an actinometer, the values of total and reflected radiation were measured with a pyranometer. These records show that the spatial characteristics of these fluxes differ significantly in a statistical sense.

Direct solar radiation displays the greatest variability, and reflected radiation shows the least variability. Statistically, these records should be interpreted as realization of $X_I(x)$ as random functions of x and only moments up to the second order should be used as a rule in the analysis. In this case the most complete characteristic of the random function $X_I(x)$ is the two-dimensional probability density $f_2(I, I', x, x')$ (Pugachev 1968), associated with the onedimensional probability density by the relation

$$f_1(I, x) = \int_0^{\infty} f_2(I, I', x, x')\mathrm{d}I'. \qquad \text{(II.1.97)}$$

Since in practice it is difficult to determine f_2 and f_1, the set of characteristics is usually limited to the following approximate characteristics: mathematical expectance $I(x)$, variance $V_I(x)$, irradiation probability density averaged over the direction x, and autocorrelation function $K_I(x, x')$ of the random function $X_I(x)$. Let us assume that the radiation field in a plant stand is horizontally homogeneous if the probability density $f_I(x_0, l, I)$ averaged over the length l in the direction x is independent either of the point x_0 position or of the selected direction of the x axis, provided that the length of average is sufficiently long, i.e.

$$\lim_{l \to \infty} f_I(x_0, l, I) = f_I(I) \qquad \text{(II.1.98)}$$

for any horizontal direction.

It follows from the probability theory that random functions which satisfy the condition (II.1.98) are stationary and ergodic.

Note that many plant stand types are not horizontally homogeneous in a statistical sense. Thus for example, dense crops growing in rows may be treated as homogeneous in directions parallel to the rows, but inhomogeneous in directions perpendicular to the rows. In the latter case $I(x)$ is a periodic function with a period equal to the spacing of rows.

II.2 Incident radiation

II.2.1 GENERAL

The incoming radiation at the upper level of a plant stand primarily depends on astronomic factors and large-scale meteorological processes, while the radiation characteristics of the ground surface determined by the reflection coefficient and relief of the surrounding ground can be considered to be of secondary importance. Therefore the incoming radiation is usually described by means of general actinoclimatological characteristics complemented with information on the surrounding ground.

The determination of incoming PAR and NIR also necessitates the use of complementary phytoactinoclimatological characteristics essential for conversion from short-wave radiation to PAR and NIR.

II.2.2 INCOMING DIRECT SOLAR RADIATION

The value of solar irradiance on a surface perpendicular to the sunrays at the atmosphere upper level is $S_{00} = 1380\ \mathrm{W m^{-2}}$. This value is called the Solar Constant. The spectral distribution of extraterrestrial solar radiation approximately corresponds to black-body radiation at an effective temperature from 5700 to 5800 K. According to the Wien law the maximum of radiation is at $\lambda = 480$ nm, i.e. in the blue-green. During its traverse through the atmosphere solar radiation is attenuated because of scattering and absorption.

Direct solar radiation is in the spectral range from 280 to 4000 nm, approximately half of it being in the region of PAR and the rest in NIR. Only one per cent of it is in the ultraviolet region of the spectrum. Since spectral selectivity is characteristic of the attenuation of radiation, not only the integral irradiance but also the spectral distribution of radiation changes. Shown in Figure 40 borrowed from the book by Kondratyev (1954) is the spectral distribution of direct solar radiation at various solar elevations. As seen from Figure 40,

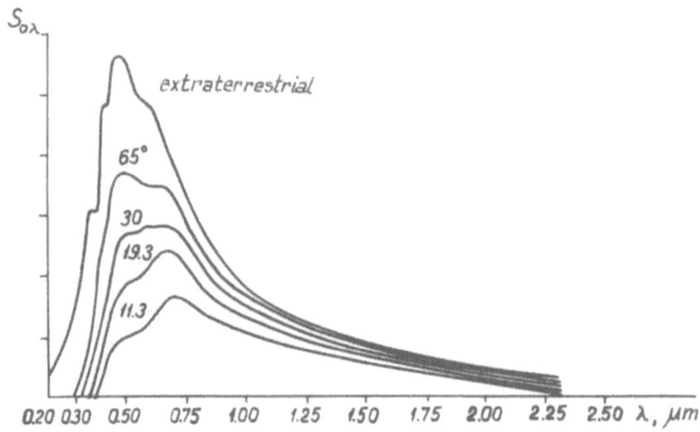

Fig. 40. Spectral distribution of direct solar radiation in dry and pure atmosphere at various solar elevations (Kondratyev 1954).

attenuation in the PAR region increases with the decreasing solar elevation considerably more rapidly than in the NIR region, where deep absorption bands occur due to water vapour in the atmosphere.

In the absence of atmosphere the irradiance of direct solar radiation on a horizontal surface depends on astronomic factors and is expressed by the formula

$$S'_{00} = \frac{S_{00}}{\rho} \sin h_{\odot}, \tag{II.2.1}$$

where

$$\sin h_{\odot} = \sin \varphi \sin \delta + \cos \varphi \cos \delta \cos \tau, \tag{II.2.2}$$

ρ is the distance between the Sun and the Earth in astronomical units, varying annually within $\pm 1.7\%$, φ is the geographical latitude of the object, and τ is the hour angle counted off from apparent noon.

The daily total of the irradiance of direct solar radiation on a horizontal surface is correspondingly expressed as

$$\sum_d S'_{00} = \frac{S_{00}}{\rho} \int_{-\tau_0}^{\tau_0} \sin h_{\odot} \mathrm{d}\tau, \tag{II.2.3}$$

where $\tau_0 = \arccos(-\tan \delta \tan \varphi)$ defines the time of sunrise (the '–' sign) and the sunset (the '+' sign).

The diurnal course of radiation as expressed by (II.2.1) is determined by the term $\cos \tau$, and the annual course of daily totals (II.2.3) by the sun declination $\delta(\tau)$ and the distance between the Sun and the Earth $\rho(\tau)$. These astronomic factors cause the diurnal and annual variations not only of direct solar radiation in the absence of atmosphere but also of all forms of radiation in a meteorological atmosphere. The effect of the atmosphere, clouds and the earth's surface is superimposed on these variations thus determining the actual radiation regime of the ground surface.

The presence of the atmosphere is the cause of the attenuation of direct solar radiation; some of the attenuated radiation is scattered resulting in diffuse sky radiation, and some is absorbed and turns into long-wave radiation of the atmosphere.

The attenuation of direct solar radiation in the atmosphere is determined by two factors:

(a) pathway length of the solar beam in the atmosphere characterized by the number of air masses $m \approx 1/\sin h_{\odot}$;

(b) mean concentration of attenuating particles in the way of the beam, which determines the atmospheric transparency.

In practical calculations of direct solar radiation the air mass is usually replaced with $1/\sin h_{\odot}$, and the formula for irradiance is written as $S_0(h_{\odot}, B) = S_{00}p(h_{\odot}, B)$, where p is a transmission function and B a certain characteristic of the atmospheric transparency. At $h_{\odot} = 50^{\circ}$ the atmosphere attenuates about 25–40% of direct solar radiation and, depending on the atmospheric transparency, the irradiance of direct radiation ranges from 0.7 to 0.98 $\mathrm{kW m^{-2}}$ during midday hours in middle latitudes.

In the presence of clouds the incoming direct solar radiation depends on whether or not the sun's disk is covered by cloud. High-level clouds covering the sun's disk decrease direct solar radiation by 10–40%, medium-level clouds decrease it 10–100 times, and below low-level clouds direct solar radiation is zero.

A quantitative account of the effect of clouds on the incoming radiation is an extremely complicated problem which is still far from having a satisfactory solution. The major difficulties are, first, the space-time variability of cloudiness, second, the lack of reliable and easily measured quantitative characteristics of cloudiness which would allow quantitative relationships to be established between clouds and incoming radiation.

II.2.3 INCOMING DIFFUSE SKY RADIATION

The part of solar radiation which reaches the ground surface from above as diffuse radiation after multiple scattering on atmospheric particles and clouds, reflection from the ground surface and additional scattering in the atmosphere is called diffuse sky radiation D_0.

The intensity of diffuse radiation depends on solar elevation and the number and size of scattering particles. Its spectral composition is different from that of direct solar radiation.

The amount of diffuse radiation depends on the part contributed by scattering to the total attenuation of solar radiation. The more the scattering the more the compensation of direct solar radiation attenuation by the additional diffuse radiation input. If all attenuation were due to scattering, i.e. in the absence of absorption, about one half of the attenuated radiation would reach the earth's surface, the rest returning to outer space. On the other hand, increasing the scattered part decreases the absorbed part correspondingly, which leads to decreasing long-wave radiation of the atmosphere. Thus an increase in the scattered part of direct solar radiation leads to increasing diffuse radiation and decreasing long-wave radiation of the atmosphere.

A second factor determining the diffuse radiation flux is solar elevation: an increase in which increases diffuse radiation. This is caused by atmospheric particles scattering radiation strongly forward along the beam path, so that the larger part of the diffuse radiation comes from the circumsolar area.

In addition, diffuse radiation depends on the ground surface reflection coefficient (albedo). The significance of multiple scattering between atmosphere and earth's surface increases with increasing albedo. The effect of albedo is particularly noticeable in the presence of a snow cover, when D_0 increases to 50%.

The spectral composition of diffuse radiation is determined by the character of scattering. Scattering below a cloudless sky occurs mainly on small particles and increases with decreasing wavelengths. Therefore the maximum of the diffuse radiation spectrum is shifted to a shorter wavelength in comparison with direct radiation and is at about $\lambda = 450$ nm. It is this circumstance that determines the blue of the clear sky. In the presence of clouds scattering takes place on large particles, those of cloud, and is less selective. Therefore, the spectrum has no maximum, the clouded sky is light-grey in colour and the diffuse radiation spectrum of the cloudy sky is nearly that of direct solar radiation. The larger the size of scattering particles, the more the importance of the red in the radiation spectrum. Therefore the diffuse radiation spectrum of a cloudless sky where blue and violet rays are dominant is most rich in PAR.

It should be noted that systematic data on the spectral composition of direct and diffuse radiation and on the limits of their variation are not sufficiently complete at present.

In addition to the spectral composition an important characteristic of diffuse radiation is its angular distribution which is determined by the form of the scattering phase function, the spatial distribution of scattering particles and the optical depth of the atmosphere.

The angular distribution of diffuse radiation is nonisotropic: under a clear sky the maximum of intensities is in the solar aureole and the minimum in the solar vertical plane opposite to the sun, at an angular distance of approximately 90° from the sun. The angular distribution of diffuse radiation intensities is symmetrical with respect to the solar vertical plane. In the case of a cloudy sky the angular distribution of diffuse radiation is less variable: the azimuthal part of the phase function is not strongly marked and a decrease in intensity from the zenith to the horizon is observed.

For practical purposes, as in studying the radiation regime in plant stands, the azimuthal part is usually neglected and only the so-called zonal brightness of the sky $D_{0z}(\theta)$ is considered. The latter is expressed

$$D_{0z}(\theta)=\int_0^{2\pi} d_0(\theta,\varphi)\cos\theta\sin\theta d\varphi. \qquad \text{(II.2.4)}$$

In calculations for purposes of comparison absolute intensities $d_0(r)$ are often replaced by the relative sky brightness in the direction

$$d_0'(r)=\frac{d_0(r)}{D_0} \qquad \text{(II.2.5)}$$

with a normalizing condition

$$\int_{\Omega_1} d_0'(r)\cos\theta d\Omega\equiv 1 \qquad \text{(II.2.6)}$$

or the relative zonal brightness of the sky

$$D_{0z}'(\theta)=\frac{D_{0z}(\theta)}{D_0}=\int_0^{2\pi} d_0'(\theta,\varphi)\cos\theta\sin\theta d\varphi \qquad \text{(II.2.7)}$$

with a normalizing condition

$$\int_0^{\pi/2} D_{0z}'(\theta)d\theta\equiv 1. \qquad \text{(II.2.8)}$$

For a uniformly bright sky

$$D_{0z}'(\theta)=2\cos\theta\sin\theta. \qquad \text{(II.2.7a)}$$

In the case of a clear sky the maximum zonal brightness is at $\theta=55\div 75°$ and does not shift to higher zones with the increasing solar elevation. Increasing turbidity decreases radiation from the higher and lower zones, but increases radiation from the middle zones.

Zonal brightness is more uniform under an overcast sky.

In the literature (Anderson 1964, 1966a; Moon & Spencer 1942) the overcast sky brightness is sometimes characterized by means of the so called *standard overcast sky*, for which brightness in the zenith is 3 times as large as brightness at the horizon. The relative zonal brightness of the standard overcast sky is given by the formula

$$D_{0z}'(\theta)=\tfrac{6}{5}(1+\sin\theta)\cos\theta\sin\theta. \qquad \text{(II.2.4a)}$$

The results of zonal brightness measurements for the amount of lower level clouds equal to 10, contributed by Põldmaa (1963), indicate that brightness in the zenith is also higher (approx. by 50%) than that at the horizon but the angular distribution is closer to uniform in comparison with the standard overcast sky.

Figure 41 shows the dependence of the relative zone brightness D_{0z}' for a uniformly bright sky (curve 1), the standard overcast sky (curve 2) and an overcast sky (data of Põldmaa, curve 3). The maximum brightness under an overcast sky is in the zone between 30 and 55°, i.e. closer to the zenith than in the case of a clear sky.

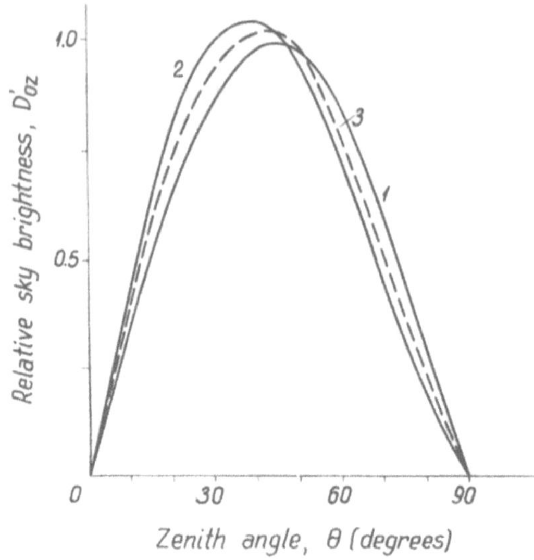

Fig. 41. Relative sky brightness D'_{0z} versus zenith angle θ.
1 – uniformly bright sky, 2 – standard overcast sky, 3 – overcast sky (after Põldmaa 1963).

The dependence of the diffuse radiation of a cloudless sky on solar elevation $h_\odot$ may be expressed by means of Kastrov's empirical formula $D_0(h_\odot) = a \sin^k h_\odot$, where a and k depend on the transparency of the atmosphere and the reflection coefficient of the ground surface.

As previously noted, the diffuse sky radiation flux D_0 decreases with the increasing transparency of the atmosphere, but this dependence is ambiguous and is determined by the part contributed by scattering to the total attenuation. The latter varies from 50 to 85% depending on the amount of water vapour, dust, water droplets and other aerosols in the atmosphere. These variations are temporal, geographical and also random in character.

The dependence of diffuse radiation on cloudiness is a complex one and involves a number of factors. It is mainly determined by the type and amount of clouds, as also their depth, optical properties and location in the sky.

The diffuse radiation of a cloudy sky can vary more than by a factor of 10. For different types of clouds these variations are not so great and do not exceed 3 to 8 times.

The irradiance of diffuse radiation of a cloudless sky at midnight is within the limits of 70–200 Wm^{-2}, in the case of low level clouds of the St and Ns type it is within the limits of 30–70 Wm^{-2}, and with middle level clouds it reaches 400–500 Wm^{-2}.

The total solar radiation Q_0 is the main component of the radiation balance; it is determined by the solar elevation, transparency of the atmosphere, reflection coefficient of the ground surface, and cloudiness.

The spectral quality of total radiation is more stable than that of direct and diffuse radiation and is practically independent of solar elevation starting from $h_\odot > 15°$. This is associated with an effect of compensation: the blue and violet content of solar radiation gradually decreases with decreasing solar elevation. At the same time the relative part contributed to total radiation by diffuse radiation, which is richer in blue and violet, increases thus practically compensating for the above-mentioned decrease in blue and violet.

The foregoing holds true for the total radiation incident on a horizontal surface; the spectral quality of the total radiation incident on an inclined surface depends on the angle of inclination and is subject to diurnal variation.

A different compensating effect is due to changes in the atmospheric transparency: the increasing transparency increases the direct solar radiation flux and decreases the diffuse radiation flux, and vice versa.

Table II.2.1 lists the values of S_{00}, S'_0, D_0 and Q_0 for various solar elevations with two extreme values of the atmospheric transparency, 0.70 and 0.83.

It will be seen from table II.2.1 that the compensation effect is not complete in summer, since Q_0 decreases a little with the decreasing atmospheric transparency.

Table II.2.1 *Short-wave radiation fluxes W m^{-2} and attenuation in the atmosphere. Tartu, 1958–1959, summertime*

	$h_\odot$			
	10°	30°	60°	90°
S'_{00}	230	665	1155	1335
$p_2 = 0.83$				
S'_{01}	110	460	915	1090
D_{01}	35	55	70	75
Q_{01}	145	515	985	1175
$S'_{01}/S'_{00}\%$	48	69	79	82
$Q_{01}/S'_{00}\%$	63	77	85	88
$p_2 = 0.70$				
S'_{02}	50	315	690	870
D_{02}	85	165	230	250
Q_{02}	135	480	920	1120
$S'_{02}/S'_{00}\%$	22	47	60	65
$Q_{02}/S'_{00}\%$	59	72	80	84
Q_{01}/Q_{02}	1.08	1.07	1.07	1.05

In winter in the presence of snow cover the diffuse radiation increases appreciably and the compensation is complete. On the whole, in the case of great solar elevations the incoming short-wave radiation under a cloudless sky is 80–90% of the incoming short-wave radiation in the absence of atmosphere.

The total irradiance of a cloudless sky Q is also called the possible total irradiance. It is calculated by the approximate formula (Barashkova 1961a)

$$Q_0(h_\odot, p, A_Q) = \frac{S_0 \sin h_\odot}{1 + f(p, A_Q) \sin h_\odot}, \qquad \text{(II.2.9)}$$

where

$$f = \begin{cases} 0.03\text{–}0.08 \text{ in winter with snow,} \\ 0.19\text{–}0.26 \text{ in summer time.} \end{cases}$$

In the presence of clouds the incoming total radiation is highly variable and depends primarily on whether or not the Sun's disk is covered by cloud.

II.2.5 INCOMING LONG-WAVE RADIATION OF THE ATMOSPHERE

Since the atmosphere ranges in temperature from 200 to 300 K it must emit infrared radiation. In accordance with Kirchhoff's law only those gases emit radiation which can absorb it. The basic gases of the atmosphere – nitrogen, oxygen and argon – absorb practically no radiation, and the main absorbers are water vapour, CO_2 and ozone. Unlike the solids in the ground surface, absorption as well as emission by these gases is not governed by the law of black body radiation and is complex and selective. Therefore, the absorption spectra of water vapour, CO_2 and ozone are essential for the mechanism of the thermal radiation of the atmosphere. The absorption bands of these gases are non-uniformly distributed over the spectrum, and absorption is comparatively weak in the region 8,500–13,000 nm, known as the atmospheric window. Most radiation is absorbed by water vapour characterized by wide and strong absorption bands with a complex spectral pattern. The experimental determination of narrow absorption spectra and their dependence on temperature and pressure is a difficult problem which has not been completely solved.

In theoretical calculations the infrared radiation of the atmosphere is assumed to be dependent on the total amount and vertical distribution of water vapour, CO_2, and ozone, and also on temperature and pressure. Calculation is reduced to an approximate solution of the radiative transfer equation for the absorbing medium, integration over wavelengths being the major difficulty (Kondratyev 1956). Because theoretical calculations are very complicated, empirical formulae and radiation nomograms for determining radiation by a graphical method (for example, (Niilisk 1960)) are widely used.

It should be noted that the infrared radiation of a cloudless atmosphere is non-isotropic: it increases by 30% from the zenith to the horizon.

Clouds increase the infrared radiation of the atmosphere, since for practical purposes a cloud several hundred metres in depth may be treated as a black body. They also increase the effective temperature and the water vapour content of the atmosphere.

Under a cloudless sky in middle latitudes the infrared radiation of the atmosphere is 0.28 to 0.35 kW m^{-2}. In the case of a cloudy sky it is 20 to 30% greater.

II.2.6 PHOTOSYNTHETICALLY ACTIVE RADIATION

So far there is no generally accepted opinion as to the definition of PAR, its limits, units and instrumentation for its measurement. In 1940 Gabrielsen suggested that PAR ranges from 400 to 700 nm. Later (Monteith 1969; Bonhomme 1969; McCree 1971, 1973) these limits became generally accepted in western countries. Soviet authors (Nichiporovich 1960) accepted the PAR interval from 380 to 710 nm divided into three subintervals: 380–535 nm, 535–610 nm and 610–710 nm. These limits of PAR are now generally accepted in the USSR and other socialist countries. In accordance with this PAR range accepted in Soviet radiation climatology, PAR distribution charts for the USSR (Moldau et al. 1963; Yefimova 1965) and for the whole Earth (Yefimova 1968) have been compiled.

It is accepted practice in phytoactinometry to measure radiation in units of energy. In western countries PAR J_P^* is expressed by the formula

$$J_P^* = \int_{400}^{700} J_\lambda \mathrm{d}\lambda, \tag{II.2.10}$$

where J_λ is the irradiance, and in the USSR and socialist countries PAR J_P is expressed by

$$J_P = \int_{380}^{710} J_\lambda \mathrm{d}\lambda. \tag{II.2.10a}$$

But since photobiological processes in plants are quantum in character, it would be more appropriate in principle to express PAR in quanta incident on, or absorbed by, the plant rather than in energy units. If J_λ is measured in Wm^{-2}(nm^{-1}) and λ in nm, then according to McCree (1973) the quantum flux density of PAR J_{QP} (einstein s^{-1} m^{-2}(nm^{-1}) is expressed by

$$J_{QP}^* = 8.36 \times 10^{-9} \int_{400}^{700} \lambda J_\lambda \mathrm{d}\lambda \tag{II.2.11}$$

or

$$J_{QP} = 8.36 \times 10^{-9} \int_{380}^{710} \lambda J_\lambda \mathrm{d}\lambda. \tag{II.2.11a}$$

The total effect of PAR on the process of photosynthesis depends not only on J_P or J_{QP} but also on the spectral curve of photosynthetic intensity P_λ expressed as moles CO_2 $s^{-1}m^{-2}(nm^{-1})$.

The spectral quantum yield of photosynthesis is defined by the relation

$$P_{Q\lambda}=\frac{P_\lambda}{8.36\times 10^{-9}\lambda\kappa_\lambda}, \tag{II.2.12}$$

where κ_λ is the leaf spectral absorption coefficient for the radiation J_λ.

The photosynthetic intensity in integral radiation is equal to

$$P^*=\int_{400}^{700} J_\lambda P_\lambda \mathrm{d}\lambda \tag{II.2.13}$$

or

$$P=\int_{380}^{710} J_\lambda P_\lambda \mathrm{d}\lambda. \tag{II.2.13a}$$

The photosynthetic efficacy of PAR (McCree 1973) is expressed

$$\varepsilon_P=\frac{P}{J_P}$$

or

$$\varepsilon_{QP}=\frac{P}{J_{QP}}$$

and it characterizes the efficiency of photosynthetic utilization of radiation averaged over the PAR region.

In his fundamental work McCree (1966; 1972a, b; 1973, 1976) investigated the limits of PAR, the functions J_λ, P_λ and the values of ε_P and ε_{QP}. Using 22 species of crop plants the action spectra, absorptance and spectral quantum yield of CO_2 uptake were measured over the wave-length range 300 to 750 nm. The following factors were varied: species, variety, age of leaf, growth conditions, temperature, CO_2 concentration and flux of J_λ. For all species and conditions the quantum yield curve had two broad maxima centred at 620 and 440 nm, with a shoulder at 670 nm.

These measurements enabled McCree to determine the functions κ_λ, P_λ and $P_{Q\lambda}$ for a 'mean' leaf in the spectral interval from 350 to 750 nm. The standard deviation in determining these functions equals 7%. Maximum quantum yield varies from 0.054 to 0.076 moles per einstein with a mean of 0.065. The part of PAR in the whole spectrum of a natural or artificial radiation source depends on the shape of the spectral curve of J_λ and is defined by the relation

$$\left.\begin{aligned} C_J^* &= \frac{\int_{400}^{700} J_\lambda \mathrm{d}\lambda}{\int_0^\infty J_\lambda \mathrm{d}\lambda} \\ &\text{or} \\ C_J &= \frac{\int_{380}^{710} J_\lambda \mathrm{d}\lambda}{\int_0^\infty J_\lambda \mathrm{d}\lambda} \end{aligned}\right\} \tag{II.2.14}$$

where C_J^* and C_J are *the PAR conversion factors for integral radiation.*

The factors C_J^* are undervalued in comparison with the factors C_J by 5–7% (Tooming & Gulyayev 1967) depending on the character of radiation J_λ. For artificial sources of radiation (various lamps) the factors C_J were calculated by Tooming and Gulyayev (1967). Values of ε_P, ε_P^*, ε_{QP} and ε_{QP}^* for total and diffuse radiation and for various artificial sources were calculated by McCree (1972b). From these calculations it follows that the quantities ε_P and ε_P^* are more variable than ε_{QP} and ε_{QP}^*. On the other hand, the quantities ε_P and ε_{QP} are more variable than ε_P^* and ε_{QP}^*. That is why McCree considers the quantum flux density J_{QP} to be more representative as a unit of PAR than the flux density J_P.

Methods of conversion from short-wave radiation to PAR were evolved at our laboratory in 1963 (Moldau et al. 1963; Tooming & Niilisk 1967) by means of conversion factors:

(a) for direct solar radiation

$$C_S = \int_{380}^{710} S_0(\lambda)\mathrm{d}\lambda \bigg/ \int_0^\infty S_0(\lambda)\mathrm{d}\lambda \tag{II.2.15}$$

(b) for diffuse radiation

$$C_D = \int_{380}^{710} D_0(\lambda)\mathrm{d}\lambda \bigg/ \int_0^\infty D_0(\lambda)\mathrm{d}\lambda \tag{II.2.16}$$

(c) for total radiation

$$C_Q = \frac{C_D + C_S \dfrac{S_0'}{D_0}}{1 + \dfrac{S_0'}{D_0}}. \tag{II.2.17}$$

The following formulae are used for conversion in radiation climatology calculations:

$$\left.\begin{aligned} \sum S_{0P}' &= \bar{C}_S \sum S_0' \\ \sum D_{0P} &= \bar{C}_D \sum D_0 \\ \sum Q_{0P} &= \bar{C}_Q \sum Q_0, \end{aligned}\right\} \tag{II.2.18}$$

where $\sum S'_{0P}$, $\sum D_{0P}$ and $\sum Q_{0P}$ are respectively the totals of direct, diffuse and total PAR, $\sum S'_0$, $\sum D_0$ and $\sum Q_0$ are respectively the totals of direct, diffuse and total radiation, and $\bar{C}_S$, $\bar{C}_D$ and $\bar{C}_Q$ are the mean factors for conversion from short-wave radiation to PAR for the given time period.

Two methods – calculation and experimental – are available for the determination of the conversion factors. The calculation method uses formulae (II.2.15)–(II.2.17), which require the knowledge of the spectral distribution of radiation either in relative units for the whole spectrum or in absolute units in the PAR region, as well as of the integral fluxes. The experimental determination of conversion factors is based on the concurrent measurement of the fluxes of integral radiation and PAR in absolute units. For this purpose actinometers with filters are usually used, for direct radiation, and various phytopyranometers and standard pyranometers, for diffuse and total radiation (for more detail see, for example, the work of Tooming and Gulyayev (1967)). The conversion factors are primarily dependent on solar elevation, turbidity of the atmosphere and cloudiness.

Figure 42 shows the dependence of C_S on solar elevation as found by Tooming and Niilisk (1967) from data measured with Yanishevsky's actinometer with filters WG7 (3 mm) and RG8 (5 mm) in Estonia, Moldavia and Tadjikistan. It will be seen that C_S increases rapidly with solar elevation increasing up to 30°. At $h_{\odot} > 30°$ C_S changes insignificantly and its mean value is 0.43. As atmospheric turbidity increases C_S slightly decreases. According to Perelyot (1970) this decrease is more pronounced at low solar elevations. Our experimental data, which agree well with both theoretical calculations and the data of Gulyayev (1965) and

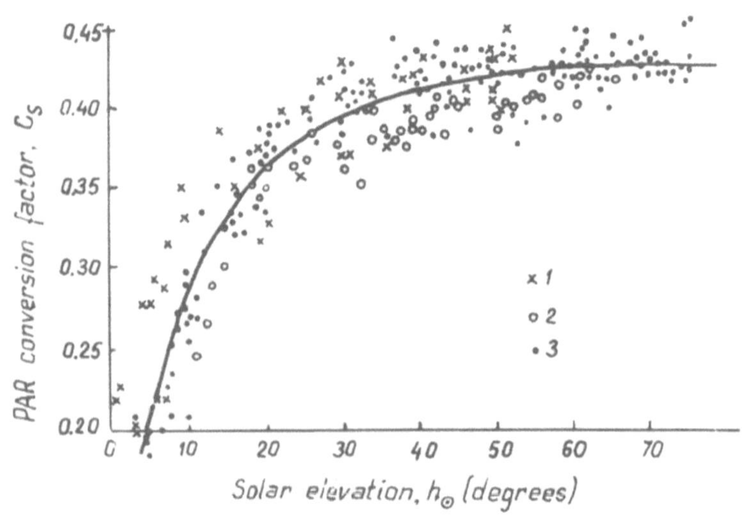

Fig. 42. PAR conversion factor C_S for direct solar radiation as a function of solar elevation (Tooming & Niilisk 1967).

1 – Jõgeva, Estonia, 1963, 2 – Moldavia, 1964, 3 – Hissar Valley, Tadjikistan, 1965.

Perelyot (1970) indicate that C_S depends but slightly on geographical locality, and the curve of Figure 42 is of a universal character.

In our climatological calculations we use $C_S=0.43$. On the basis of their experimental results Goisa and Perelyot tabulated C_S as a function of the solar elevation $h_\odot$ and of the transmissivity p_2 (Goisa & Perelyot, 1977). In the table $h_\odot$ varies from 5° to 65° and p_2 from 0.60 to 0.80; C_S varies from 0.12 to 0.46 and on the whole coincides with the data of Figure 42. These authors have compiled a procedure for the calculation of direct, diffuse and total PAR (Goisa & Perelyot, 1977).

According to measurements performed by Bedenko and Fedyushin (1971) and Fedyushin and Bedenko (1974) C_S in the foothills of Kazakhstan is diminished because of the dry atmosphere and great turbidity caused by the presence of dry aerosols. Instead of $C_S=0.43$ generally accepted for climatological calculations these authors use $C_S=0.41$. Instead of the generally accepted PAR conversion factor value $C_D=0.57$ they use $C_D=0.62$, noting that in the case of Cirrus C_D varies from 0.45 to 0.72 under a clear sky and in good atmospheric transparency. Between altitudes 500 to 1400 m above sea level the conversion factors vary but insignificantly.

The most detailed data for C_D were obtained by Perelyot (1970), and less systematic data by our laboratory (Moldau et al. 1963; Tooming & Niilisk 1967) and Gulyayev (1965).

PAR conversion factors for diffuse radiation C_D exceed the values of C_S for direct radiation. Perelyot (1970) obtained the dependence of C_D on the solar elevation $h_\odot$ (Figure 43) under a cloudless sky. The plot exhibits a great spread of data, which is due to measurement errors and atmospheric turbidity variation. The values of C_D vary between 0.55 and 0.80. At low solar elevations $C_D \approx 0.65$, at $h_\odot = 50$–60° $C_D \approx 0.75$. The upper portion of the curve of $C_D(h_\odot)$ agrees well with experimental data, and the lower portion with the theoretical calculations of

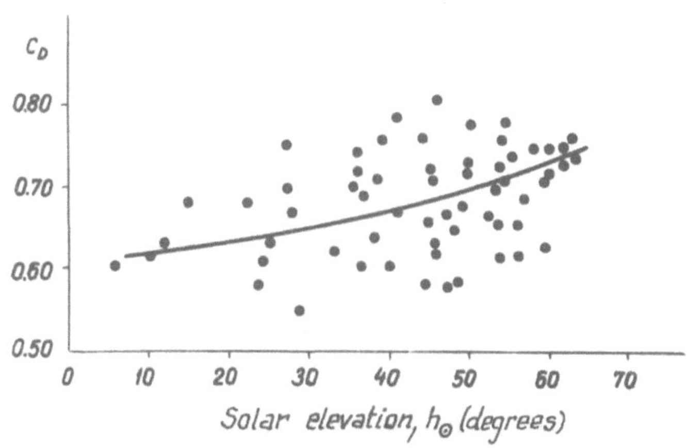

Fig. 43. PAR conversion factor C_D for diffuse solar radiation of cloudless sky as a function of solar elevation (Perelyot 1970).

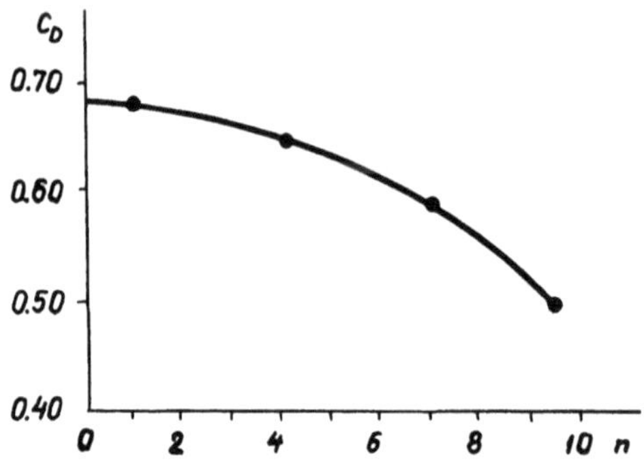

Fig. 44. PAR conversion factor for diffuse solar radiation C_D as a function of the amount of clouds n (Perelyot, 1970).

Tooming and Niilisk (1967). These data do not agree with the $C_D(h_{\odot})$ values calculated by Monteith (1969) from data of Robinson and showing that C_D increases as $h_{\odot}$ decreases.

Perelyot (1970) also studied the dependence of C_D on cloudiness (Figure 44). The $C_D(n)$ curve was obtained by averaging over 450 points. It indicates that C_D decreases from 0.69 under a clear sky to 0.50 under an overcast sky. For Cirrus, C_D is more dependent on cloud density than on the amount of cloud. For thin Cirrus ($S'/D = 1.0 - 1.5$) the values of C_D are close to those under a clear sky. In the case of dense Cirrus ($S'/D \ll 1$)C_D assumes values close to those under an overcast sky.

In our phytoactinoclimatological calculations we assumed the value of C_D to be constant and equal to 0.57. For an approximate calculation of total PAR $C_Q = 0.52$.

Speaking of works published abroad, Williams (1976) calculated the factors C_S^*, C_D^* and C_Q^* by the formula $S_0(\lambda) = S_{00}(\lambda)t_\lambda^m$, where m is the air mass and t_λ is the spectral transmittance of the atmosphere. Data contributed by Thekaekara were used for determining the spectral distribution of direct solar radiation outside the atmosphere $S_{00}(\lambda)$. Williams' data of C_S^* for mid-latitude atmosphere will agree well with the exponential curve of Figure 42. The values of C_S^* for tropical atmosphere are 0.07–0.03 higher. In Williams' work the C_D^* factor for a clear sky varies between 0.57 and 0.63 and is slightly undervalued in comparison with the experimental results of Perelyot given in Figure 43. As noted by Williams, the error made in determining S_{0P} does not exceed 5 mW cm^{-2}, and that made in determining D_{0P} of a clear sky does not exceed 2.5 mW cm^{-2}.

Measurements performed by McCree (1976) have shown that as $h_{\odot}$ changes from 15° to 70° C_D^* for a clear sky changes from 0.51 to 0.47, C_D^* of a clear sky equals 0.69, C_D^* of an overcast sky varies from 0.54 to 0.65. For high-level clouds, $C_Q^* = 0.50$ to 0.52. These data are in good agreement with those obtained by Soviet authors.

According to data of Britton and Dodd (1976) the annual variation of the monthly mean values of C_Q^* in Texas is from 0.46 to 0.48. The dependence of daily values of $\overline{C_Q^*}$ on $\sum Q_0$ is given by the table below

$\sum Q_0$ (MJ m^{-2}d^{-1})	0–5	5–10	10–15	15–20	20–25	25–30
$\overline{C_Q^*}$	0.58	0.55	0.50	0.48	0.48	0.47

This dependence includes not only the influence of solar elevation and atmospheric transparency but also that of cloudiness, which is the most important factor.

Monteni and Gosse (1978) have proposed the following formula, based on their experiments, for the calculation of total PAR:

$$Q_{OP}=\alpha\frac{S_{OOP}}{S_{00}}Q_0,$$

where S_{OOP} and S_{00} are the PAR and short-wave radiation of the sun outside the atmosphere. The coefficient α is equal to 1.19 for a clear atmosphere and 0.92 for a turbid atmosphere.

In the case of a cloudless sky

$$D_{OP}=0.97\alpha\frac{S_{OOP}}{S_{00}}D_0.$$

Experiments performed by Stanhill and Fuchs (1977) in the USA and Israel show that at $h_\odot>10°$ C_Q^* is constant and equal to 0.49 ± 0.05. For daily totals $C_Q^*=0.47\pm0.07$, and for monthly totals $\overline{C_Q^*}=0.49\pm0.02$.

However, Stanhill and Fuchs' statement that $C_Q^*=C_D^*$ and Szeicz's (1966) proposal to use the value $C_Q^*=0.40\pm0.03$ do not agree with the results obtained by other authors. The dependences $C_S^*(h_\odot)$ and $C_D^*(h_\odot)$ found by Bonhomme (1969) are less pronounced than those found by other authors.

Slomka and Slomka (1973) propose to take account of the effect of clouds in calculating the sum totals of total PAR $\sum Q_{OP}$ by means of an empirical function f_C and the formula

$$\sum Q_{OP}=\overline{C_Q}f_C\left(\frac{\sum S_0}{\sum S_{00}}\right)\sum Q_0,$$

where the function of cloudiness f_C is a linear function of the relative totals of direct solar radiation $\sum S_0/\sum S_{00}$. In this formula $f_C=-0.195\sum S_0/\sum S_{00}+1.195$ and $\sum S_{00}$ is the total of direct solar radiation in the absence of clouds. For C_Q Slomka and Slomka use a value of 0.5. A table of $\overline{C_Q f_C}$ has been compiled, in which $\sum S_0/\sum S_{00}$ varies from 0.10 to 1.11 and $\overline{C_Q f_C}$ accordingly from 0.59 to 0.48. Gorska and Slomka (1976) determined the monthly values of $\overline{C_Q f_C}$ and $\sum Q_{OP}$ for the Belsk observatory and the industrial city of Pulawy in Poland. The value of $\overline{C_Q f_C}$ determined for summer months at Belsk is equal to 0.525, and at Pulawy 0.412. The authors attribute this discrepancy partly to air pollution at Pulawy and partly to differences in measuring techniques.

It should be noted that experimental data of C_S, D_D and C_Q contributed by different authors are not always comparable because of the differences in radiation detectors, methods of calibration and in measuring techniques.

A fairly good stability and a universal geographical applicability of the conversion factors C_S, C_D and C_Q allow the solution of an important practical problem: there is no need in setting up a special network of phytoactinometric stations to determine the incoming PAR, as this can be done with the available actinoclimatoligical data if the conversion factors are used.

PAR distribution charts for the European part of the Soviet Union were compiled by Moldau et al. (1963), and Yefimova (1965, 1969) compiled charts for the entire Soviet Union. More accurate charts were compiled for the Ukraine by Perelyot (1971), for Byelorussia by Savikovsky and Kasperovich (1973) and for mountain slopes in Georgia by Dolidze (1974). Our method was also used by Lingova (1970) for the calculation of incoming PAR in Bulgaria and by Felmery for Hungary (1974). Eiyubov and Ragimov (1973) used our conversion factors to calculate the dependence of Q_{OP} on altitude for the mountainous districts of the Caucasus. It was found that Q_{OP} decreases as altitude increases from sea level to 0.5–1.0 km, and then increases with the increasing altitude. The said authors used this dependence to compile PAR distribution charts for Azerbaijan.

II.3 Optical properties of phytoelements

II.3.1 GENERAL

The scattering phase function $(1/2\pi)\gamma_L(r', r)$ in the short-wave region of the spectrum and the emission coefficient $\eta(r_L)$ in the long-wave region are the main radiation characteristics of phytoelements. They are given by formulae (II.1.23) and (II.1.32), respectively. This chapter deals mainly with the scattering phase function $(1/\pi)\gamma_L$ and also with more integral optical characteristics of phytoelements (leaves), viz. the coefficients of reflection R_L, transmission T_L, and absorption κ_L.

A vast (though not sufficiently comprehensive) body of experimental data relating to the optical properties of phytoelements have been accumulated by now. These data are systematized and generalized in monographs (Brandt & Tageyeva 1967; Shulgin 1963) and reviewed in articles (Gates et al. 1965; Birkebak & Birkebak 1964; Gausman et al. 1971; Gausman & Allen 1973). Problems of method and instrumentation are dealt with in Rvachev's monograph (1966). Work on the development of the optical model of the leaf has been initiated (Guminetsky & Rvachev 1966; Moldau 1967; Gausman et al. 1971).

The optical properties of leaves are determined by the following conditions:

(1) The surface roughness characteristics and the refractive index of the cuticular wax of the upper epidermis control the reflectance from this surface.

(2) The composition, amount and distribution of pigments determine the absorption of radiation in the ultraviolet and visible ranges.

(3 The internal leaf structure, arrangement of tissues and cells cause the internal scattering of incident radiation. Leaf absorptance, reflectance and transmittance are largely determined by the mean optical pathlength.

(4) The concentration and distribution of leaf water determine the absorption of IR radiation.

The interaction between solar radiation and leaves was first explained by Willstätter and Stoll (1918).

One part of incident solar radiation normal to the epidermis is reflected from the platelike epidermal cells. The other part enters the densely packed cylindrical palisade parenchyma cells arranged roughly parallel to the incident radiation. The radiation passes through this cell layer into the spongy mesophyll. Here the cells are not densely packed and are interspersed with intercellular air spaces. In the absence of absorption the spongy mesophyll causes internal scattering of the incident radiation due to multiple reflections from the cells and refractions caused by refractive index differences between the hydrated cell walls and the intercellular air spaces. This nature of the spongy mesophyll causes the scattering of the radiation. A portion of this scattered radiation, diffuse in nature, escapes through the lower epidermis and is designated as transmitted radiation. The other fraction diffuses upward and escapes through the upper epidermis and is considered as a part of reflected radiation.

Later Sinclair et al. (1973) demonstrated that the high level of reflectance in the 720–1300 nm spectral region cannot be explained by the laws of reflection and transmission accepted in geometrical optics, which describe the radiative transfer at the cell wall - air interfaces in the spongy mesophyll tissue. The assumption that transmission and reflection from the cell walls is diffuse as defined by Lambert's cosine law will be in better agreement with experimental data.

In terms of optics the plant leaf is a non-uniform plane-parallel turbid medium defined on both sides by rough surfaces which reflect and transmit radiation. Since the optical properties and morphological structure of the external and internal layers of the leaf are different, it is wise to distinguish between scattering from the surface and from the internal layers when constructing its optical model.

II.3.2 OPTICAL MODELS OF THE LEAF

Several optical models of the plant leaf have been proposed. In *the model of Allen, Gausman et al.* (1969), a compact plant leaf is considered as a transparent plate with a rough plane-parallel surface. The transmissivity at an interface between air and leaf is designated T_{12}, the corresponding reflectivity is given by $R_{12}=1-T_{12}$.

The transmissivity T_{12} is calculated from Fresnel's equations. If n is the relative index of refraction between air and leaf (or cell) and τ is the transmissivity of the plate, then the leaf reflectance is given by

$$R_L=(1-T_{12})+\frac{\tau^2 T_{12}^2(n^2-T_{12})}{n^4+\tau^2(n^2-T_{12})^2}$$

and leaf transmittance

$$T_L=\frac{\tau n^2 T_{12}^2}{n^4-\tau^2(n^2-T_{12})^2}.$$

Later Allen et al. (1970) generalized their model to explain the variation of optical properties caused by the development of intercellular air spaces in the leaf mesophyll. A leaf represented by a plate of thickness D is subdivided conceptually into N compact layers of individual thickness D/N. The thickness of the air spaces that separate the layers is taken as infinitesimal. With this model as a basis, formulae for the calculation of leaf transmissivity, reflectivity and scattering coefficient have been derived. The plate theory explains the approximate equality of leaf reflectance and transmittance on the basis of the scattering of diffuse light within the leaf by critical internal reflections.

The model proposed by Kumar and Silva (1973) does not consider the absorption of radiation and is applicable in the spectral region from 700 to 1300 nm. The leaf is assumed to consist of homogeneous and isotropic media – cell wall, chloroplasts, cell sap and air. The light ray traversing through a leaf cross section is treated by geometrical optics and Fresnel equations. The Rayleigh and Mie scattering is neglected. The two-dimensional cross section of a leaf is used for predicting the reflectance from the leaf. The angle of incidence of the radiation beam is equal to 5°. A total of eight optical interfaces in the leaf are considered. When tested on the soybean leaf this model has shown good agreement with experimental results.

The Moldau (1967) *model* was worked out at our laboratory. In this model reflection and transmission at an interface between air and leaf in the PAR region are practically nonselective and are described by the laws of geometrical optics. The rough leaf surface is divided into elements the orientation of whose normals is given by a distribution function. Reflection from each element is described by the Fresnel law and depends on the refractive index of the leaf's surface layer relative to the refractive index of air and on the angles of incidence and reflection. Radiative transfer inside a leaf is described in this model by a transfer equation for a horizontally homogeneous turbid medium. This is a very simplified approach, since a leaf's internal structure is very complicated, absorption is nonselective and is determined by the properties of substances in cell walls and protoplasm, the amounts of water, enzymes, numerous yellow and green pigments, etc.

Moldau's optical model allows a qualitative and, in some cases, a quantitative explanation of the dependence of scattering on the conditions of radiation incident on the leaf, the thickness and surface shape of the leaf, the pigments it contains, and other anatomical and physiological characteristics of the leaf. The reflection coefficient R_L is mainly dependent on the leaf surface structure and does not exceed 5 to 7% at angles of incidence $\vartheta' < 60°$.

Intraleaf scattering is highly selective spectrally. The reflection (transmission) coefficient does not exceed 2–3% in the absorption bands and reaches 40 to 50% outside them. This explains why the surface component of reflected radiation is dominant in absorption bands, while the inside component dominates outside them.

The model proposed by Tucker and Garrat (1977) proceeds from the leaf anatomy and the optical properties of individual structures. The leaf is considered to consist of a reflecting and transmitting surface (the cuticle) and layers of polisade parenchyma and spongy mesophyll. Four radiation states (solar, reflected, absorbed and transmitted) are represented by six compartments: the solar input, initial reflection from the cuticle, the scattering and absorption in the palisade parenchyma, the scattering and absorption in the spongy mesophyll, diffuse reflected radiation and diffuse transmitted radiation. The model accounts for all incident electromagnetic irradiance (400–2500 nm), the totality being divided among reflecting, transmitting or absorbing radiation states. The processes of scattering and absorption were principal driving processes in the model, and only 1% of the incident radiation is initially reflected from the cuticle. The intraleaf scattering refers to the combined effect of reflective-refractive differences between or among cells, air spaces, cellular constituents and the Mie scattering which occurs. The strong absorption in the spectral regions 350–700 nm is caused by chlorophyll a and b and by carotenoids. Absorptance in the middle IR region (1300–2500 nm) of the spectrum is caused by absorption by leaf water molecules. The stochastic model of radiative transfer in the leaf is treated as a Markov chain, stationary in time, and the leaf scattering is proportional to cellular density in the palisade parenchyma and spongy mesophyll. Comparison of the spectral reflectance modelled for black maple and measured values for silver maple demonstrates a close agreement in the 700–2500 nm region of the spectrum and higher values for the modelled leaf spectral reflectance and transmittance in the 500–700 nm interval.

II.3.3 SCATTERING PHASE FUNCTION OF THE LEAF

Sufficiently reliable and complete experimental data on the angular dependence of the scattering phase function are scarce.

The reflection function of a large class of mesophytic, succulent and xerophilous plants in the PAR and NIR region was investigated by Shulgin (1960) by means of a goniophotometer with an integrating sphere.

Xeromorphic leaves with a glossy surface have an elongated scattering phase function because of the specular reflection component of the leaf. A waxy film on the glossy surface of the leaf causes strong reflection both at small angles (up to 15°) and at large angles (up to 60–80°). At a constant angle the variation of reflection is largely determined by the leaf surface structure, but at large angles (60–80°) reflection increases also with an increasing leaf thickness.

The angular dependence of the reflection function is less pronounced in the group of succulents. A waxy film increases reflection. As in the case of xeromorphic leaves a specular reflection exists in the presence of a glossy surface.

A large variety in absolute values, as well as in the angular dependence of the reflection function is characteristic of mesophytic plants. Typical of wrinkled and mat leaves is a flat scattering phase function with a maximum of reflection at 70 to 80°. Thin and wrinkled leaves show weak and isotropic reflection. Typical of mesophytic plants with drooping leaves is a maximum reflection at angles between 10–15 and 60–70°.

The scattering phase function of the leaf was measured both in reflected and transmitted beams by Moldau at our laboratory. A spectrogoniograph was designed for this purpose (Moldau 1964), which enables the determination of the angular distribution of spectral brightness coefficients for various flat objects in reflected and transmitted beams on different planes with independently variable radiation beam angles and apertures. The instrument is useful for measuring in natural and polarized light in a spectral range 400–800 nm. The use of a polarized beam permits to discriminate between reflection from the leaf surface and that from the inside layers (Moldau 1965, Shulgin 1963).

According to the data of Moldau (1965) the scattering phase functions of an etiolated maize leaf are weakly elongated in reflected beams, and are strongly elongated towards direct transmission in transmitted beams. The brightness coefficients at $\lambda = 495$ nm are small, which indicates considerable absorption in this region.

Upon unfurling of the leaf its brightness coefficients decrease rapidly in the absorption bands, especially at $\lambda = 673$ nm. They are least variable at $\lambda = 730$ nm. The scattering phase functions extend in the incident beam direction, and the protrusion in the direction of direct transmission rapidly disappears in transmitted beams.

As the plant continues to grow these changes slow down gradually, and about the second decade of July the scattering phase functions become practically stable in shape stabilizing at $\lambda = 552$ nm. Scattering phase functions remain stable up to the ear milk ripeness stage (the period of maize ageing was not investigated, because early frosts in Estonia set in during the milk ripeness stage).

Figure 45 drawn from data of Moldau shows the angular dependence (in the plane of incidence) of the radiation $(1/\pi)\gamma_L(\vartheta) \cos \theta' \cos \vartheta$ in the blue ($\lambda = 485$ nm) and nearinfrared ($\lambda = 731$ nm) scattered from a maize leaf unit area at angles of incidence equal to 0 and 60° to the leaf upper side. It will be seen from the diagram that in the case of a normal incidence of radiation a leaf unit area scatters predominantly forwards and backwards and that the angular dependence is the same in form in the region of PAR and NIR, but scattering is considerably stronger in the latter region. The curves are symmetrical in relation to the normal.

There is a considerable amount of specular reflection at the angle of incidence equal to 60°. This is particularly felt in the PAR region.

When radiation is transmitted the greater part of it is also scattered downwards, but the symmetry relative to the vertical is disturbed.

Analogous data were obtained by Moldau for birch leaves, in which the effect of specular reflection is still more pronounced. At $\vartheta = 60°$ and for the PAR region brightness in the direction of the specular beam of PAR is 1.5 times as high as that in the direction of the incident beam.

Investigations performed by Moldau (1965) show that the shape of the scattering phase function and its values for transmitted beams are subject to greater variation during the growing season than for reflected beams; these variations are smaller in the case of weak absorption than in that of strong absorption. A living

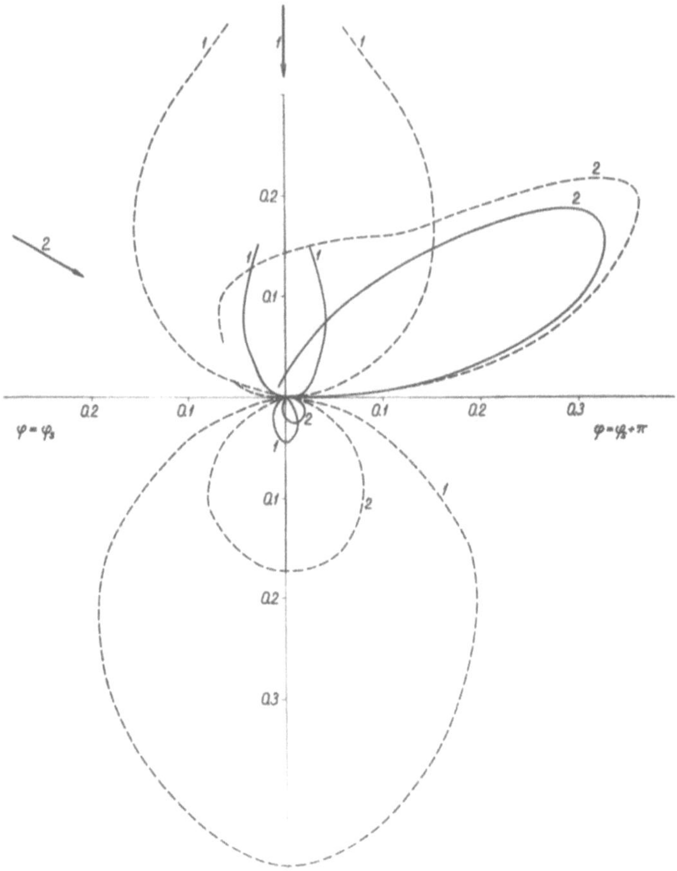

Fig. 45. Angular distribution of radiation scattered by a unit leaf area of maize in the blue (solid lines) and in the near infrared (dashed lines) spectral regions at two angles of incidence 0° and 60° (1 and 2, respectively) (according to data of Moldau).

leaf has a stable scattering phase function during the greater part of its life. This considerably simplifies the account of the angular distribution of scattered radiation in calculations of radiation inside plant stands.

However, investigations also show that leaves of most plant species are not mat. The greatest deviations from a mat surface are observed in glossy leaves having a specular reflection effect and in thin and young leaves showing a noticeable forward transmission. It is also stressed in works of Shulgin (1963) and Brandt and Tageyeva (1967) that the Lambert law is not valid in this case.

A detailed investigation of the angular dependence of the radiation diffused by soybean and maize leaves in the spectral interval from 375 to 1000 nm by means of a spectrogoniometer was performed by Breece III and Holmes (1971), and the angular dependence of the reflected radiation of leaves of white philodendron, soybean and maize was investigated at 450 and 800 nm by Woolley (1971). The results obtained by these authors show that the general character of the angular distribution of diffuse radiation changes only slightly over the spectrum. The glossy leaves of philodendron appear to be ten times as bright in the direction of specular reflection as in other directions. In the visible spectrum region the peak of reflection at the specular reflection direction equals more than 50% of the whole amount of reflected radiation. Maize leaf surface irregularities extend mostly in a direction parallel to the veins so that the leaf reflects light more diffusely in a plane perpendicular to the veins than in a plane parallel to the veins. No leaves tested showed retroreflection.

II.3.4 SPECTRAL OPTICAL PROPERTIES OF PHYTOELEMENTS

Since a leaf is a complex optical system, it is characterized by a pronounced spectral selectivity in absorbing, reflecting and scattering radiation.

The effects of various factors on leaf spectral reflectance and transmittance are reviewed by Woolley (1971).

In spite of an extremely great variability in size, form and structure of phytoelements in the plant covering of the earth, the spectral optical properties have universal specific features which permit the concept of 'a mean' living leaf (Shulgin 1963, Monteith 1965b).

Shulgin and Moldau estimated the range of variation of the curves of absorption κ_L, reflection R_L and transmission T_L of living leaves for various species (Figures 46 and 47).

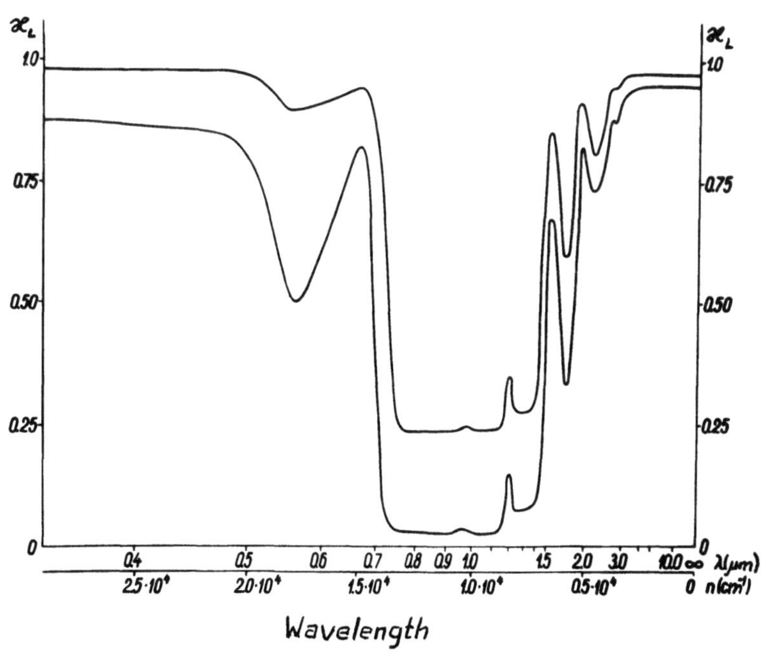

Fig. 46. Variation of spectral absorption coefficient κ_L of green leaves (data of Moldau).

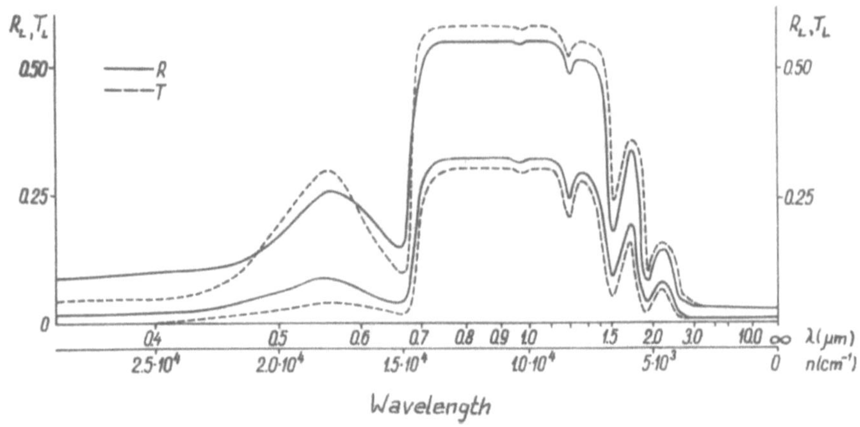

Fig. 47. Variation of spectral coefficients of reflection R_L and transmission T_L of green leaves (data of Moldau).

Absorption in the ultraviolet region of the spectrum (300–380 nm) reaches 90–99%. This is due to the optical properties of protoplasm and cell walls. The effect of specific and morphophysiological properties of leaves on absorption in the ultraviolet is very slight.

Generally absorption in the PAR region is strong, and is due, mainly, to the pigment system – chlorophyll, and carotenoid pigments. It has two distinct maxima in the blue-violet (380–480 nm) and orange-red (620–689 nm) regions of the spectrum and is equal to about 90%. It depends comparatively slightly on specific, ecological and other factors. Between these maxima, in the green portion of the spectrum is an absorption minimum which determines the green colour of phytoelements. In this region the absorption coefficient varies in a wide range – from 0.50 to 0.90 – depending on numerous intrinsic and environmental factors.

Absorption decreases drastically starting from 690 nm, and there is the deepest absorption minimum, 5–25%, in the NIR region (750–1200 nm). Absorption of radiation increases again starting from 1200 nm and reaches its maximum value in the absorption bands of liquid water, 1500 and 2000 nm, respectively. It is seen from Figures 46 and 47 that spectral reflection and transmission vary similarly, inversely with absorption. On an average the integral absorption of solar radiation by green parts of plants is 40–75%, absorption in the PAR region is 75–90%, and in the NIR 5–24%.

According to data of Shulgin (1963) and also Gates (Gates et al. 1965) spectral transmission in thin leaves exceeds reflection in the green and NIR regions. Reflection is dominant in other cases.

The spectral properties of leaves are most variable in the green and NIR. According to Shulgin (1963) absorption in young bright coloured maize leaves amounts to 50–60% at $\lambda = 540$ nm and increases to 80% in mature dark-green leaves. Then absorption decreases owing to the yellowing of leaves. This dependence of PAR absorption on leaf age determines both the dependence of absorption on the leaf ordinal number and different optical properties of leaves in different layers of a stand.

Rvachev (Rvachev et al. 1965) discovered that the absorption of NIR (740–1150 nm) by leaves of evergreen and needles of coniferous plants has an annual variation. This absorption is small (5–7%) in summer, increases in autumn and reaches its maximum value 20–30%, then decreases in winter and spring down to the summertime values. Large variations of the optical properties of plant parts due to difference in species are also observed in the NIR region.

The determination of the spectral properties of needles involves serious technical difficulties. Investigations by Gates et al. (1965) show that among green parts of plants needles have the highest absorption coefficient: assuming $T_L = 0$, it is equal to 97% for PAR and 59% for NIR.

In recent years a group of American workers (Gausman et al. 1971, 1973; Gausman & Allen 1973) carried out fundamental research on the spectral optical properties of plant leaves. They measured the spectral coefficients of reflection and transmission of thirty plant species in the spectral interval 550 to 2200 nm. They discovered that the bottom leaf surfaces of dorsiventral leaves had higher reflectance values than the top leaf surfaces, indicating that the spongy parenchyma contributes more to light scattering than the palisade parenchyma or the leaf mesophyll. Thick oleander and ficus leaves had the lowest light transmittance, 1% at 550 nm, and begonia the highest, 21%. The difference in transmittance was caused by greater light diffusion by top leaf surfaces. Youngest leaves of cotton had approximately 10% lower reflectance over the 750 to 1350 nm wave-length interval, reflectance of the older leaves slowly but progressively increased from 2 to 3% going from nodes 3 down to 6.

According to Gausman et al. (1972) maize leaf water reaction within growth stages had little influence on near infrared light (750–1350 nm wavelength interval) reflectance because leaf mesophylls remained relatively compact. Cotton leaf mesophylls, however, developed many air spaces with maturation, and mature leaves had higher nearinfrared light reflectance than young leaves. The largest differences among reflectances of maize leaves of different age within growth stages occurred when tassels were appearing in the leaf whorls.

Shulgin (1963) points out that the universal character of the spectral optical properties of green plant parts reflects the common laws of adaptation of plants to direct and diffuse radiation. Also spectral optical properties change in growing plants under the effect of radiation and other environmental factors. Thus specific and ontogenetic variations of optical characteristics may be attributed to the response of plants to environmental conditions.

II.3.5 INTEGRAL COEFFICIENTS OF LEAF REFLECTION, TRANSMISSION AND ABSORPTION FOR SHORT-WAVE RADIATION AND PAR

Optical properties of leaves in the PAR region depend on many factors: conditions of irradiation, species, leaf thickness and surface structure, amounts of chlorophyll and carotenoid pigments contained in the leaf, dry matter content per leaf unit area, and leaf's internal structure (number of cells per unit area, number of layers of the palisade and spongy parenchyma, etc.).

Since the spectral composition and angular distribution of direct and diffuse solar radiation change diurnally, this must influence the absorption coefficient κ_L according to formulae (II.1.27) – (II.1.30). Shulgin (1963) evaluated the variation of κ_L of direct solar PAR (400–720 nm) due to the changing spectral composition of radiation as a function of solar elevation (Table II.3.1).

Table II.3.1 *Absorption coefficient of 'mean' green leaf in direct solar PAR as a function of solar elevation* $h_\odot$

$h_\odot$	11.3°	19°	30°	65°	Total radiation
Absorption coefficient	0.785	0.80	0.81	0.795	0.798

As is seen from Table II.3.1 the absorption of PAR by a leaf depends on solar elevation but slightly and is practically equal to the absorption of total radiation below a clear sky.

The difference between absorption coefficients is more considerable below an overcast sky. According to data of Gates (Gates et al. 1965) the absorption coefficient of mesophytic and xerophilous plants under an overcast sky is on a average 0.10 higher than below a clear sky. This is due to the difference in the spectral quality of the incident radiation, because the diffuse radiation spectrum is poorer in NIR. In succulents and coniferous plants the absorption coefficient is practically independent of irradiation conditions. The highest absorption coefficient is possessed by the needles of coniferous plants. According to Shulgin (1963) the highest absorption is characteristic of xerophilous plants. Shaded leaves absorb less than sunlit ones in the same species. Within a species absorption increases from North to South.

The dependence of absorption on the angle of incidence has been studied in most detail experimentally by Brandt and Tageyeva (1967). Table II.3.2 based on their data gives values of the transmission T_L, reflection R_L and absorption κ_L coefficients as a function of the angle of incidence θ' for two plant species irradiated with white light from a 500-W film projector lamp. The leaves of lettuce (*Lactuca sativa*) were thin, pale-green in colour, those of Chinese hibiscus (*Hibiscus rosa sinensis*) were thick, with a glossy surface.

Table II.3.2 *Transmission, reflection, and absorption coefficients for leaves as a function of the angle of incidence of white light*

Coefficient	Angle of incidence θ'						
	0	30	40	50	60	70	75
Lettuce							
T_L	17.5	17	16	14.5	12	11	10
R_L	14	14.5	15.5	16.5	19	23	25
κ_L	68.5	68.5	68.5	69	69	66	66
Chinese hibiscus							
T_L	6	5.5	5	4.5	4	4	4
R_L	6.5	7.5	8	8.5	9.5	11	13
κ_L	87.5	87	87	87	86.5	85	83

According to Table II.3.2 transmission through a leaf decreases with the increasing angle of incidence, but reflection increases. Because of their mutual compensation absorption is more stable; up to $\theta' = 60°$ the absorption coefficient is practically independent of the angle but it will somewhat decrease with the further increase of θ'.

The transmission coefficient for PAR of mature African crop plant leaves (Farrar & Mapunda 1977) ranges from 0.043 (maize) to 0.093 (common beans), and that of young leaves from 0.097 (*Sesame indicum*) to 0.394 (*Eleusine corocana*).

Coefficients of reflection and transmission for both leaf sides in a large number of woody plants are determined in the work of Birkebak and Birkebak (1964). According to these data reflection, as a rule, is stronger on the downward side of a leaf; no such dependence is observed for transmission. Therefore, the absorption of radiation incident on the downward side of a leaf is less than the absorption of radiation incident on the upward side. On an average the coefficient of absorption of integral direct radiation for the leaves of woody plants is 0.40–0.50, the coefficient of reflection is 0.25–0.35, and the coefficient of transmission is 0.18–0.24. According to Shulgin (1963) the PAR absorption coefficient for the leaves of grassy plants is 5–10% lower than for woody plants. This is due to a larger transmission by the leaves of grassy plants which are thinner in comparison with those of woody plants.

According to Thomas et al. (1971) the reflectance of cotton leaves in the 1450 and 1930 nm water absorption bands is correlated ($r = -0.90$) with relative turgidity when the relative turgidity is below 70% and leaf wilting is visible on plants. Within the relative turgidity range 70 to 80% reflectance changes are small and may not be predictable because of variation among leaves. Nevertheless, Carlson et al. (1971) found in leaves of maize, soybean and sorghum a close relation between the relative water content and the area under the reflectivity curves in the wavelength interval from 1000 to 2500 nm.

In his works Shulgin (1960, 1963, 1973, 1967) gives a detailed analysis of the optical properties of plants as a function of species, degree of morphological adaptation to light, structural characteristics, etc.

According to data of Shulgin, chlorophyll in plant leaves is one of the most important factors which determine the high absorptive capacity of the leaf in the PAR region. The chlorophyll '*a*' content varies from species to species to a greater extent than the content of chlorophyll '*b*'. The maximum occurrence of species with chlorophyll '$a+b$' 3.5–4.0 mg cm^{-2} coincides with the beginning of the absorption plateau, and the amount of chlorophyll is abundant in many species.

In recent years Shulgin has been evolving a concept according to which leaves having numerous pigments and plastids, i.e. leaves with a considerable photoactive area (10–50 cm^2/cm^2 of leaf) may in an optico-structural sense be treated as a cenosis having a vertical radiation profile.

The absorption coefficient of pine-tree shoots for PAR (Molchanov & Khazanov 1975) varies from 0.79 to 0.72 with the angle of incidence increasing from zero to 90° and increases from 0.54 to 0.78 with the leaf area index of a shoot with needles increasing from 3.7 to 6.5. The mean value of κ_L at $L_0=5$ is equal to 0.66.

Further calculations in this book will use a mat leaf. Both of its sides have equal optical characteristics:

$T_L=0.06$ and $R_L=0.09$ in the PAR region, and $\omega_L=0.85$ in the NIR region.

II.3.6 OPTICAL PROPERTIES OF PHYTOELEMENTS IN THE LONG-WAVE SPECTRAL REGION

No trustworthy data on the scattering phase function $(1/\pi)\gamma_L$ of a leaf in the long-wave spectral region are available, but it is known that absorption in this region is large and is close to black body absorption. Therefore it is common practice nowadays to treat plant leaves as grey bodies whose radiation obeys Kirchhoff's laws. Consequently, scattering and absorption in the leaf are isotropic, its absorption coefficient κ_L and emissivity δ_L are equal, and radiation from the leaf is determined by its temperature in accordance with formula (II.1.31). The reflection coefficient $R_L=1-\kappa_L$ has been studied experimentally in the work of Gates and Tantraporn (1952), where it is shown that at the angle of incidence $\theta'=65°$ $R_L<5\%$ and at $\theta'=20°$ $R_L=3\%$. The upward side of a leaf reflects more strongly than the downward side, old leaves reflect more strongly than young ones, and shaded leaves reflect more strongly than sunny ones. The emissivity of fresh green leaves of various trees and grasses, according to measurements by Pak Sen Su (1957), is between 0.95–0.98.

The most detailed data of the emissivity δ_L, obtained by measurement of δ_L for 34 species, are presented in the work of Idso et al. (1969). The least values of δ_L are possessed by the leaves of *Phaseolus vulgaris cv. Bountiful* (0.938 ± 0.008) and *Zea mays cv. Mexican June* (0.944 ± 0.004), the highest values of δ_L were obtained for *Rosa* (0.993 ± 0.006) and *Saccharum officinarum* (0.995 ± 0.004).

The later data show that in different plant species δ_L varies in a comparatively wide range, though the data of all three authors coincide in the essential.

With an equal amount of infrared radiation $E_T = 0.453$ W m^{-2} the leaf temperature of *Phaseolus vulgaris* is 29.7 °C, and that of *Saccharum officinarum* 25.1 °C, i.e. depending on emissivity δ_L leaf temperature may differ by 4.6 °C at the same value of infrared radiation. The model leaf emissivity assumed for further calculations in this book is $\delta_L = 0.95$, i.e. in the long-wave spectral region $R_L = 0.05$, $T_L = 0$ and $\kappa_L = 0.95$.

II.4 Penetration of direct solar radiation into a plant stand

II.4.1 GENERAL

In terms of energy, direct solar radiation is the main component of the total radiation field in a stand, especially in localities where cloudless days predominate during the growing season. Also it is the most variable component of the radiation field, thus requiring a large scale of average. The latter circumstance creates serious difficulties both in the way of theoretical treatment of the problem and of obtaining reliable experimental data. Therefore, the problem is far from being solved and still is one of the most difficult problems in phytoactinometry. The present state of this problem will be discussed below on the basis of data available in the literature as well as investigations by the author and his colleagues.

II.4.2 STATISTICAL CHARACTER OF PENETRATION OF DIRECT RADIATION IN A PLANT STAND. PENUMBRAS

The irradiance of direct solar radiation in a plant stand is subject to considerable changes (see Figure 39). In sunflacks it is close to the values of S_0 above the stand, and is zero in shaded areas. Neglecting penumbras, i.e. if the Sun is treated as a point source of radiation, a record of direct solar radiation, when considered statistically, represents the realization of a random function $X_S(x)$, having only two values

$$X_S(x) = \begin{cases} 0 & \text{in shaded area} \\ S_0(r_S) & \text{in sunflecks.} \end{cases}$$

The mathematical expectance of X_S is

$$S = \lim_{l \to \infty} \frac{l_S}{l} S_0,$$

where l_S is the total length of gaps on the pathlength l.

Practically,

$$\lim_{l \to \infty} \frac{l_S}{l} = a_S$$

i.e. equal to the probability of solar radiation S_0 on the pathlength l. Consequently, $1-a_S$ may be treated as the probability of shaded areas on the pathlength l. Thus the probability density of direct solar radiation $f_S(I)$ averaged over the length l has only two values and may be written using Dirac's delta-function

$$f_S(I) = [a_S\delta(I - S_0) + (1 - a_S)\delta(I - 0)] \tag{II.4.1}$$

and the mathematical expectance of the penetration function of direct solar radiation is equal to $a_S(L, r_S)$, the variance $V_S = S_0^2 a_S(1-a_S)$ and the relative variance $V_S' = V_S/S = S_0(1-a_S)$. The autocorrelation function $K_S(x, x')$ depends on the dimensions and distribution of sunflecks on a horizontal plane and can be found from experimental data.

Thus, neglecting penumbras, direct solar radiation is described statistically by means of the penetration function $a_S(L, r_S)$ and autocorrelation function $K_S(x, x')$.

The random function $X_S(x)$ becomes more complicated when penumbras are taken into account. In order to understand the peculiar character of the variation of direct solar radiation intensity in penumbra areas, consider Figure 48 which shows the shadow effect of a circular disk at different distances above the ground. On the earth's surface the Sun is viewed at an angle $\alpha = 32'$ which defines a cone $D_1A_0D_2$ at the point A_0. At a height h the apparent disk area of the Sun is $S_I = \pi r_I^2$ where $r_I = h \tan \alpha/2 = 0.0047h$.

Consider the shadow effect of a circular disk having a radius r, placed successively at four distances above the ground. In position 1 the disk is just above the ground. The variation of the irradiance of direct solar radiation corresponding to this position, caused by translating a point detector along the straight line C_1C_4, is represented by curve 1. It will be observed that no penumbra exists in this case, and curve 1 has pronounced jumps. In disk position 2 corresponding to the height $h = h_0/2$, where $h_0 = r/0.0047$, shading sets in at the point A_1. As the radiation detector moves from the point A_1 to A_2 the whole solar disk gradually becomes shaded by the screen disk and the irradiance (curve 2) decreases from S_0 to 0. As the detector moves from A_2 to A_3 the Sun is in the shade and the irradiance is zero. Between the points A_3 and A_4 the Sun emerges from under the screen disk and irradiance increases gradually to the maximum S_0. It is seen from Figure 48 that in the penumbra areas A_1A_2 and A_3A_4 the irradiance has all values from 0 to S_0. In position 3 of the screen disk, i.e. at $h = h_0$ shading sets in at the point B_1, i.e. sooner than in position 2. Complete shading takes place only at the central point A_0. The variation of irradiance is described by curve 3. In position 4 of the disk $h = 2h_0$ and shading sets in at the point C_1 and is complete at the point C_2, where the whole screen disk is posed against the background of the apparent solar disk, but only a part of the solar disk is obstructed, because of the small size of the screen disk. As the detector moves on from the point C_1 to C_3 the irradiance does not change (curve 4), because the screen disk is moving against the background of the apparent solar disk (assuming the same brightness at all points of the solar disk). Attenuation of irradiance within C_2C_3 is defined by the ratio of the screen disk area $S = \pi r^2$ to the apparent solar disk area $S_I = \pi r_I^2$ i.e. $S(C_2) = S_0(1 - S/S_I) = \frac{3}{4}S_0$. Consequently, if the angular size of the Sun exceeds that of the screen disk ($r < r_I$), there is no full shading and the shadow effect is reduced to a penumbra area. The irradiance at the centre of this area is determined by the ratio of the angular size of the screen disk to that of the solar disk. Thus the structure and size of the penumbra area and the irradiance of direct solar radiation are determined, first, by the size of plant parts and, second, by the their distance from the point considered. It is also seen from Figure 48 that the penumbra area and its irradiance increase with the increasing depth of the plant stand. For example, a screen disk 2 cm in diameter will produce no full shade when placed at a height not less than 2 m, and $S \geqslant 0.8S_0$ throughout the penumbra area when the height is 5 m.

The above examples show that taking account of the penumbra effect considerably complicates the description of the direct solar radiation field in the plant stand, a statistical treatment of the field requires more than the two values of the random function $X_S(x)$ and all irradiance values from zero to S_0 must be represented in the realization of this function. From general considerations it follows that the probability density of direct solar radiation $f_S(I)$ is U-shaped (Figure 49). The first maximum corresponding to $I = 0$ is proportional to the shade area a_{SS}, and the second at $I = S_0$ is proportional to the sunfleck area a_S.

The probability density $f_S(I)$ can be represented as the sum of three terms, the first two terms defining the direct radiation in the full shade area and the third term $f_{SP}(I)$ characterizing the distribution of radiation in the penumbra area, i.e.

$$f_S(I) = a_{SS}\delta(I - 0) + a_S\delta(I - S_0) + f_{SP}(I), \tag{II.4.1a}$$

where

$$\int_0^\infty f_{SP}(I)\mathrm{d}I = 1 - a_S - a_{SS}$$

being the part contributed by the penumbra area.

A statistical treatment of direct solar radiation in plant stands using Dirac's delta-function was first proposed in a work of Isobe (1962a). General observations on the role of penumbras in the radiation regime in plant stands are contained in the work of Anderson and Miller (1974). A considerable advance in the theory of penumbra and sunflecks has been made by Norman with coauthors (Norman et al. 1971; Miller & Norman 1971a, b) who proposed a new theory which takes account of the statistics of sunfleck dimensions and the intensity distribution due to penumbra in the sunflecks. Variants for a random and non-random distribution of phytoelements have been worked out. Comparison of this theory with experimental results obtained in a stand of Sitka spruce

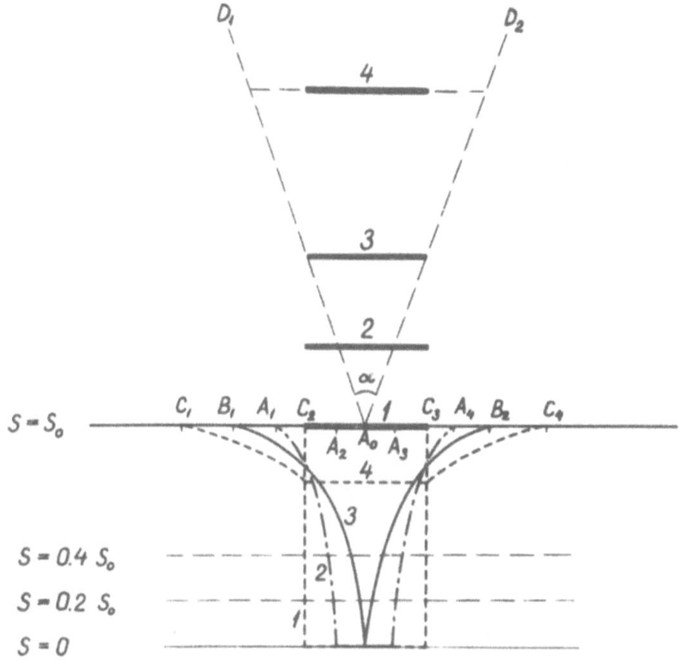

Fig. 48. Penumbra effect due to a round disk placed between Sun and detector.
Curves 1–4, corresponding to disk positions 1–4, approximate the variation of irradiance due to detector position changing in direction C_1C_4. See text.

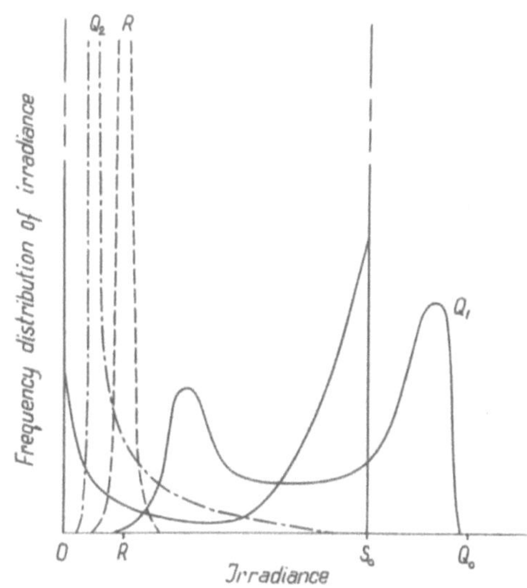

Fig. 49. Distribution functions of radiative fluxes in a horizontal direction within a plant stand. S – direct solar radiation with penumbra taken into account, Q_1 and Q_2 – total radiation respectively in upper and deeper layers in a dense stand, R – reflected radiation.

(Norman & Jarvis 1974, 1975) has shown good agreement in the case of a clumped distribution of phytoelements.

II.4.3 THEORETICAL EXPRESSIONS FOR DIRECT SOLAR RADIATION PENETRATION

Exponential formula. The general case was discussed in the work of Ross and Nilson (1965). Integrating Equation (II.1.61) with the boundary condition (II.1.61a) gives the following expression for the penetration function of direct solar radiation

$$a_S(z, r_S) = \exp\left[-\operatorname{cosec} h_\odot \sum_k \int_z^{z_0} u_k(z') G_k(z', r_S) \mathrm{d}z'\right]. \qquad \text{(II.4.2)}$$

The index k implies summation either over individual plant organs or over individual species in plant stands of a complex structure.

Consider some special cases of formula (II.4.2) which are important for practical application.

1. Attenuation of radiation is dominated by leaves of one species ($k=1$).

$$a_S(z, r_S) = \exp\left(\frac{-1}{\sin h_\odot} \int_z^{z_0} G(z', r_S) u(z') \mathrm{d}z'\right). \qquad \text{(II.4.3)}$$

2. The distribution function of foliage area orientation is independent of height. Then $G(z, r_S) = G(r_S)$, and taking into account (I.2.32) we obtain

$$a_S(L, r_S) = \exp\left[-\frac{G(r_S)}{\sin h_\odot} L\right]. \qquad \text{(II.4.4)}$$

3. The stand consists of leaves and vertical cylndrical stems, for which

$$\frac{1}{2\pi} g_S = \delta\left(\vartheta_S - \frac{\pi}{2}\right)$$

and according to (I.2.18d),

$$G_C(r_S) = \frac{1}{\pi} \cos h_\odot$$

$$a_S(z, r_S) = \exp\left[\frac{G_L(r_S) L}{\sin h_\odot} - \frac{L_C}{\pi} \cot h_\odot\right], \qquad \text{(II.4.5)}$$

where

$$L_C = \int_z^{z_{0C}} u_C(z') \mathrm{d}z'$$

and z_{0C} is the uppermost level of stems in the stand.

Denoting

$$K = \frac{G(r_S)}{\sin h_\odot}$$

we obtain the Monsi and Saeki (1953) formula from formula (II.4.4).

4. The stand consists of horizontal leaves, i.e., in accordance with (I.2.18c) $G_L = \sin h_\odot$
and

$$a_S(L) = \exp(-L). \tag{II.4.6}$$

5. The stand consists of leaves with normals oriented in all directions, with equal probability (spherical orientation according to Nichiporovich, 1961). Then according to (I.2.18b) $G_L = \frac{1}{2}$
and

$$a_S(L, r_S) = \exp\left(\frac{-L}{2\sin h_\odot}\right) \tag{II.4.7}$$

6. The stand consists of vertical leaves distributed uniformly in azimuth. Then, according to (I.2.18d)

$$G_L = \frac{2}{\pi}\cos \mathrm{h}_\odot \quad \text{and}$$

$$a_S(L, r_S) = \exp\left(-\frac{2}{\pi} L \cot h_\odot\right). \tag{II.4.8}$$

7. All leaves have a constant angle of inclination $\vartheta_L = \theta$ and a uniform azimuthal distribution. In this case

$$a_S = \exp(-KL), \tag{II.4.9}$$

where $K(\theta, h_\odot)$ is defined by the formula

$$K(\theta, h_\odot) = \begin{cases} \cos\theta & \theta \leqslant h_\odot \\ \cos\theta\left[1 + \frac{2}{\pi}(\tan\varphi_0 - \varphi_0)\right] & \theta > h_\odot \end{cases} \tag{II.4.10}$$

and $\cos\varphi_0 = \tan h_\odot \cot\theta$.

This special case was discussed in detail in works of Warren Wilson (1960), Anderson (1966a), Duncan et al. (1967). Setting $\theta = 0$ and $\pi/2$ in (II.4.10) gives special cases 4 and 6, respectively.

Figure 50 borrowed from the work of Anderson (1966a) shows the attenuation coefficient K as a function of solar elevation for different leaf inclination angles. We have supplemented this diagram with the curve of $K = \frac{1}{2}\sin \mathrm{h}_\odot$ (shown dashed) corresponding to a uniform orientation of leaves. It is seen from the diagram that at $h_\odot > 40°$ the attenuation coefficient depends but slightly on $h_\odot$ if the leaf inclination angle is θ small or the leaves are uniformly oriented. The dependence of K on $h_\odot$ is closer for a vertical leaf orientation.

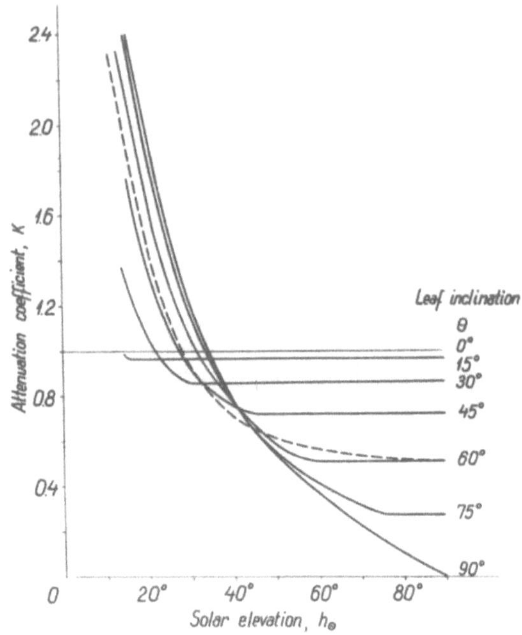

Fig. 50. Attenuation coefficient K as a function of solar elevation $h_{\odot}$ for different leaf inclinations θ (Anderson 1966a).

Binomial formula. Let us derive a binomial formula in its general form. Divide a horizontally uniform stand of a depth L_0 consisting of leaves into N_0 layers. Suppose that ΔL_k is the depth of the k-th layer,

$$L_0 = \sum_{k=1}^{N_0} \Delta L_k$$

and that the direct solar radiation $S_0(r_S)$ is incident on the stand. Determine the shade area at the lower level of the first layer ΔL_1. Assuming that no mutual shading of leaves in the direction of sunrays exists within the layer, we obtain the expression $G_1(r_S)\Delta L_1$ for the leaf area projected on the direction of sunrays, and $G_1 \Delta L_1/\sin h_{\odot}$ for the same area projected on a horizontal surface. Upon the passage of radiation through the first layer the unshaded horizontal surface area will consequently be $1 - G_1 \Delta L_1/\sin h_{\odot}$. Analogously we obtain $1 - G_2 \Delta L_2/\sin h_{\odot}$ for the second layer, etc.

Assuming a random distribution of leaves in individual layers and in the direction of the Sun, i.e. that distribution in the k-th layer is independent of leaf distribution in other layers, we multiply the probabilities of independent random quantities to derive an expression for the gap area which exists upon the passage through layers, i.e. the generalized binomial formula

$$a_S(L, r_S) = \prod_{k=1}^{N} \left[1 - \frac{\Delta L_k G_k(h_\odot)}{\sin h_\odot} \right] =$$

$$= \exp \left[\sum_{k=1}^{N} \ln \left(1 - \frac{\Delta L_k G_k}{\sin h_\odot} \right] . \quad \text{(II.4.11)}$$

Taking into account that $\ln(1-a) \approx -a - a^2/2$ gives the following apexpression for the binomial formula

$$a_S(L, r_S) \approx \exp \left[\frac{-1}{\sin h_\odot} \sum_{k=1}^{N} \Delta L_k G_k \left(1 + \frac{\Delta L_k G_k}{2 \sin h_\odot} \right) \right] . \quad \text{(II.4.12)}$$

At the limit $\Delta L_k \to 0$ or $N \to \infty$ formulae (II.4.11) and (II.4.12) are reduced to the exponential formula (II.4.2).

Thus the exponential formula (II.4.2) is the extreme case of the binomial one and implies two essential assumptions: (1) the distribution of leaves in individual layers is random, (2) the stand may be divided into any large number of horizontal layers. As compared with the exponential formula, the binomial one contains an additional parameter N, the number of 'elemental' layers (or depth of 'elemental' layer ΔL).

In a special case where $\Delta L_k = \Delta L$ and $G_k(r_S) = G(r_S)$, i.e. orientation of leaves is the same in all layers, we derive the following respectively from (II.4.11) and (II.4.12)

$$a_S(L, r_S) = \left[1 - \Delta L \frac{G(r_S)}{\sin h_\odot} \right]^{L/\Delta L} \quad \text{(II.4.13)}$$

and

$$a_S(L, r_S) \approx \exp \left[-\frac{LG}{\sin h_\odot} \left(1 + \frac{G \Delta L}{2 \sin h_\odot} \right) \right] . \quad \text{(II.4.14)}$$

It is to be observed from (II.4.14) that the binomial distribution of penetration can be described approximately by means of an exponential function whose exponent contains an additional term, $G\Delta L/\sin h_\odot$, which characterizes the effect which the finite depth of an 'elemental' layer ΔL has on the penetration.

A binomial formula to describe the attenuation of solar radiation in a plant stand was first proposed by Monsi and Saeki (1953) in the form $J = J_0(1 - \bar{s}/S)^n$, where n is the maximum possible order of mutual shading, S is the basal area of the vertical column considered, $\bar{s}$ is the projection of leaves in the column upon S in the sunrays. In our notation $n = L/\Delta L$, $\bar{s}/S = G\Delta L/\sin h_\odot$ and $J/J_0 = a_S$.

A binomial formula in the form (II.4.13) was first derived by de Wit (1965).

Mototani (1968) proposed a binomial formula for negative values of $N = L/\Delta L$. In this case instead of (II.4.13) and (II.4.14), respectively, we obtain

$$a_S(L, r_S) = \left[1 + L \frac{G(r_S)}{\sin h_\odot} \right]^{-L/\Delta L} \quad \text{(II.4.15)}$$

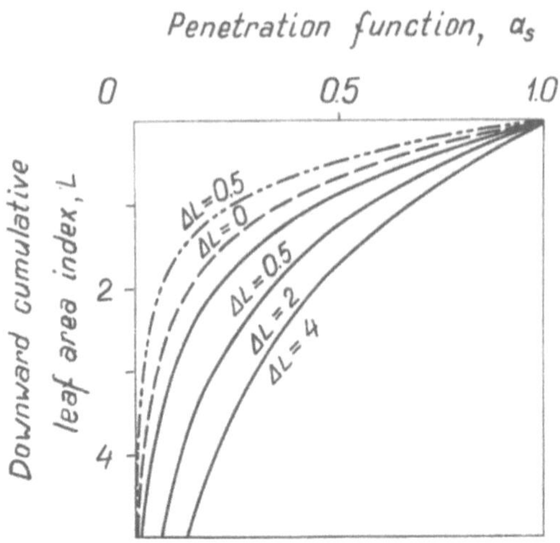

Fig. 51. Vertical profiles of penetration function of direct solar radiation a_s, given by the positive binomial ($\triangle L = 0.5$), exponential ($\triangle L = 0$) and negative binomial (solid lines) formula (Nilson 1971a).

and

$$a_S(L, r_S) \approx \exp\left[-\frac{LG(r_S)}{\sin h_\odot}\left(1+\frac{G\Delta L}{2\sin h_\odot}\right)\right]. \tag{II.4.16}$$

Analyses of formulae (II.4.4), (II.4.13) and (II.4.15) indicate that the positive binomial formula underestimates and the negative binomial formula overestimates values of penetration in comparison with the exponential formula, the degree of deviation being determined by the number of layers $N = L/\Delta L$.

Figure 51 taken from Nilson's work (1971a) illustrates the course of the exponential, positive binomial and negative binomial penetration functions for $G/\sin h_\odot = 1$ and $\Delta L = -0.5$; 0.0; 0.5; 2.0 and 4.0. It follows from the curves that differences between the three functions become immaterial when $\Delta L < 0.2$. In a special case when $L = \Delta L$, we obtain

$$a_S(L, r_S) = 1 - L\frac{G(r_S)}{\sin h_\odot}$$

for the positive binomial formula and

$$a_S(L, r_S) = \frac{1}{1 + LG(r_S)/\sin h_\odot}$$

for the negative binomial formula. In the latter case the penetration function for the first layer is not determined by the projection of a foliage area on a horizontal surface in the sunrays.

Mototani (1968) proposed the negative binomial formula as an empirical one without a theoretical substantiation. Nilson in his work (1971a) pointed out that this formula could be interpreted theoretically. Let us discuss this problem in more detail.

Since $1-a_S=\sigma_S$ is the part of the shaded area, then upon penetrating through the first 'elemental' layer $\sigma_S^+=G\Delta L/\sin h_\odot=\tau$ for the positive binomial formula, and $\sigma_S^-=\tau/(1+\tau)$ for the negative binomial formula. The amount of shading is less in the latter case. The difference $\sigma_S^+-\sigma_S^-=\tau^2/(1+\tau)>0$ indicates that in the case of the negative binomial relation there exists a mutual shading of leaves, which constitutes a fraction $(\sigma_S^+-\sigma_S^-)/\sigma_S^+=\tau/(1+\tau)$ of their effective projection, in the 'elemental' layer.

The relative variance V' of the distribution of phytoelements is equal to unity for an exponential penetration (the Poisson distribution); and for the binomial formula

$$V'=1\pm\frac{G\Delta L}{\sin h_\odot}, \tag{II.4.17}$$

where the minus sign corresponds to the positive binomial formula, and the plus sign to the negative binomial formula.

Acock et al. (1970) have analyzed and compared the three types of formula, exponential, positive and negative binomial ones, and found that failing to take account of the deviation from a random distribution of phytoelements in real plant stand types results in a 5–15% error in penetration values. However, this is an underestimate, as will be made obvious in II.4.7.

Nilson's formula. A new theoretical formula for the penetration function a_S based on the Markov chain has been derived by Nilson (1971a).

Consider a horizontally uniform plant stand. Divide it into $N=L/\Delta L$ equal horizontal 'elemental' layers so that no mutual shading exists within each layer. Assume that the conditional probability of observing a shaded area in an 'elemental' layer is $\lambda_0(G\Delta L/\sin h_\odot)\leqslant 1$ in the absence of a shading area in the overlying layer in the direction of the Sun r_S, and $\lambda(\Delta LG/\sin h_\odot)\leqslant 1$ if such an area exists. In contrast to the binomial model in which there is no mutual dependence of the positions of shaded areas in adjacent areas, the Markov model treats the positions of shaded areas in a given layer as dependent on the arrangement of shaded areas in the overlying layer. If $\lambda_0=\lambda=1$, the binomial model of penetration is derived as a special case from the Markov model.

The probability of shading in the overlying layer is $\tau=G\Delta L/\sin h_\odot$ and that of no shading is $1-\tau$. In this case the total probability of shading in the given layer, which is also equal to τ consists of two conditional probabilities $\lambda\tau$ and $\lambda_0\tau$. The probability of shaded areas occuring in two layers in the direction of the Sun is equal to the product of the corresponding probabilities for both layers, i.e. $\lambda\tau\cdot\tau$; and the probability of a shaded area in a given layer with an unshaded area occuring concurrently in the overlying layer is $\lambda_0\tau(1-\tau)$. The total probability of a shaded area occuring in a given layer independently of the situation in the overlying layer is $\lambda_0\tau(1-\tau)+\lambda\tau^2=1$, whence

$$\lambda_0 = \frac{1-\lambda\tau}{1-\tau} = \frac{1-\lambda\dfrac{G\Delta L}{\sin h_\odot}}{1-\dfrac{G\Delta L}{\sin h_\odot}}.$$

Since $\lambda_0 \geqslant 0$ and $\lambda \geqslant 0$, the upper limit of λ_0 is

$$\lambda_0 \leqslant \frac{1}{1-\tau}$$

But since, on the other hand, $\lambda_0 \tau \leqslant 1$, max λ_0 can be realized under the condition $\tau < \frac{1}{2}$.

The probability of the penetration of direct solar radiation through the first layer is equal to $1-\tau$, and through each succeeding layer $1-\lambda_0\tau$. Consequently, after penetrating N layers, the probability of a sunfleck, i.e. the part of the unshaded horizontal area is

$$a_S(L, r_S) = \begin{cases} 1 - \dfrac{LG}{\sin h_\odot}, & 0 < L \leqslant \Delta L \\ \left(1 - \dfrac{G\Delta L}{\sin h_\odot}\right)\left(1 - \lambda_0 \dfrac{G\Delta L}{\sin h_\odot}\right)^{N-1}, & \Delta L < L \end{cases} \quad \text{(II.4.18)}$$

According to Nilson (1971a) the relative variance is expressed by

$$V' = (1-\tau)\{1 + 2[(N-1)\varepsilon + (N-2)\varepsilon^2 + \ldots + \varepsilon^{N-1}]/N\}, \quad \text{(II.4.19)}$$

where

$$\varepsilon = 1 - \lambda_0.$$

It is seen from (II.4.19) that V' is not constant, as in the case of a binomial distribution, but changes with the depth L.

Let us discuss some special cases of formula (II.4.18).

1. $\lambda_0 = 0$. Then (II.4.17) gives $\lambda\tau = 1$. That implies that a shaded area can only be present in a given layer where there is also a shaded area above it in the overlying layer, i.e. shading elements in all layers must be arranged over one another. Such an arrangement of leaves provides a maximum mutual shading and, consequently, a maximum penetration. From (II.4.18) we obtain the following for $\lambda_0 = 0$

$$a_S(L, r_S) = \begin{cases} 1 - \dfrac{GL}{\sin h_\odot}, & \text{if } 0 < L \leqslant \Delta L, \\ 1 - \dfrac{G\Delta L}{\sin h_\odot}, & \text{if } \Delta L < L. \end{cases} \quad \text{(II.4.20)}$$

It follows that the penetration of radiation is determined by the arrangement and number of leaves in the first layer and that a further increase in the number of leaves will not diminish the penetration because leaves in lower layers are located under those (in the direction of sunrays) of the first layer. The relative variance in this case is $V'=(1-\tau)N$.

2. $\lambda_0=1$. From (II.4.17) we obtain $\lambda=1$ and formula (II.4.18) changes to the binomial one (II.4.13). In this case the relative variance is $V'=1-\tau$.

3. Max $\lambda_0=1/(1-\tau)$. From (II.4.17) we obtain $\lambda=0$; from the conditions $0\leqslant\lambda_0\leqslant 1$ we obtain $0<\tau\leqslant\frac{1}{2}$, and from (II.4.18)

$$a_S(L, r_S)=\begin{cases} 1-\dfrac{GL}{\sin h_{\odot}}, & \text{if } 0\leqslant L\leqslant \Delta L \\ \dfrac{(1-2\tau)^{N-1}}{(1-\tau)^{N-1}}, & \text{if } \Delta L<L, \end{cases}$$

for $0<\tau\leqslant\frac{1}{2}$.

4. $\lambda_0=1/\tau$. Expression (II.4.17) gives $\lambda=(2\tau-1)/\tau^2$, and it follows from the conditions $0\leqslant\lambda\tau\leqslant\lambda$ that $\tau/2\leqslant\tau\leqslant 1$. Consequently, upon the penetration of the first layer $a_S\leqslant\frac{1}{2}$. From the condition $\lambda_0\tau=1$ we can conclude that shading can only exist in a given layer when there is no shading in the overlying layer. Consequently, leaves in the first layer must cast no shade on leaves in the second.

Thus this special case relates to a situation where no mutual shading exists in the first and second layers, but the foliage area is so large that the penetration below the second layer is zero, i.e.

$$\begin{cases} 1-\dfrac{GL}{\sin h_{\odot}}, & \text{if } 0\leqslant L\leqslant\dfrac{\sin h_{\odot}}{G} \\ 0, & \text{if } \dfrac{\sin h_{\odot}}{G}<L. \end{cases} \tag{II.4.21}$$

5. The number of layers is infinitely large. $N\to\infty$ or $\Delta L\to 0$. Since $\lim_{\Delta L\to 0}\ln(1-\lambda_0\tau)(L/\Delta L-1)=-\lambda_0\tau$, (II.4.17) gives

$$a_S(L, r_S)=\exp\left(-\lambda_0\frac{GL}{\sin h_{\odot}}\right) \tag{II.4.22}$$

and (II.4.19) gives $V'=2/\lambda_0-1$.

The expression for a relative variance shows that in the case of $0\leqslant\lambda_0\leqslant 1$ $V'>1$ and we have a clumped distribution of leaves in the stand.

The first of the above special cases, i.e. $\lambda_0=0$, corresponds to the highest degree of clumping of leaves, when leaves are arranged one over another, and the fourth case $\lambda_0=1/\tau$ corresponds to the highest regularity, when no mutual shading exists as long as $a_S\to 0$.

For example, shown in Figure 52 borrowed from the work of Nilson (1971a) are curves of a_S corresponding to the Markov model of penetration for $G(r_S)/\sin h_{\odot}=1$. Curve 1 ($\lambda_0=1.3$ and $\Delta L\to 0$) corresponds to a semiregular distribution of leaves; curve 4 ($\lambda_0=0.7$, $\Delta L\to 0$), to a clumped distribution; and curve 2 ($\lambda_0=0.7$, $\Delta L=0.3$), to a semi-regular distribution in the upper and a clumped one in the lower layers. Curve 3 corresponds to the exponential formula (a Poisson distribution for $\lambda_0=1$ and $\Delta L\to 0$.

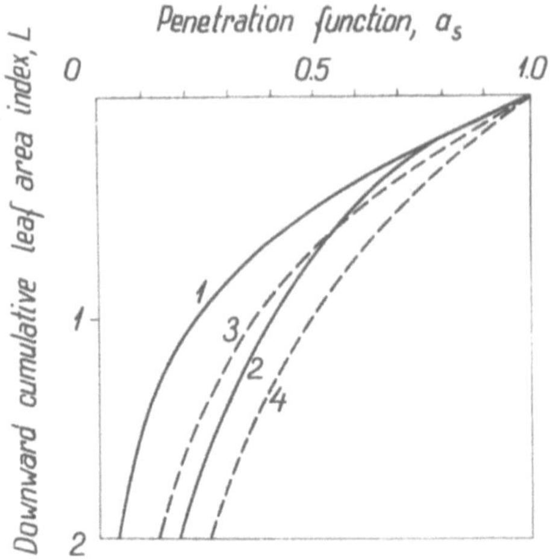

Fig. 52. Vertical profiles of penetration function of direct solar radiation a_s, given by Nilson's formula (Nilson 1971a). See text.

II.4.4 PENETRATION THEORY FOR DIRECT SOLAR RADIATION IN HORIZONTALLY INHOMOGENEOUS PLANT STANDS

There are two ways of taking account of the horizontal inhomogeneity of a stand.

First, the stand may be treated as a turbid medium in which the distribution functions of phytoelements $u_k(P)$ do not only depend on the height z but also on the horizontal coordinates x and y. In the case of a regular plant distribution the functions u_k, being periodic, depend either on one horizontal coordinate (a row-sown crop) or on both coordinates (a square-sown crop). This approach is suitable for a closed stand with no considerable gaps between plants. Of course, this requirement limits the application of the radiative transfer theory as regards many types of plant stand.

The other way, which is more general, though more complicated, is not to treat the stand as a homogeneous turbid medium but rather as a totality of separate plants or small plant communities arranged in a certain way on the ground. This approach to the penetration theory necessitates more detailed data on the stand structure, for example, the regularity of the distribution of plants in the stand, three-dimensional distribution of phytoelements for a single plant, distribution of plants according to height and diameter, character and degree of foliage overlapping for neighbouring plants, etc.

Generalized exponential formula. This was derived jointly by Nilson and the author.

Prescribing $\eta_L = \Gamma_L = 0$ in the transfer equation (II.1.38a) and assuming that G_L does not depend on P, we obtain the transfer equation for direct solar radiation in the form

$$\sin\vartheta_\odot \cos\varphi_\odot \frac{\partial S}{\partial x}+\sin\vartheta_\odot \sin\varphi_\odot \frac{\partial S}{\partial y}+\cos\vartheta_\odot \frac{\partial S}{\partial z}=$$

$$=-G(\vartheta_\odot,\varphi_\odot)u_L(x,y,z)S(x,y,z,\vartheta_\odot,\varphi_\odot) \quad \text{(II.4.23)}$$

which when solved, with the boundary conditions

$$S(x,y,z_0,\vartheta_\odot,\varphi_\odot)=S_0(\vartheta_\odot,\varphi_\odot)$$

gives this expression for the penetration function

$$a_S(x,y,z,\vartheta_\odot,\varphi_\odot)=\exp[-G(r_S)\tau(P,r_S)], \quad \text{(II.4.24)}$$

where

$$\tau(P,r_S)=\frac{1}{\cos\vartheta_\odot}\int_z^{z_0} u_L(z', x-(z_0-z-z')\tan\vartheta_\odot \cos\varphi_\odot, y-$$

$$-(z_0-z-z')\tan\vartheta_\odot \sin\varphi_\odot)dz' \quad \text{(II.4.25)}$$

is the pathway length of the solar beam from the upper level of the plant canopy to the point $P=(x,y,z)$ in the direction $r_S=(\vartheta_\odot,\varphi_\odot)$.

For square-sown crops $u_L(x,y,z)$ is a periodic function of x and y with the periods of x_0 and y_0 corresponding to the distance between rows.

Suppose that the separation of the variables governing the foliage area density u_L is permissible, i.e. $u_L(x,y,z)$ can be presented as (I.2.36) with the normalizing relations (I.2.37). The horizontal variability of a_S is defined by the functions $u_2(y)$ and $u_3(x)$. Some of these were discussed in I.5.3. In an actual stand the function $u_L(x,y,z)$ is random or discontinuous.

Note that the horizontal inhomogeneity in a plant stand always leads to an increase in the average penetration, i.e. the penetration function averaged over a horizontal surface

$$\overline{a_S(x,y,z)}=\frac{1}{x_0 y_0}\int_0^{x_0}\int_0^{y_0} a_S(x,y,z)dydx \quad \text{(II.4.26)}$$

is always larger than the penetration function in the case of the averaged leaf area density

$$\overline{u_L(x,y,z)}=\frac{1}{x_0 y_0}\int_0^{x_0}\int_0^{y_0} u_L(x,y,z)dydx. \quad \text{(II.4.27)}$$

This is due to the non-linearity of penetration.

The author's theory. The turbid medium model of the plant stand is but an idealization, which fails to take into account a number of factors. The most important factors are differences in species and age, the distribution of plants over the ground surface, and particularly the habitus of individual plant species.

Let us neglect differences in species and age for the sake of simplicity and consider a pure stand uniform in species and age.

Let us model plants as vertical columns of the same height and a constant horizontal cross-section area. Let S_{1Pk} be the horizontal cross-section area of the

Fig. 53. Irregular distribution of plants.

k-th plant. Practically S_{1Pk} may be defined as the vertical projection of a plant crown on the ground surface and S_{1Pk} can have various contours. There may be gaps within S_{1Pk} depending on the plant architecture. The vertical column used as a plant model determines the living space of the plant. The coverage of the ground surface by plants will be characterized by means of the overlapping index η given by the formula (I.2.27). Let N_P plants be distributed in a certain fashion over a unit ground area (Figure 53). Divide the ground area into three parts: ω_0 the part of the ground area without plants; ω_1 the part of the ground area occupied by a single plant; ω_2 the part of the ground area formed by the overlapping of neighbouring plants, so that

$$\omega_0 + \omega_1 + \omega_2 = 1. \tag{II.4.28}$$

The average leaf area index is different for each of these areas. Area ω_0 has no plants, consequently, the leaf area index is zero. In area ω_1 the leaf area index is determined by the leaf area of a single plant, i.e. $L_1 = L^*$. Let the leaf area index of area ω_2 be L_2. Because of overlapping, $L_2 > L^*$. What is the leaf area index averaged over the whole ground surface? The foliage area of the k-th plant is $L^* S_{1Pk}$, and taking into account (I.2.27) the foliage area of all N_P plants is

$$L^* \sum_{k=1}^{N_P} S_{1Pk} = \eta L^*.$$

Consequently,

$$L = \eta L^* = L_1 \omega_1 + L_2 \omega_2, \tag{II.4.29}$$

whence

$$L_2 = \frac{\eta - \omega_1}{\omega_2} L^* = H L^* = \frac{H}{\eta} L, \tag{II.4.30}$$

where

$$H(\eta) = \frac{\eta - \omega_1}{1 - \omega_0 - \omega_1} \tag{II.4.30a}$$

is the so-called *foliage thickening function.* The function H indicates the factor whereby the leaf area in the thickened area of the stand exceeds that of the area containing a single plant.

Since each part ω_0, ω_1 and ω_2 has a different foliage area, the penetration of radiation is also different. Obviously, the penetration function for part ω_0 is equal to unity.

Let a_{S1} and a_{S2} define the penetration functions for the overlapping regions above ω_1 and ω_2. Then the penetration function a_S averaged over the whole ground surface is

$$a_S=\omega_0+\omega_1 a_{S1}+\omega_2 a_{S2} \tag{II.4.31}$$

and the determination of the average penetration function is reduced to that of the areas ω_0, ω_1 and ω_2 and the penetration functions a_{S1} and a_{S2}.

The areas ω_0, ω_1 and ω_2 depend on the distribution of plants over the ground surface, the contours of the cross-section area of plants S_{1P} and the overlapping index η. For instance, in a 39-year old Scots pine forest planted in rows (Ford 1976) $\omega_0=0.175$, $\omega_1=0.703$, the parts of double and triple overlapping are equal to 0.119 and 0.003, respectively, i.e. $\omega_2=0.122$.

Let us discuss two special cases: a random and a regular plant distribution.

A random plant distribution is described using the Poisson distribution:

$\omega_{0C}=e^{-\eta}$ is the probability of occurrence of no plant projection at a given point,

$\omega_{1C}=\eta e^{-\eta}$ is the probability of occurrence of a single plant projection at a given point,

$\omega_{2C}=1-(1+\eta)e^{-\eta}$ is the probability of occurrence of more than one plant projection at a given point.

The foliage thickening function for a random plant distribution is, according to (II.4.30),

$$H_C(\eta)=\eta\frac{1-e^{-\eta}}{1-(1+\eta)e^{-\eta}}=\frac{\eta}{1-\dfrac{\eta e^{-\eta}}{1-e^{-\eta}}}. \tag{II.4.32}$$

For large values of η, $H_C\approx\eta$.

In the case of a regular plant distribution (square-sown plant stand, with a square contour of cross-section S_{1P}),

$$\left.\begin{aligned}
\omega_{0P}&=\begin{cases}1-\eta, & \text{if } 0<\eta\leqslant 1,\\ 0, & \text{if } \eta>1,\end{cases}\\
\omega_{1P}&=\begin{cases}\eta, & \text{if } 0<\eta\leqslant 1,\\ (2-\sqrt{\eta})^2, & \text{if } 1<\eta\leqslant 4,\\ 0, & \text{if } \eta>4,\end{cases}\\
\omega_{2P}&=\begin{cases}0, & \text{if } 0<\eta\leqslant 1,\\ 1-(2-\sqrt{\eta})^2, & \text{if } 1<\eta\leqslant 4,\\ 1, & \text{if } \eta>4,\end{cases}
\end{aligned}\right\} \tag{II.4.33}$$

$$H_P(\eta)=\begin{cases}\dfrac{\eta-(2-\sqrt{\eta})^2}{1-(2-\sqrt{\eta})^2}=\dfrac{1}{1-\dfrac{1+\sqrt{\eta}}{4}}, & 1<\eta\leqslant 4\\ \eta, & \eta>4.\end{cases} \tag{II.4.34}$$

The functions $\omega_0, \omega_1, \omega_2$ and H are shown in Figure 54, which demonstrates that foliage thickening is always greater for a random plant distribution. But at large values of η the foliage thickening function depends but slightly on the type of plant distribution. The areas ω_0 and ω_1 dominate at low values of η, and the area of multiple overlapping ω_2 dominates at high values. The penetration of radiation is most complex in character in the range $\eta=0.5$–1.5, where all components, ω_0, ω_1 and ω_2 should be taken into account. This range is frequent in nature.

Let us proceed to the penetration functions a_{S1} and a_{S2}. Shown in Figure 55 is a plant having a stem with leaves on it. Let z_{1L} be the uppermost foliage level of a plant in its natural position, n_L the number of leaves, S_{Lj} the area of the j-th leaf, z_{Lj} the distance between the $(j-1)$-th and j-th leaves. Suppose that the leaves are arranged on the stem following the genetic spiral (Sinnot 1960) so that i successive leaves complete a full cycle, i.e. the first, $(i+1)$-th, $(2i+1)$-th, etc., leaves are placed one over another. Divide the plant foliage into $n_k=n_L/i$ 'elemental' layers. Then each layer will have i leaves casting no shade on one another. The depth of each layer is

$$z_k=\sum_{j=ik+1}^{ik+i} z_{Lj},$$

the foliage area is $\sum_{j=ik+1}^{ik+i} S_{Lj}$, and the leaf area index in the k-th 'elemental' layer is

$$\Delta L_k^*=\frac{\sum_{j=ik+1}^{ik+i} S_{Lj}}{S_{1Pk}}.$$

The foliage area in a unit volume of the k-th layer of plant foliage is

$$u_k^*=\frac{\Delta L_k^*}{z_k}=\frac{1}{z_k S_{1Pk}}\sum_{j=ik+1}^{ik+i} S_{Lj}.$$

The effective projection of leaves on the downward surface of the k-th layer in the direction of sunrays is defined by the function $K_k(r_S)=G_k(r_S)/\sin h_{\odot}$. Then the part of the shaded area in the k-th layer is equal to $K_k\Delta L_k^*$ and the part of the gap area in the k-th layer is $1-K_k\Delta L_k^*$.

Assuming that the arrangement of leaves in the k-th layer does not depend on that in other layers, the penetration of radiation may be described by means of the binomial formula (II.4.11) according to which the penetration function for a single plant after passing N layers is

$$a_{P1}=\prod_{k=1}^{N}(1-K_k\Delta L_k^*). \tag{II.4.35}$$

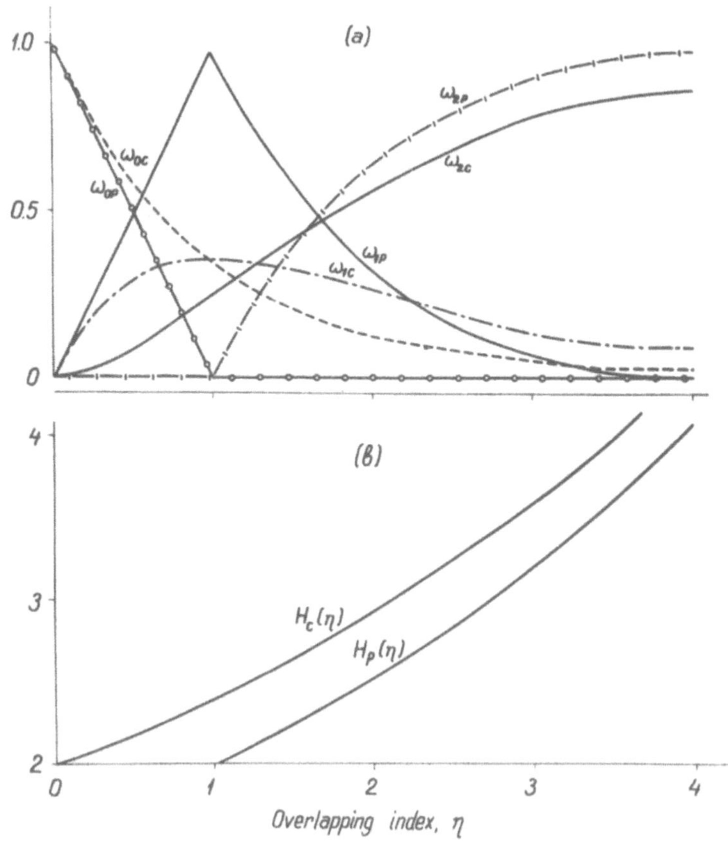

Fig. 54. Dependence of (a) functions ω_0, ω_1, ω_2 and (b) foliage thickening function $H(\eta)$ on overlapping index η for random and regular distribution of plants.

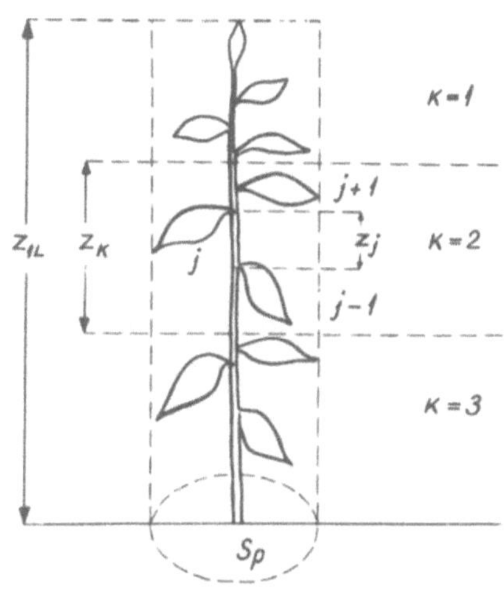

Fig. 55. Illustration of 'elemental' layer of plant. See text.

In the special case in which the orientation and area of leaves are the same for all layers

$$a_{P1} = (1 - K\Delta L^*)^N \tag{II.4.36}$$

with

$$L^* = N\Delta L^*. \tag{II.4.37}$$

It should be emposized that formulae (II.4.35) and (II.4.36) are not valid for low solar elevations.

A more complicated situation exists above area ω_2 because of the interlacing branches of neighbouring plants. The foliage area of an elemental layer is not equal to ΔL^*, but $\Delta L_2 = H\Delta L^*$ and is determined by the plant distribution. Depending on the overlapping of neighbouring plants, a mutual shading of leaves in the elemental layer can occur. No experimental data on overlapping are available at this stage. We shall, therefore, discuss two theoretical cases, where either a random or a regular overlapping exists.

Random overlapping. In this case the occurrence of a new leaf at a point over the area ω_2 does not depend on whether or not any leaves have been present in the vicinity of this point. No mutual shading of leaves existed in the elemental layer prior to intrusion. Because of the overlapping, the foliage area increased by a factor of H. A random overlapping gives rise to mutual shading. Suppose that the maximum order of leaf shading above the area ω_2 is equal to the order of foliage thickening H. Let us divide the elemental layer into H sub-layers, the foliage area of each being equal to $\Delta L_2/H$. In the case of a random overlapping the distribution of leaves in a sub-layer does not depend on leaf distribution in other sub-layers, and the penetration of radiation in an elemental layer can be described using a binomial expression $(1 - K\Delta L_2/H)^H$. Since we still have an independent distribution of leaves in the elemental layers, the penetration of radiation above the area ω_2, for a random overlapping, can described by a binomial expression

$$a_{S2}^C = \left(1 - \frac{K\Delta L_2}{H}\right)^{HN} \tag{II.4.38}$$

with

$$L_2 = N\Delta L_2 = HN\Delta L^*. \tag{II.4.39}$$

Since the form of the foliage thickening function depends on the type of plant distribution, we obtain the penetration function for a regular distribution of plants and a random overlapping by taking into account formulae (II.4.32), (II.4.34) and (II.4.39)

$$a_{S2}^{PC} = \left(1 - \frac{K\Delta L_2^P}{H_P}\right)^{NH_P} = (1 - K\Delta L^*)^{NH_P} =$$

$$= \left(1 - \frac{KL}{N\eta}\right)^{NH_P} = (1 - K\Delta L^*)^{(L/\Delta L^*)(H_P/\eta)} \tag{II.4.40}$$

and for a random distribution and random overlapping

$$a_{S2}^{CC}=\left(1-\frac{K\Delta L_2^C}{H_C}\right)^{NHc}=(1-K\Delta L^*)^{NHc}=$$

$$=\left(1-\frac{KL}{N\eta}\right)^{NHc}=(1-K\Delta L^*)^{(L/\Delta L*)(Hc/\eta)}. \tag{II.4.41}$$

Regular overlapping. In this case foliage in an 'elemental' layer is so thickened by the leaves of neighbouring plants that an unshaded foliage area is constantly maintained. Shading in the elemental layer is equal to $K\Delta L_2$, and the penetration is $1-K\Delta L_2$. After passing N independent elemental layers the penetration is defined by a binomial relation in the form of $(1-K\Delta L_2)^N$.

In the case of a regular plant distribution and regular overlapping the radiation penetration function for the part ω_2 is

$$a_{S2}^{PP}=(1-K\Delta L_2^P)^N=(1-KH_P\Delta L^*)^N=$$

$$=\left(1-\frac{KL}{N}\frac{H_P}{\eta}\right)^N=(1-KH_P\Delta L^*)^{L/\eta\Delta L*} \tag{II.4.42}$$

and in the case of a random distribution and regular overlapping

$$a_{S2}^{CP}=(1-K\Delta L_2^C)^N=(1-KH_C\Delta L^*)^N=$$

$$=\left(1-\frac{KL}{N}\frac{H_C}{\eta}\right)^N=(1-K\Delta L^* H_C)^{L/\eta\Delta L*}. \tag{II.4.43}$$

Taking into account (II.4.33), (II.4.40)–(II.4.43) we finally obtain the following functions for the penetration of direct solar radiation in a plant stand from (II.4.31).

1. Random plant distribution and random overlapping

$$a_S^{CC}=\begin{cases} e^{-\eta}+\eta e^{-\eta}(1-\Delta L^* K)^{L/\eta\Delta L*} \\ +(1-e^{-\eta})(1-K\Delta L^*)^{(L/\Delta L*)(Hc/\eta)}, \\ \qquad\qquad\qquad 0<K\Delta L^*<1, \\ e^{-\eta}, \qquad\qquad 1\leqslant K\Delta L^* \end{cases} \tag{II.4.44}$$

2. Random plant distribution and regular overlapping:

$$a_S^{CP}=\begin{cases} e^{-\eta}+\eta e^{-\eta}(1-K\Delta L^*)^{L/\eta\Delta L*} \\ +(1-e^{-\eta}-\eta e^{-\eta})(1-KH_C\Delta L^*)^{L/\eta\Delta L*}, \\ \qquad\qquad\qquad 0\leqslant K\Delta L^*<\dfrac{1}{H_C}, \\ e^{-\eta}+\eta e^{-\eta}(1-K\Delta L^*)^{L/\eta\Delta L*}, \quad \dfrac{1}{H_C}\leqslant K\Delta L^*<1, \\ e^{-\eta}, \qquad\qquad 1<K\Delta L^* \end{cases} \tag{II.4.45}$$

3. Regular plant distribution and random overlapping:

$$0 \leqslant \eta < 1$$

$$a_S^{PC} = \begin{cases} 1-\eta+\eta(1-K\Delta L^*)^{L/\eta\Delta L^*}, & 0 \leqslant K\Delta L^* < 1 \\ 1-\eta, & 1 \leqslant K\Delta L^* \end{cases} \tag{II.4.46}$$

$$1 \leqslant \eta < 4$$

$$a_S^{PC} = \begin{cases} (2-\sqrt{\eta})^2(1-K\Delta L^*)^{L/\eta\Delta L^*}+[1-(2-\sqrt{\eta})^2]\times \\ \times(1-K\Delta L^*)^{(L/\Delta L^*)(H_P/\eta)}, & 0 \leqslant K\Delta L^* < 1 \\ 0, & 1 \leqslant K\Delta L^*, \end{cases} \tag{II.4.47}$$

$$4 \leqslant \eta$$

$$a_S^{PC} = \begin{cases} (1-K\Delta L^*)^{(L/\Delta L^*)(H_P/\eta)}, & 0 \leqslant K\Delta L^* < 1, \\ 0, & 1 \leqslant K\Delta L^*. \end{cases} \tag{II.4.48}$$

4. Regular plant distribution and regular overlapping:

$$0 \leqslant \eta < 1 \qquad a_S^{PP} = a_S^{PS}$$

$$1 \leqslant \eta < 4$$

$$a_S^{PP} = \begin{cases} (2-\sqrt{\eta})^2(1-K\Delta L^*)^{L/\eta\Delta L^*}+[1-(2-\sqrt{\eta})^2]\times \\ \times(1-KH_P\Delta L^*)^{L/\eta\Delta L^*}, & 0 \leqslant K\Delta^* < \dfrac{1}{H_P}, \\ (2-\sqrt{\eta})^2(1-K\Delta L^*)^{L/\eta\Delta L^*}, & \dfrac{1}{H_P} < K\Delta L^* < 1, \\ 0, & 1 \leqslant K\Delta L^* \end{cases} \tag{II.4.49}$$

$$4 < \eta$$

$$a_S^{PP} = \begin{cases} (1-K\eta\Delta L^*)^{L/\eta\Delta L^*}, & 0 \leqslant K\Delta L^* < \dfrac{1}{\eta}, \\ 0, & \dfrac{1}{\eta} \leqslant K\Delta L^*. \end{cases} \tag{II.4.50}$$

Formulae (II.4.44)–(II.4.50) characterize the penetration of direct solar radiation as dependent on the following plant stand characteristics:

(a) total leaf area defined by the quantity L;

(b) shadow effect of foliage area described by the quantity $K(r_S) = G_L(r_S)/\sin h_{\odot}$;

(c) individual plant structure characterized by the quantities ΔL^* and η;

(d) number of plants on the test area, defined by the quantity η.

In addition, the above formulae take into account both the type of plant distribution and the type of overlapping. A number of special cases can be derived from these formulae. Let us describe some of them.

1. In a sparse plant stand the value of η is small,

$$\omega_2 \approx 0, \ e^{-\eta} \approx 1-\eta$$

and

$$a_S^{CC} = a_S^{CP} = a_S^{PC} = a_S^{PP} = 1-\eta+\eta(1-K\Delta L^*)^{L/\eta\Delta L^*},$$

i.e. the penetration is independent of the type of plant distribution and overlapping.

Since $L/\eta\Delta L^*$ is large,

$$a_{P_1} = (1 - K\Delta L^*)^{L/\eta\Delta L^*}$$

decreases rapidly, and starting from a certain depth

$$a_S = 1 - \eta.$$

2. In a dense stand the value of η is large,
$H_P = \eta \approx H_C$, $e^{-\eta}$ and $\eta e^{-\eta}$ are small and

$$a_S^{CC} = a_S^{PS} = (1 - K\Delta L^*)^{L/\Delta L^*},$$

$$a_S^{PP} = a_S^{CP} = (1 - K\eta\Delta L^*)^{L/\eta\Delta L^*},$$

i.e. the penetration does not depend on the type of plant distribution but on the type of overlapping, a greater penetration corresponding to a random overlapping.

3. For plants having a long stem and a large number of 'elemental' layers (grasses, reeds, coniferous trees, etc.) we can assume $N \to \infty$ and

$$a^{PP} = a_S^{PC} = \begin{cases} 1 - \eta + \eta\, e^{-LK/\eta}, & 0 \leqslant \eta < 1, \\ (2 - \sqrt{\eta})^2 e^{-LK/\eta} + [1 - (2 - \sqrt{\eta})^2]\, e^{-LKH_P/\eta}, & 1 \leqslant \eta < 4, \\ e^{-LK}, & 4 \leqslant \eta, \end{cases}$$

$$a_S^{CC} = a_S^{CP} = e^{-\eta} + \eta\, e^{-\eta - LK/\eta} + (1 - e^{-\eta} - \eta\, e^{-\eta})\, e^{-KL(H_C/\eta)}.$$

i.e. the penetration does not depend on the type of overlapping but on the type of plant distribution.

4. Plants low in height and with broad leaves have a small number of 'elemental' layers N.

For $N = 1$

$0 \leqslant \eta < 1$

$$a_S^{PP} = a_S^{PC} = \begin{cases} 1 - LK, & 0 \leqslant LK < \eta, \\ 1 - \eta, & \eta \leqslant LK, \end{cases}$$

$4 < \eta$

$$a_S^{PP} = \begin{cases} 1 - LK, & 0 \leqslant LK < 1, \\ 0, & 1 \leqslant LK, \end{cases}$$

$$a_S^{PC} = \begin{cases} \left(1 - \dfrac{LK}{\eta}\right)^{H_P}, & 0 \leqslant LK < \eta, \\ 0, & \eta \leqslant LK, \end{cases}$$

$1 \leqslant \eta \leqslant 4$

$$a_S^{PP} = \begin{cases} 1 - LK, & 0 \leqslant LK < \dfrac{\eta}{H_P}, \\ (2 - \sqrt{\eta})^2 \left(1 - \dfrac{LK}{\eta}\right), & \dfrac{\eta}{H_P} \leqslant LK < \eta, \\ 0, & \eta \leqslant LK, \end{cases}$$

$$a_S^{PC}=\begin{cases}(2-\sqrt{\eta})^2\left(1-\dfrac{LK}{\eta}\right)+[1-(2-\sqrt{\eta})^2]\left(1-\dfrac{LK}{\eta}\right)^{H_P}, & \\ & 0\leqslant LK<\eta,\\ 0, & \eta\leqslant LK\end{cases}$$

and

$$a_S^{CP}=\begin{cases}1-LK, & 0\leqslant LK<\eta/H_C,\\ (1+\eta-LK)e^{-\eta}, & \dfrac{\eta}{H_C}\leqslant LK<\eta\\ e^{-\eta}, & \eta\leqslant LK\end{cases}$$

$$a_S^{CC}=\begin{cases}(1+\eta-LK)e^{-\eta}+(1-e^{-\eta}-\eta e^{-\eta})\left(1-\dfrac{LK}{\eta}\right)^{H_C}, & \\ & 0\leqslant LK<\eta\\ e^{-\eta}, & \eta\leqslant LK.\end{cases}$$

It is seen from the special cases discussed above that the usual binomial relation of penetration corresponds to a regular plant distribution and a random overlapping with a high degree of coverage. Hence we obtain the usual exponential relation for penetration when the number of penetration layers is large.

Row crop. This type of plant distribution is frequent among pure culture stands. Let y_0 be the distance between rows, $a<y_0$ the mean distance between plants in a row. Then $N_1=1/y_0$ is the number of rows per metre length, $N_2=1/a$ is the number of plants in a row per metre length, and $N_P=1/ay_0$ is the number of plants per square metre. Let us model plants as vertical cylinders of the radius r. Then $S_{1P}=\pi=r^2$ and $\eta r^2/ay_0$, whence $r=(\eta a y_0/\pi)^{\frac{1}{2}}$.

If $0<r\leqslant a/2$, no mutual overlapping of plants exists and a_S is defined by formula (II.4.36). If $a/2\leqslant r\leqslant y_0/2$, plants overlap along rows, but as yet no overlapping exists between rows. Let us divide the test area into two parts: ω_0 the part of area without plants and $\omega_1=1-\omega_0$ the part of area occupied by plants, and the order of overlapping will not be taken into account over this ground area. Let us find the average relative area for the part ω_1. Let us approximate

$$\omega_1=2rN_1=2\frac{r}{y_0}=\frac{\sqrt{\eta}}{C_1}, \tag{II.4.51}$$

where

$$C_1=\left(\frac{\pi}{4}\frac{y_0}{a}\right)^{\frac{1}{2}}. \tag{II.4.52}$$

Since for the total test area $L=\eta L^*$, then $L_1=L_L/\omega_1=HL^*$ where the foliage thickening function is

$$H=C_1\sqrt{\eta}. \tag{II.4.53}$$

When $y_0/2<r$ overlapping also sets in between rows and $\omega_0\approx 0$. Taking into account (II.4.51)–(II.4.53) we finally obtain the formulae for the penetration of direct solar radiation in a row crop.

1. Regular overlapping:

$$a_S^P = \begin{cases} 1-\eta+\eta(1-\Delta L^* K)^{L/\eta\Delta L*}, \\ \qquad\qquad 0 \leqslant \eta < \left(\dfrac{\pi}{4}\dfrac{a}{y_0}\right)^{\frac{1}{2}}, \\ \dfrac{\sqrt{\eta}}{C_1} + \left(1-\dfrac{\sqrt{\eta}}{C_1}\right)(1-KC_1\Delta L^*)^{L/\eta\Delta L*}, \\ \qquad\qquad \left(\dfrac{\pi}{4}\dfrac{a}{y_0}\right)^{\frac{1}{2}} \leqslant \eta < \left(\dfrac{\pi}{C_1 a}\right)^{\frac{1}{2}}, \\ (1-K\eta\Delta L^*)^{L/\eta\Delta L*}, \\ \qquad\qquad \left(\dfrac{\pi}{C_1 a}\right)^{\frac{1}{2}} \leqslant \eta. \end{cases} \tag{II.4.54}$$

2. Random overlapping:

$$a_S^C = \begin{cases} 1-\eta+\eta(1-K\Delta L^*)^{L/\eta\Delta L*}, \quad 0 \leqslant \eta < \left(\dfrac{\pi}{4}\dfrac{a}{y_0}\right)^{\frac{1}{2}}, \\ \dfrac{\sqrt{\eta}}{C_1} + \left(1-\dfrac{\sqrt{\eta}}{C_1}\right)(1-K\Delta L^*)^{(L/\Delta L*)(C_1/\sqrt{\eta}}, \\ \qquad\qquad \left(\dfrac{\pi}{4}\dfrac{a}{y_0}\right)^{\frac{1}{2}} \leqslant \eta < \left(\dfrac{\pi}{aC_1}\right)^{\frac{1}{2}}, \\ (1-K\Delta L^*)^{L/\Delta L*}, \qquad \left(\dfrac{\pi}{aC_1}\right)^{\frac{1}{2}} \leqslant 1. \end{cases} \tag{II.4.55}$$

3. Special case $\Delta L^* \to 0$.

$$a_S^C = a_S^P = \begin{cases} 1-\eta+\eta\,\mathrm{e}^{-KL/\eta}, & 0 \leqslant \eta < \left(\dfrac{\pi a}{4 y_0}\right)^{\frac{1}{2}}, \\ \dfrac{\sqrt{\eta}}{C_1} + \left(1-\dfrac{\sqrt{\eta}}{C_1}\right)\mathrm{e}^{-KLC_1/\sqrt{\eta}}, & \left(\dfrac{\pi}{4}\dfrac{a_1}{y_0}\right)^{\frac{1}{2}} \leqslant \eta < \left(\dfrac{\pi}{aC_1}\right)^{\frac{1}{2}}, \\ \mathrm{e}^{-KL}, & \left(\dfrac{\pi}{aC_1}\right)^{\frac{1}{2}} \leqslant \eta \end{cases} \tag{II.4.56}$$

Calculations for $\Delta L^* = 0.1$ and $K = 1$ have been performed using formulae (II.4.44)–(II.4.50) and (II.4.54)–(II.4.56). Analyses of the results allows to find out to what degree the penetration of radiation into a plant stand is influenced by individual parameters.

Figure 56 shows the dependence of the penetration function a_S^{CP} on the depth of a stand for different values of η. When L is constant, low values of η correspond to a high degree of clumping: all vegetation is concentrated in a narrow area.

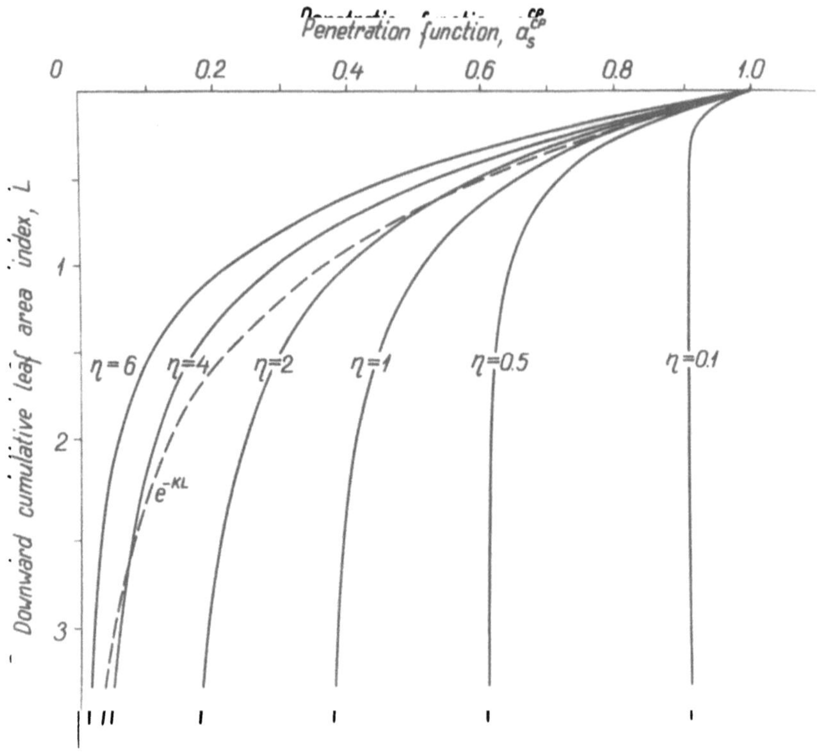

Fig. 56. Vertical profiles of penetration function a_S^{CP} at different values of overlapping index η. Random distribution and regular overlapping of plants.

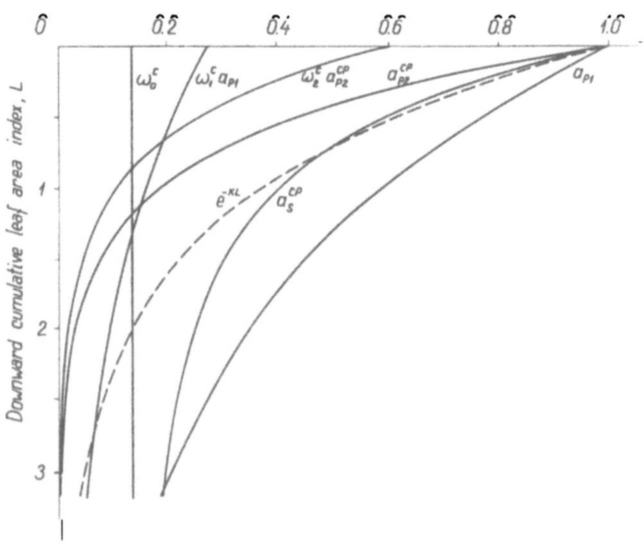

Fig. 57. Vertical profiles of components of penetration function. Random distribution and regular overlapping of plants.

Curves: ω_0^C – part of ground area without plants, a_{P1} – penetration function for single plant, a_{P2}^{CP} – penetration in the overlapping region, $\omega_1^C a_{P1}$ – part of area ω_1^C in total penetration, $\omega_2^C a_{P2}^{CP}$ – parts of areas ω_2^C in total penetration, a_S^{CP} – total penetration function, e^{-KL} – exponential penetration.

Therefore, as the average value of L in the parts ω_1 and ω_2 increases, the gap area rapidly decreases to zero, and then the penetration is determined only by the part ω_0. As η increases the area of the parts ω_1 and ω_2 increases, the phytoarea density in these areas decreases, while the penetration and the role of the parts ω_1 and ω_2 increase, but when $\eta<1$ the penetration a_S^{CP}, practically for all L values, is larger than in the case of an exponential penetration. For medium values of η the penetration a_S^{CP} deviates from the exponential penetration towards regularity in the upper layers of the stand and towards clumping in its lower layers. This situation occurs in a maize stand. At large values of η the penetration a_S^{CP} decreases rapidly and approaches zero even in the upper layers of the stand.

A further increase in η is accompanied by a multiple overlapping of neighbouring plants, and as a result the fluctuation of density decreases. This determines a decrease in the mean penetration.

The dependence of individual components of the penetration function $a_S^{CP}=\omega_{0C}+\omega_{1C}a_{P1}+\omega_{2C}a_{P2}^{CP}$ on the depth of stand L at $\eta=2$ is shown in Figure 57. It is to be observed that a_{P2}^{CP} decreases more rapidly than a_{P1}. This is due to a higher foliage area density in the multiple overlapping region ω_{2C} and the regular type of overlapping. The largest parts of the total penetration are constituted by the mutual overlapping regions in the upper layers, and by the gaps between

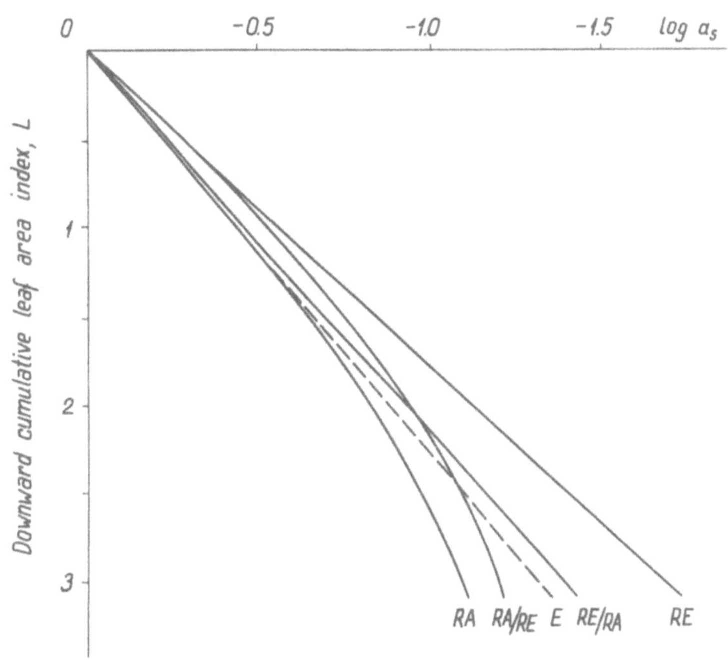

Fig. 58. Vertical profiles of log a_S for different plant stand models.
Curves: E – exponential model, RE – regular distribution and overlapping of plants, RA – random distribution and overlapping, RE/RA – regular distribution and random overlapping, RA/RE – random distribution and regular overlapping.

plants in the lower layers. At the depth $L \approx 1$ all the three components contribute more or less equally to the total penetration.

Figure 58 shows the variation of $\log a_S$ with L for all the four cases of plant distribution and overlapping at $\eta = 4$. It is seen that in upper layers a_S^{CC} and a_S^{PC} deviate but slightly from the exponential expression. The functions a_S^{PP} and a_S^{PC} deviate from $\exp(-KL)$ towards regularity at all values of L. The functions $\log a_S^{CC}$ and $\log a_S^{CP}$ are non-linear in form. In upper layers they deviate from $\exp(-KL)$ towards regularity, but in lower layers they deviate towards clumping.

The above examples show that this new theory covers a wider field of application than just a horizontally uniform plant stand and that the formulae derived contain some parameters having a concrete meaning. It will be necessary in the future to determine the parameters ΔL^* and η experimentally and test the derived formulae against experimental data.

Nilson's silhouette theory. For calculating the penetration function $a_S(z, r)$ in a stand consisting of separate trees, with a mean density N_P and a given plant distribution law, Nilson (1977a) worked out a so called silhouette theory, in which the main factor is the tree area $S(z, r)$ projected in the direction r on a horizontal surface at a height z. Let $P_0(S)$ denote the probability of no tree trunks occurring on S; $P_1(S)$, the probability of one trunk on S; $P_2(S)$, the probability of two trunks, etc. when $\sum_i P_i(S) = 1$. Then let $a'_P(z, r)$ denote the penetration of radiation in the direction r through one crown; and $a''_P(z, r)$, the penetration of radiation simultaneously through two crowns in the path of the beam coming in the direction r, etc. Since $P_0(S)$ characterizes the part of gaps between the tree crowns, $a_P^{\circ} = 1$. The average penetration function for a stand of trees is then defined by the series

$$a_S(z, r) = P_0(S) + a'_P P_1(S) + a''_P P_2(S) + \dots \qquad \text{(II.4.57)}$$

The number of essential terms in the series depends largely on the direction and increases with the increasing inclination angle ϑ.

Nilson's silhouette theory involves a number of plant stand architecture characteristics in addition to the height z and direction r:

(a) the shape and dimensions of the crown of a separate tree, which determine the area and shape of the projection S;

(b) the number of trees per unit area and the character of distribution of trees expressed by the probability $P_i(S)$;

(c) the penetrability of the crown of a separate tree, which determines the value of $a'_P(z, r)$;

(d) the character of mutual shading of several crowns, which determines the values of a''_P, a'''_P, etc.

This theory treats gaps between crowns separately from those within crowns, and its success depends on the concrete determination of the quantities S, $P_i(S)$, and $a^i_P(S)$. Nilson has derived formulae for S in the case of the following model figures for crowns: a cone, a cylinder, an ellipsoid of revolution, and a truncated paraboloid of revolution. The probabilities $P_i(S)$ are discussed for three types of the distribution of trees:

the Poisson distribution

$$P_i(S)=\frac{(N_P S)^i}{i!}\exp(N_P S), \qquad \text{(II.4.58)}$$

the positive binomial distribution

$$P_i(S)=\frac{N!}{i(N-i)!}(1-N_P S)^{N-i}(N_P S)^i \qquad \text{(II.4.59)}$$

and the negative binomial distribution

$$P_i(S)=\frac{(N-1+i)!}{i(N-1)!} \qquad \text{(II.4.60)}$$

When calculating a'_P the exponential or binomial law is usually used for the attenuation of radiation (Charles-Edwards & Thornley 1973; Tung Liang & Wing Ming Chan 1977), but the calculation of a^i_P is laborious. In the case of a vertical line of sight some of these parameters may be treated as is customary in forestry. For example, $N_P S(0)$ is the relative closure of crowns, $P[S(0)]$ is the relative closure of the tree canopy, $a_S(z, 0)$ is the vertical penetrability of the canopy.

II.4.5 CALCULATED PENETRATION OF DIRECT SOLAR RADIATION AND ITS DEPENDENCE ON VARIOUS FACTORS

Influence of the vertical distribution of phytoarea. The exponential penetration function (II.4.4) determines the vertical profile of direct solar radiation expressed in units of the downward cumulative leaf area index $L=\int_z^{z_0} u_L(z)\mathrm{d}z$. In actual crop stands, however, radiation profiles must be expressed in units of the height z rather than in units of L. A radiation profile expressed in L can have different realizations over z depending on the form of the function $u_L(z)$. To estimate the influence of $u_L(z)$ on the penetration function $a_S(z)$, let us consider four model functions $u_L(z)$ shown in Figure 59B. In actual stands the vertical distribution of phytoarea is between these extreme cases and the influence of different model functions $u_L(z)$ on $a_S(z)$ allows estimating the limits of maximum variability of $a_S(z)$ due to the variations of $u_L(z)$ in real stands. All model functions correspond to stands having the same height $z=200$ cm and depth $L_0=6$. Calculations of $a_S(z)$ have been made using formula (II.4.7) at $h_\odot=60°$ and $G_L(r_S)=\frac{1}{2}$. Figure 59A shows the functions $a_S(z)$ corresponding to the model functions $u_L(z)$. It will be seen from Figure 59 that the most uniform distribution of radiation inside a stand is provided by a vertical leaf distribution of type 1. Among actual plant species grasses (see Figure 14) and young conifers are closest to this type. Distribution type 2 is very frequent in nature (maize, clover, lucerne, trees, etc.) and also results in a comparatively uniform attenuation of radiation. Distribution type 4 is the

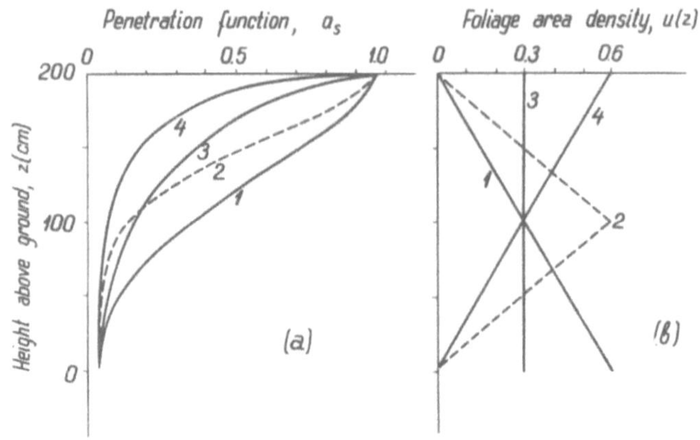

Fig. 59. Different models (1–4) of foliage area vertical distribution (b) and corresponding penetration functions $a_S(z)$.

least favourable one as regards a uniform distribution of radiation over the phytoarea, since most attenuation of radiation occurs in the top layers. Among real plants, horse beans (see Figure 18) correspond to this type. Differences between the separate curves are large, particularly in the upper and middle layers. This indicates that the vertical distribution of phytoarea is an important factor of the radiation regime in plant stands.

Influence of solar elevation. The influence of solar elevation $h_\odot$ on the penetration a_S is defined by the function $K(h_\odot)=G(h_\odot)/\sin h_\odot$. The relation $G(h_\odot)$ for some model crop stands were shown in Figure 50.

Figure 60 shows a_S as a function of $h_\odot$ in the upper ($L=0.5$) and middle ($L=2.0$) layers for some model leaf orientations, as well as for the two most different types of orientation in real plants (white clover, ryegrass). The G-functions for these plants were shown in Figure 32 (curves 1, 4). An analysis of Figure 60 shows that the dependence of a_S on $h_\odot$ increases as leaf orientation approaches the vertical one. Differences between the types of orientation are less pronounced in the upper layers of the stand ($L=0.5$) than in the middle layers ($L=2.0$). For actual stands, solar elevations from 30° to 40° constitute a neutral region where the penetration is practically independent of the distribution function of foliage area orientation $(1/2\pi)g_L$.

Penetration in maize stands. Phytometric data of $u_L(z)$ and $G(z, r_S)$ obtained for maize stands at Väimela (experiment No. 3) and Jõgeva (experiment No. 4) were used for calculating the penetration and estimating the azimuthal effect of the orientation of leaves. The results of calculation show that the azimuthal effect, which increases with the Sun declining to the horizon, is more pronounced in the upper layers and is practically insignificant in the lower ones. A 10–15% error results in calculating a_S of individual layers when the azimuthal effect is neglected.

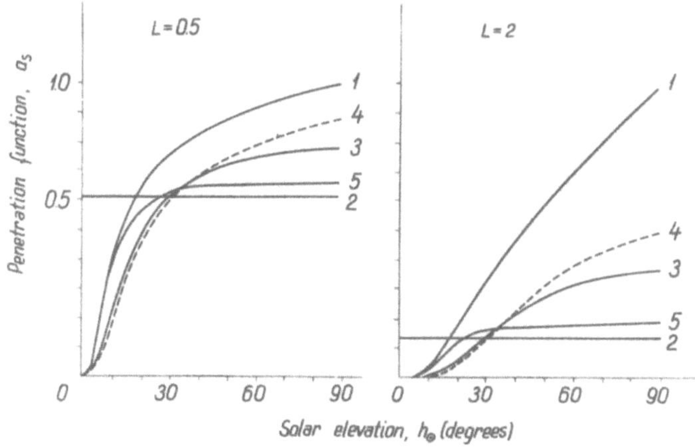

Fig. 60. Penetration function a_S versus solar elevation for different foliage orientations. (a) $L=0.5$, (b) $L=2$; 1 – vertical leaves, 2 – horizontal leaves, 3 – uniform orientation of leaves, 4 – real crop with most vertical leaves (ryegrass, after Warren Wilson 1959b), 5 – ditto, with most horizontal leaves (white clover, after Warren Wilson 1959b).

Most attenuation of radiation occurs in the top layers of the stand. It is so strong at low solar elevations that the topmost layer, 20–30 cm in depth, practically accounts for the entire attenuation of radiation. The penetration is maximum when the Sun is in the zenith, but even in this case only 2% of direct solar radiation reaches the ground surface.

Since the difference between a_S and its values calculated on the assumption of a uniform leaf orientation ($g'_L=1$ and $G=1$) is insignificant, it can be concluded that the spatial orientation of leaves in a maize stand may be assumed to be uniform. This assumption leads only to a few percent error in calculations of the penetration of direct solar radiation because deviations from a uniform orientation in individual layers can be opposite to one another, thus resulting in mutual compensation. The effect of stalks on the total penetration in a maize stand is insignificant and may be neglected.

As much as 85% of radiation is attenuated in the first 40-cm layer containing about 26% of the total foliage area of the stand. The upper 80-cm layer containing 60% of the foliage area attenuates practically the whole of direct solar radiation. Since early in September in Estonia max $h_⊙=38°$, about 40% of the total foliage area does not receive any direct solar radiation. The lower layers receive some PAR as diffuse sky radiation, but this component is only 5–10% of the total radiation. Consequently, 40% of the maize leaves located in the lower layers receive so little PAR that their photosynthetic activity is hardly enough to compensate for respiration, and their importance as assimilating plant parts is nil. That is why photosynthetic activity in individual layers varies in a wide range depending on the radiation regime, the topmost layers being the most important ones in this respect. Of course, the importance of middle and lower layers increases in sparse crops.

Effect of stems and spikelets on the penetration of direct solar radiation in cereal crops. It is usual practice in studying the radiation regime and photosynthetic activity of stands to consider leaves alone. This, however, cannot be permitted in the case of cereals where the green area of stems and spikelets equals or even exceeds that of leaves. In order to estimate the effect of stems and spikelets on penetration we performed calculations using the exponential formula (II.4.5) for a stand of barley (cv. Domain) at the milk ripeness stage (19.07.66, experiment No. 23). The height of stand was 90 cm. The vertical distribution of the downward cumulative leaf area index $L(z)$ as well as that of the effective area index of stems $L_{EC}(z)=(1/\pi)L_c(z)$ and of spikelets $L_{EF}(z)=(1/\pi)L_F(z)$ is given in Figure 61A. The stems and spikelets were modelled as vertical cylinders. The function $G_L(h_\odot)=0.49+0.02\sin h_\odot$ was taken from our work (Mägi & Ross 1969b). The vertical penetration profiles calculated separately for stems, spikelets and leaves at two solar elevations, 10° and 50°, are shown in Figure 61B. It will be seen from Figure 61 that the total effect of stems and spikelets increases with depth and with decreasing solar elevation. At $h_\odot=10°$ most shading occurs in the upper layers and is due to spikelets. The importance of leaves in total shading is minimal. At $h_\odot=50°$ most shading is due to both leaves and stems. At $h_\odot=50°$ only 6.5% of sunrays reach the ground surface below a barley stand, while at $h_\odot=10°$ direct solar radiation penetrates only to a depth of 40 cm and the lower 50-cm layer is fully shaded.

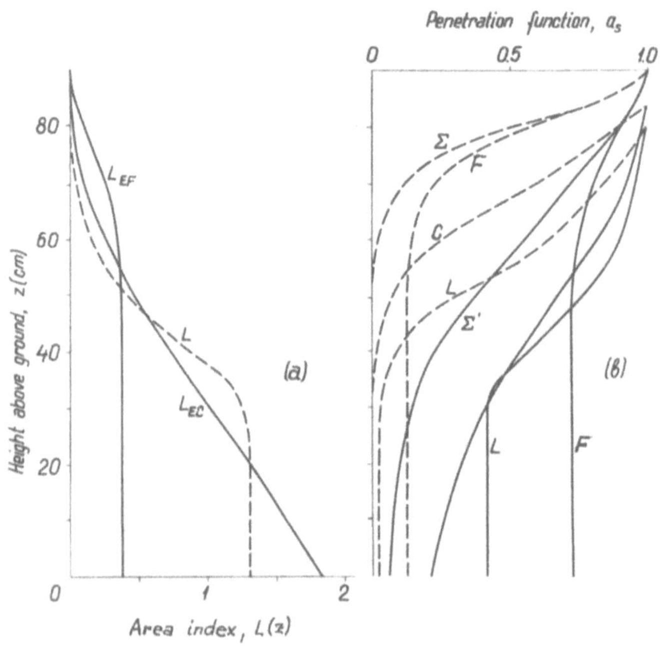

Fig. 61. Vertical profiles of phytoarea $L(z)$ (a) in a barley stand (experiment No. 23, Estonia, 1966) and calculated penetration function of plant organs for the same stand (b).
(a) L – foliage area, L_{EC} – effective stem area, L_{EF} – same for ears; (b) L – leaves, C – stem, F – ears, $\sum$ – their total; dashed lines – $h_\odot=10°$, solid lines – $h_\odot=50°$.

Thus, in cereal crops at medium solar elevations the shading effect of stems and spikelets is comparable to that of leaves, and at low solar elevations most of the shading is due to spikelets. Therefore, neglecting the effect of stems and spikelets in calculations of the penetration in cereal crops results in major errors and is not to be permitted.

Horizontally nonhomogeneous stand. Azimuthal effect of plant distribution. Calculations of $a_S(z, x, y, \vartheta, \varphi)$ were performed by the author jointly with Nilson using formula (II.4.24) on the basis of experimental data of the function $u_L(x, y, z)$ for a square-sown maize stand in Moldavia (experiment No. 11). The function $u_L(x, y, z)$ was presented in the form (I.2.36) with the normalizing relations (I.2.37). Figure 11 (curve 4) shows $u_L(z)$ and Table I.5.2 gives the functions $u_2(y)$ and $u_3(x)$ approximated respectively in the form

$$u_2(y) = \frac{1}{y_0} + f_2\left(2\pi \frac{y}{y_0}\right)$$

$$u_3(x) = \frac{1}{x_0} + f_3\left(2\pi \frac{x}{x_0}\right),$$

where f_2 and f_3 are certain periodic functions conveniently approximated as a Fourier cosine series.

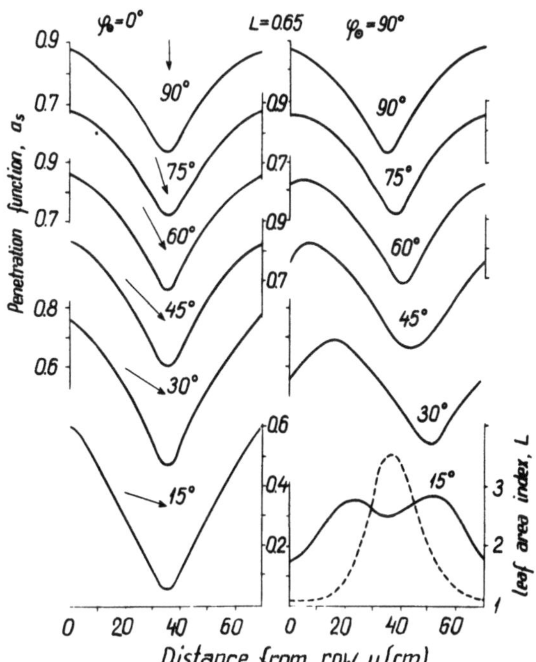

Fig. 62. Change of penetration function a_S in horizontal direction y perpendicular to rows at different solar elevations.
$\varphi_{\odot} = 0$ – sunbeams are parallel to rows, $\varphi_{\odot} = 90°$ – sunbeams are perpendicular to rows. Depth of stand $L = 0.65$. Dashed line follows the variation of the leaf area index $L(y)$ in direction y (maize stand, experiment No. 11, Moldavia, 1964, as calculated).

In modelling a row crop we assumed u_3=const, i.e. the direction of rows coincides with that of the x-axis.

Figure 62 shows the horizontal variation of penetration across rows at the depth $L=0.65$ at various solar elevations for two values of azimuth $\varphi_{\odot}$. Marked on the y-axis is the distance from the plane equidistant from the rows. Since in this case the distance between rows is 70 cm, the centre of the row corresponding to the maximum leaf area index $L(y)$ corresponds to $y=35$. It is seen from Figure 62 that because of the wavelike variation of the phytoarea density $u_2(y)$, penetration varies in a wide range, greater changes occurring with decreasing solar elevation in the case of $\varphi_{\odot}=0$. For example, at $h_{\odot}=30°$ penetration at the centre of a row is three times as large as between rows. The curve of penetration $a_S(y)$ is symmetrical about the row centre if $\varphi_{\odot}=0$, i.e. if the plane of incidence of sunrays is parallel to the direction of rows, but is asymmetrical if $\varphi_{\odot}=90°$. It is also seen that in the latter case the range of variation is less than at $\varphi_{\odot}=0°$.

Calculations show that horizontal nonuniformities of a plant stand are responsible for considerable fluctuations of the penetration functions at separate points in the stand. These fluctuations are determined both by the properties of the horizontal nonuniformity of phytoarea distribution and by the solar elevation and azimuth. This indicates the importance of an appropriate horizontal averaging during the actinometric measurements in stands.

The regularity of horizontal nonuniformities in plant stands (row crops, square-sown crops, etc.) creates an additional effect, i.e. dependence of the penetration function on the azimuth of the row direction and that of the Sun.

When comparing the azimuthal effects either due to the azimuthal dependence of leaf orientation $(1/2\pi)g''_L(\varphi_L)$ or to the regular distribution of plants it will be observed that the latter exceeds the former, sometimes by several orders of magnitude.

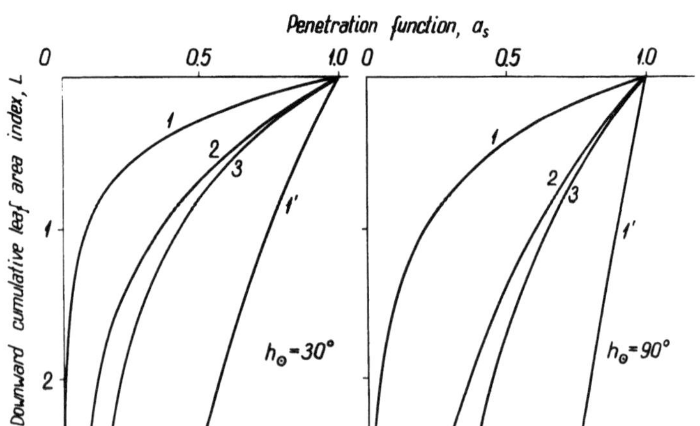

Fig. 63. Vertical profiles of penetration function a_S, as calculated for a square-sown maize stand (experiment No. 4, Moldavia, 1964). Sun beams are parallel to rows. See text.

Figure 63 presents the vertical profiles of the penetration function in the case of a square-sown maize crop at $h_{\odot}=30°$ and $90°$ and $\varphi_{\odot}=0°$. The extreme curves 1 and 1′ define the limits of minimum and maximum penetration onto a horizontal surface. Curves 2 correspond to the penetration averaged over the entire xy-plane, i.e.

$$\overline{a_S(x, y)} = \frac{1}{x_0 y_0} \int_0^{x_0} \int_0^{y_0} a(x, y) dx dy,$$

and curves 3 to the penetration for the foliage area averaged over the xy-plane, i.e. where the horizontal nonuniformity of the stand may be neglected.

It can be concluded from the data of Figure 63 that the presence of horizontal nonuniformities in a plant stand tends to increase penetration, this tendency increasing with depth. Since variations of $u_L(x, y)$ in the case of a square-sown plant distribution are larger in comparison with the distribution of plants in a row crop, variations of $a_S(x, y)$ are also larger. Thus, for example, at a depth of $L = 2$ and $h_{\odot} = 90°$ the maximum and minimum values of a_S differ by about a factor of 20.

II.4.6 METHODS OF EXPERIMENTAL INVESTIGATION

An experimental determination of the penetration of direct solar radiation in a plant stand is reduced to that of the gap area at different levels of the stand. Therefore, all instruments and methods discussed below are suited for the measurement of the horizontal sunlit area between plant parts.

Direct determination of gap area. The contours of sunflecks on a sheet of paper placed horizontally between plants are either traced in pencil or photographed. The area covered by sunflecks on the paper is determined by means of planimetry. Then the paper sheet is displaced horizontally and the operation repeated several times. This procedure is carried out on different levels, and vertical profiles of penetration are thus obtained.

This method is only historic at present in view of its essential drawbacks.

Inclined point quadrats method. This method described in greater detail under I.3.3 allows determining the penetration in any direction and not only in that of the sunrays. Its disadvantage when used to determine penetration is a low efficiency, and its advantage is a higher accuracy and representativeness in comparison with various rods.

Lopukhin rod. Lopukhin (1948a) proposed determining gap area by means of a rod, 250 cm long, with black dots marked at 5 cm intervals (Figure 64). The sunlit dots were counted in a horizontal position of the rod.

The method of measurement was improved by Budagovsky at the Institute of Geography, Academy of Sciences of the USSR. Instead of the radial shifting of the rod it was placed consecutively in 10 parallel positions at equal intervals, so that the area under study was, as it were, covered with a grid of squares whose

Fig. 64. Lopukhin rod.

nodes coincided with the marks on the rod. In this case each mark on the rod characterizes the area of a square the side of which is equal to the distance between neighbouring marks on the rod. The rod length and the number of marks on it are chosen to correspond to the stand structure.

In their work Goisa and Rogachenko (1969) estimated the accuracy of determining a_S depending on the number of dot marks and found that at large values of a_S 100 dot marks determine a_S with an error of 20%, when $a_S \approx 0.5$ the error is 10%, and when $a_S \approx 0.1$ the error is only a few percent.

A similar rod and method of measurement were worked out afterwards in Japan (Horie 1966) independently of the Soviet authors.

Niilisk rod. In this rod (Figure 65) designed and constructed by H. Niilisk (1963) at our laboratory the length exposed to direct solar radiation was measured by means of a photocell or a photodiode. In the initial device the photocell was driven along the 240 cm rod by means of a crank handle, disk and rope. The photocell voltage was transmitted through sliding contacts and wires extended along the rod to the box containing the electronics. As the photocell moves along the rod electric pulses are generated after every 0.5 cm and are registered by the counter if the photocell is illuminated by direct solar radiation concurrently with the pulse. If $N_0 = 480$ is the number of pulses for a complete photocell travel and N is the number of pulses counted, the ratio N/N_0 gives the sunlit part of the rod length, i.e. the penetration coefficient of direct solar radiation averaged over the rod length. In 1962–1963 the rod was used in stands of maize and horse beans in Estonia. This experience revealed some design deficiencies, which were eliminated in the subsequent design. The method of measurement is dealt with in our work (Ross & Nilson 1965).

While two observers have to be engaged in measurements using the Lopukhin rod, the Niilisk rod can be handled by a single observer. Also, the Niilisk rod gives a more accurate determination of the sunlit part of the rod, since its spacing is 0.5 cm, and the Lopukhin rod has 1–5 cm spacings. However, the Niilisk rod is

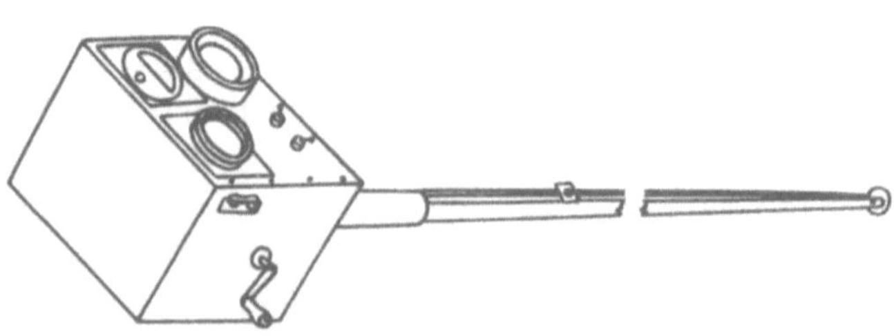

Fig. 65. Niilisk rod.

cumbersome and less efficient in comparison with the Lopukhin rod. That is why the Niilisk rod was superseded by the Lopukhin rod in our investigations starting from 1965.

'Mouse' device designed by Laisk. In localities where cloudiness dominates most times, the use of the above rods is hampered by the scarcity of sufficiently long sunlit periods. Besides that, in order to determine the penetration of diffuse sky radiation into a plant stand one must know the coefficient of penetration not only in the direction of sunrays but in all directions. That is why we require an instrument capable of distinguishing between phytoelements and the sky and thus independent of the presence of direct solar radiation. Such a device called 'Mouse' was designed and constructed by Laisk (1969) at our laboratory. It consists of a metallic pipe, 40 mm in diameter and 380 cm in length (Figure 66); the pipe wall is perforated with 360 drilled holes, 5 mm in diameter, spaced at 10 mm. A type FEU-33 photomultiplier sensitive in the blue spectral region and fitted with an optical system can travel lengthwise inside the pipe. The system 'peeps' through each hole as it moves. This equipment registers pulses only when the sky is visible through the window. The field of vision is 18′. The optical system is capable of discriminating an object 1 cm in size placed at a distance not more than 2 m. The equipment allows measuring the penetration in any direction within a plant stand having comparatively large leaves. In 1965 it was employed in Tadjikistan for measurements of the penetration of radiation into row crops of

Fig. 66. 'Mouse' device designed and constructed by Laisk.

maize, sorghum, sunflower and cotton and in 1966 in a stand of maize in Estonia. Its advantages are operability in the absence of direct solar radiation, a high resolution, reliability and accuracy.

Errors due to the use of measuring rods. A number of circumstances conducive to various errors should be taken into consideration in estimating the reliability of experimental data obtained by means of rods.

First, because of its finite dimensions, each rod requires an adequate free space inside the stand. When the rod is placed parallel to rows, it cannot be placed, for example, in the centre of a row because of the stem dimensions and the irregular distribution of plants in the row, therefore, the rod will always be displaced some distance from the centre. If the rod is to be positioned within plants having large leaves (maize, sunflower, etc.), the leaves will have to be shifted aside. If the rod is placed perpendicular to rows, some plants will have to be shifted aside. That is why the natural structure of the stand is disturbed to some extent in positioning the rod, or else the rod is placed where illumination exceeds the average value for the whole stand. For this reason the experimental data obtained by means of rods are systematically overvalued.

Second, the attenuation of direct solar radiation inside a stand is determined by the total shading area in the path of sunrays. The length and direction of this path change considerably as the Sun travels over the sky in the course of a day. Ideally, therefore, for the exact determination of the penetration function a_S for a given leaf area index L the latter should be specified exactly at the measurement point and in the direction of sunrays r_S. But practically it is impossible. In our measurements in 1962–1964 $L(z)$ was determined by the stratifying clip method on a separate plot and the results thus obtained were compared with the data obtained with all rods which were used in the same stand. Visually, all crops appeared to be similar. However, the statistical characteristics of plant heights in a maize stand (experiment No. 4, 1963, Estonia) (Ross & Vlasova 1967) showed differences between plots. Starting from 1965 $L(z)$ values have been calculated using the statistical sampling method (see I.3.6) from the distribution function of the foliage uppermost level determined in the location where measurements were performed with the rod.

Thus comparison of the data of a_S and $L(z)$ is always unrepresentative and leads to errors hard to estimate.

Third, if the shading phytoelements are remote enough from the detector, the borders between sunflecks and shaded areas are indistinct because of the finite size of the solar disk, penumbras result between sunfleck and shadow. The performance of the rod in penumbra areas depends on the sensitivity adjustment of the detecting system prior to measurement. The sensitivity of the Lopukhin rod is determined visually; that of the Niilisk rod and the 'Mouse' device, by the sensitivity of the measuring system. The latter is adjustable within certain limits. In order to estimate the importance of sensitivity, consider Figure 48 and suppose that the threshold of response of the instrument is $0.4S_0$. Then as the detector moves from point C_1 to point C_4 the rod is absolutely incapable to measure the area of disk, only 7% of disk 3 is measured, $\frac{2}{3}$ of disk 2, and disk 1 is fully measured. If sensitivity is equal to $0.2S_0$, the area measured will be different. The diagram shows that the ability of the rod to measure shading elements is reduced as the distance between rod and elements increases, and that the instrument indications become more overvalued with the increasing pathlength of sunrays in the stand. Regrettably, no measurements to compare the sensitivity of rods of different designs have been conducted. In view of the above circumstances experimental data should be treated with caution.

Rods cannot be used for determining the penetration of direct solar radiation in stands of coniferous and small-leaved trees.

When estimating the representativeness of data obtained with rods of different designs it should be taken into account that they have different scales of average. The largest was for the Lopukhin rod. It was equal to 4 m^2 for maize and other tall crops, and 1 m^2 for grasses. Three positions in a row crop and six positions in a square-sown crop give the respective scales of average, 1 and 1.8 m^2 for the Niilisk rod. The rod of the 'Mouse' device records only one position, and averaging was made over a 5 m length.

Data processing methods. Horizontally averaged coefficients of the penetration of direct solar radiation a_S as functions of the solar elevation $h_\odot$ and height z inside a crop stand, were obtained from data of direct measurements performed with rods. In order to refer the data of a_S to a constant solar elevation, the curves of a_S as a function of $h_\odot$ for each height were constructed separately for periods before noon and afternoon. The latter division permitted judging whether any azimuthal dependence of a_S existed and whether the stand structure was subject to any diurnal changes. By way of example Figure 67A shows the dependence of

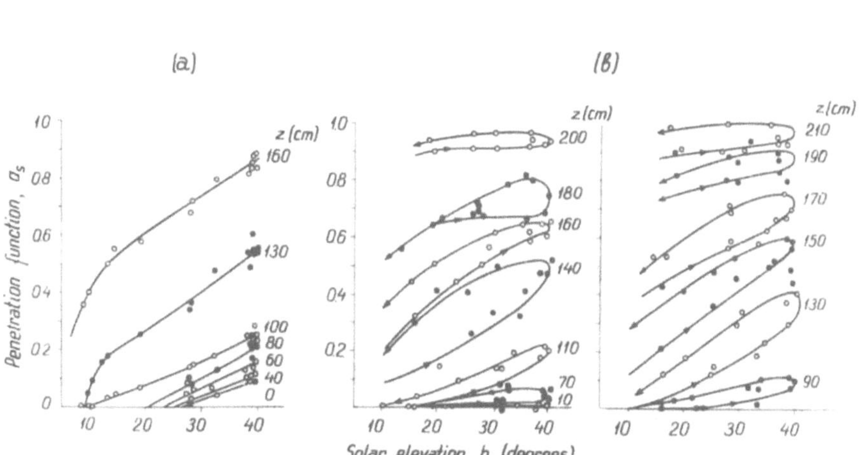

Fig. 67. Variation of penetration function a_S with solar elevation at different heights z above ground. Maize stand, experiment No. 6, Estonia, September 1–2, 1963.
(a) – Lopukhin rod data, (b) – Niilisk rod data.

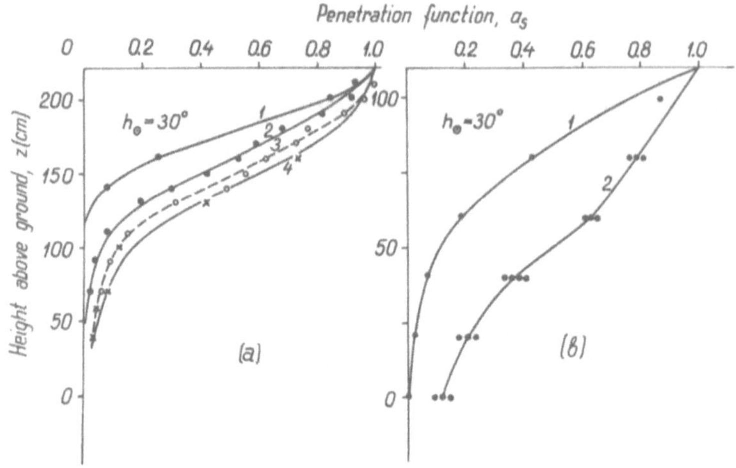

Fig. 68. Vertical profile of penetration function a_S.
(a) – maize stand, experiment No. 6, Estonia: 1 – calculated from geometrical structure using Eq. II.4.4., 2 – Niilisk rod, measurements before noon, 3 – ditto, after noon, 4 – Lopukhin rod (measurements of Budagovsky); (b) – winter wheat crop, experiment No. 7, Estonia, July 17, 1963: 1 – calculated from geometrical structure using Eq. II.4.5., 2 – Lopukhin rod, $h_\odot = 30°$.

a_S on $h_\odot$ in a row crop (experiment No. 6, Estonia, Sept. 1–2, 1913) in accordance with data obtained using the Lopukhin rod, and Figure 67B shows the same obtained for the same stand but in accordance with data obtained with the Niilisk rod.

Figure 68A shows the curves of penetration as a function of height obtained by calculation (curve 1), with the Niilisk rod (curve 2 – before noon, curve 3 – after noon) and the Lopukhin rod (curve 4). It is seen that all methods for determining a_S show systematic deviations. The difference between a_S values obtained before noon and after noon, determined using the Niilisk rod, is apparently due to the insufficiently large number of rod positions which did not provide an adequate scale of average. The Niilisk rod data overestimate values in comparison with the Lopukhin rod data. Figure 68 demonstrates a large discrepancy between the calculated curve and the data obtained with the Lopukhin rod (measurements of Budagovsky) for a winter wheat crop. The possible causes of this discrepancy were discussed in the previous section, however, it is characteristic of all types of stands that calculations using the exponential formula (II.4.4) underestimate values a_S in comparison with experimental data. We shall dwell on this question further below.

On the basis of the averaged curves of $a_S(h_\odot)$ constructed for each height z and the curve of $L(z)$ obtained from phytometric measurements in accordance with the method described in Chapter I.3 data of a_S and $L(z)$ were compared in a logorithmically plotted graph. The main question was whether the dependence between $-\ln a_S$ and $L(z)$ may be approximated with a straight line, i.e. $-\ln a_S = k_S L$, where k_S is an empirical attenuation coefficient for direct solar radiation.

It should be noted that at $G(L)=\text{const.}$ all theroretical formulae – exponential (II.4.4), binomial (II.4.13) and (II.4.15) and the Nilson formula (II.4.22) – are represented by straight lines but having different angles in the graph of $-\ln a_S$, L. In the case of the exponential formula this inclination is equal to $G/\sin h_{\odot}$, for the Nilson formula – $\lambda_0 G/\sin h_{\odot}$, and for binomial formulae $\tilde{C}_S G/\sin h_{\odot}$, where

$$\tilde{C}_S=\frac{\ln V'}{1-V'}, \tag{II.4.61}$$

where $\tilde{C}_S$ is *the phytoelements distribution coefficient* and the relative variance V' is given by formula (II.4.17). $V'=\tilde{C}_S=1$ corresponds to a random distribution and the exponential formula, $V'<1$ and $\tilde{C}_S<1$ to a semiregular distribution and the binomial formula, and $V'>1$ and $\tilde{C}_S>1$ to a clumped distribution and the negative binomial formula. Formally and numerically λ_0 in the Nilson formula and $\tilde{C}_S$ in the binomial formula coincide, but they are different in meaning. When G and $\sin h_{\odot}$ are known, the values of $\tilde{C}_S$ and V' (or λ_0) can be determined from the inclination. However, the experimental data on the graphs $-\ln a_S$, L are not always located on a straight line. This may be due to the following causes: a_S and L are not sufficiently averaged, a_S and L are not representative, a_S measurements contain random and systematic errors, G is not constant with the changing height, or else the penetration in the stand does not correspond to the above theoretical models. So if the coefficient $\tilde{C}_S$, which characterizes the relative variance V' in the binomial formula, is determined experimentally from a_S measurements, it is semi-empirical and contains all dependences not provided by the formula as well as the measuring errors.

Our attention in analyzing experimental data was mainly concentrated on $\tilde{C}_S$.

Where in addition to leaves, other plant organs dominated having a vertical cylindrical structure (tassels in the case of maize and sorghum, stems and spikelets in the case of grasses) the initial formula was

$$-\ln a_S=\tilde{C}_S\sum_j k_j, \tag{II.4.62}$$

where $\sum_j k_j=kL+\sum_j L_{Ej}\coth_{\odot}$, L_{Ej} is vertical distribution of the effective area of the j-th plant organ.

In the work of Savina (1968) experimental data of penetration in a maize stand are also approximated by formula (II.4.60) where k_S has the form $k_S=0.19+0.23\sin h_{\odot}$.

II.4.7 EXPERIMENTAL DATA ON PENETRATION

Maize. A sparse stand. Tadjikistan, experiment No. 14. Penetration was determined using the Lopukhin rod and the 'Mouse' device. Figure 69A gives, by way of example, the dependence of a_S on L at the milk ripeness stage, obtained with the Lopukhin rod. It is evident that in this case the dependence can be approximated by a straight line with a satisfactory accuracy. The systematic underestimates of a_S at small depths L can be attributed to the disregard of the tassel effective area in the upper layers of the stand when plotting $L(z)$, i.e. the values of $L(z)$ plotted on the graph correspond to the effective area of the leaves alone and are, therefore, slightly underestimated. It also follows from the figure that with greater solar elevations the inclination of straight lines increases, and with it the penetration.

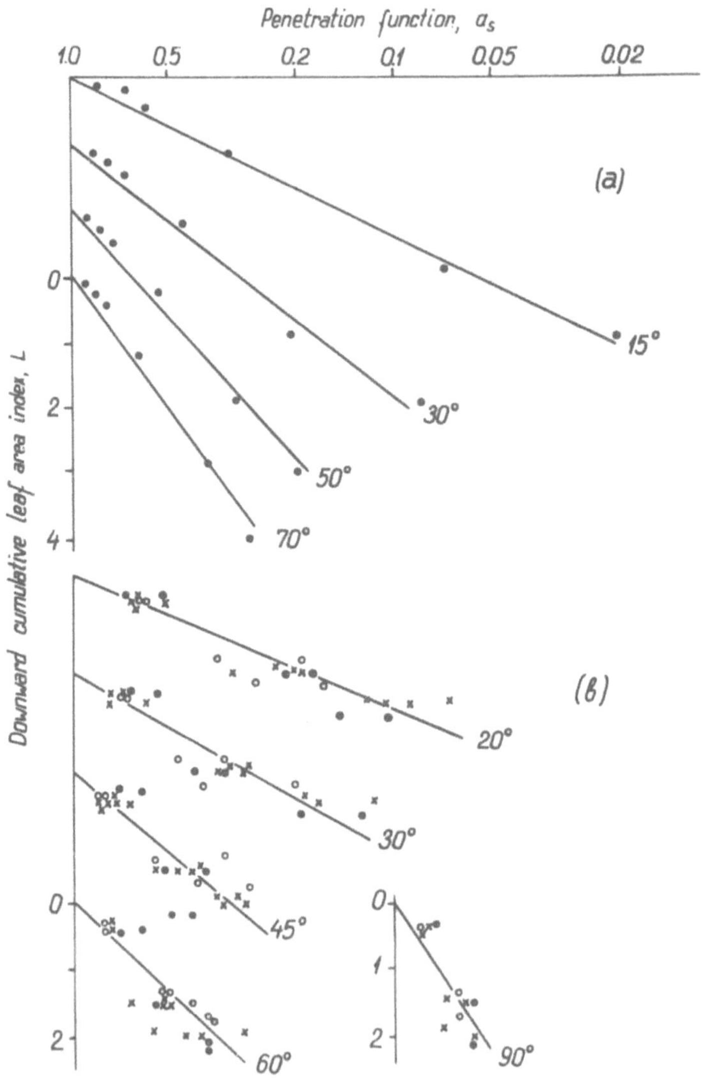

Fig. 69. Penetration function a_S plotted logarithmically as a function of depth L at different solar elevations. Maize stand, experiment No. 14, Tadjikistan.
(a) – August 16, 1965, wax ripeness, Lopukhin rod, (b) – June 28, 1965, tasseling, 'Mouse' device, measurements of Laisk. Instrument is placed parallel to rows (•), perpendicular to rows (○), or angle between rows and instrument is 45°(x).

Figure 69B depicts penetration into the same stand at tasseling, as measured with the 'Mouse' device. To every point in the graph there is one position of the rod. The scatter of points in this graph is much greater but random and independent of the rod position. The scatter proves also that one rod position is not enough to ensure a sufficient scale of average.

Anything that is specific about the results obtained with either of the instruments and presented in Figure 69 is present in nearly all the measurements performed during the growing season. Similar graphs plotted for all the vertical penetration profiles were used to determine the values of the factor $\tilde{C}_S$. In calculating $G(h_\odot)$ the data on the spatial orientation of leaves presented in Chapter I.6 were also used.

Analysis of the data showed that the factor $\tilde{C}_S$ depends upon solar elevation. Figure 70 depicts this dependence at the time of 15th leaf emergence (25, 28 and 30.06) and tassel flowering (19.07), obtained with the 'Mouse' device. The following conclusions can be drawn from Figure 70.

1. The factor $\tilde{C}_S$ is found to be dependent on solar elevation: the maximum of $\tilde{C}_S$ is found between 40 and 60°, but $\tilde{C}_S$ decreases with increasing or decreasing solar elevation.

The revealed dependence indicates that in a sparse stand there exists an additional penetration anisotropy caused not by the leaf orientation (which is taken account of by the G-function), but by a nonrandom dispersion of foliage in

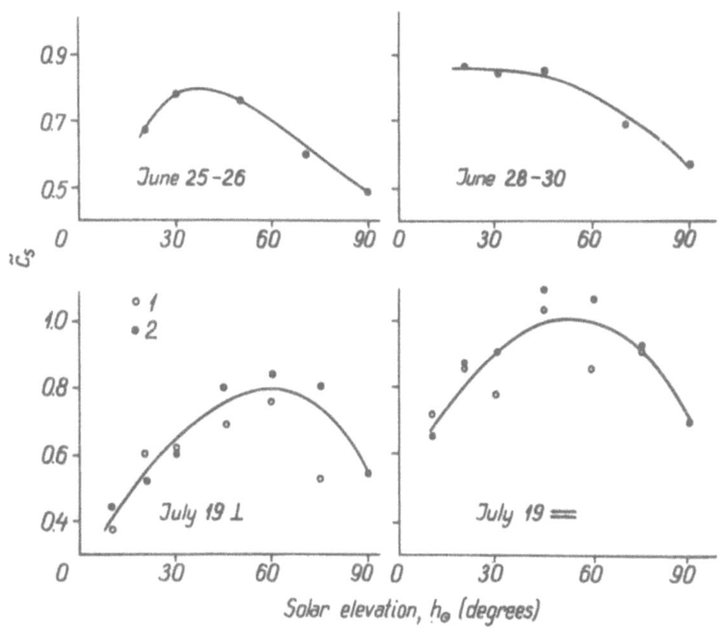

Fig. 70. Dependence of coefficient $\tilde{C}_S$ on solar elevation at various times. Maize stand, experiment No. 14, Tadjikistan, 1965. On July 19 instrument was placed perpendicular (⊥) or parallel (=) to rows to measure sunny side (o) and opposite side (•).

Measurements of Laisk using the 'Mouse' device.

the stand which, in turn, is due to both the habitus of individual plants and their row type arrangement. The decrease of $\tilde{C}_S$ at greater solar elevations points to the presence of gaps between plants, i.e. to the growing importance of the region ω_0 at large values of $h_\odot$. The diminishing of $\tilde{C}_S$ at low solar elevations is apparently due to the nonuniform distribution of leaves in horizontal layers.

2. The data of 19.07 point clearly to the difference that the two positions of the 'Mouse' device rod (Figure 70) imply. If in the first position the rod is parallel to rows and halfway between them, turning it about its axis causes 'solar elevation' to change and the beam incidence plane to follow the detector as it is perpendicular to the rows. Averaging in this instance was along the x-axis, $y = y_0/2$ and $\varphi_\odot = 90$ and $270°$. In the second position the rod is perpendicular to the rows (position 1 ⊥), therefore, the beam incidence plane is parallel to the rows and follows the detector movement. Averaging is done along the y-axis and $\varphi_\odot = 0$ and $180°$. Because of the row pattern of the stand there are more leaves in the way of the beam in the latter case than with the position of the rod, consequently ⊥ is greater and penetration is less.

Experimental data also point to variation of the factor over the growing season.

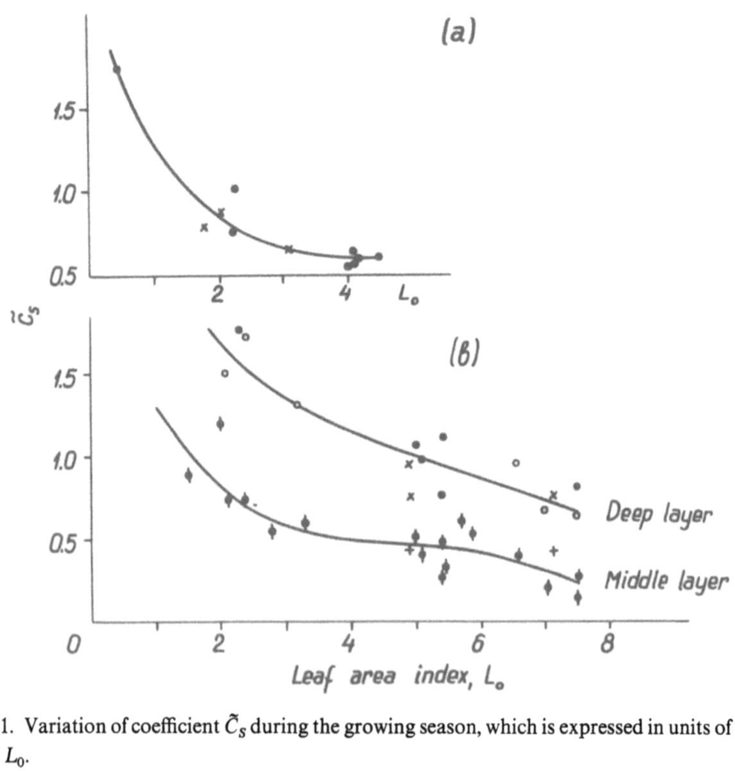

Fig. 71. Variation of coefficient $\tilde{C}_S$ during the growing season, which is expressed in units of leaf area index L_0.
(a) – sparse maize crop, experiment No. 14, Tadjikistan, 1965; (b) – dense stand, Estonia, 1962, 1963, 1966. Measured with the help of 'Mouse' device (×), Lopukhin rod (•) and Niilisk rod (○).

Figure 71 shows the variation of $\tilde{C}_S$ as a function of the stand leaf area index L_0 at a solar elevation $h_\odot = 30°$. An important conclusion can be made from the data of this drawing, viz. in the early stage of plant development (before the emergence of the 12th leaf) leaves of the stand are arranged mosaically and the mutual shading among them is not pronounced. Their distribution tends from random to regular. As the plants continue to grow, the regular leaf distribution gives way to a random (at the time of 13–15th leaf emergence) and then to a clumped one. Leaf location practically does not change following the tassel

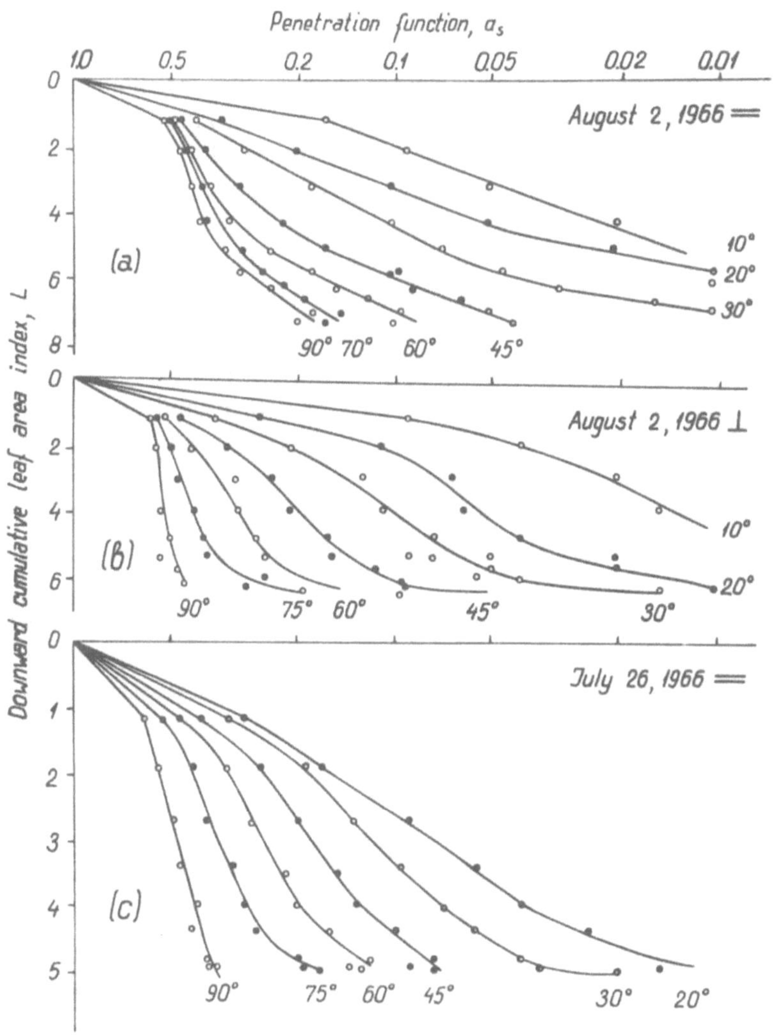

Fig. 72. Vertical profiles of penetration function a_S plotted logarithmically at different solar elevations. Dense maize stand in Estonia. Measurements of Laisk using the 'Mouse' device.
(a) – August 2, 1966, rod is parallel to rows, (b) – August 2, 1966, rod is perpendicular to rows, (c) – July 26, 1967, rod is parallel to rows.

flowering stage. The value of the factor $\tilde{C}_S$ drops from 1.7 at the 12th leaf emergence stage to 0.6 at the ear flowering stage.

A dense stand. In the early stages of development, when the stand has not closed, the attenuation of radiation is governed by the same principles as in a sparse stand and the dependence $-\ln a_S(L)$ can be approximated by a straight line. But as the stand closes its structural arrangement undergoes changes. Some plants start to overlap one another in the middle layers and the radiation attenuation principles can change. The dependence $-\ln a_S(L)$ may be approximated by a straight line but only as a first approximation or if the number of heights taken to determine the vertical attenuation profile is small. More detailed measurements, however, reveal quite a different situation.

Figure 72 shows the dependence $a_S(L)$ in a maize stand (experiment No. 19, Estonia, 26.07.66, 13th leaf emergence, and on 02.08.66, beginning of tasseling) obtained using the 'Mouse' device with the rod parallel to rows. Measurements were taken every 10 cm of height in the upper layer and 20 cm in lower layers. The points in Figure 72 correspond to previously smoothed values of $a_S(h_\odot)$. Because of the scatter of initial data, the curves are not very accurate, but their specific S-shape is still recognizable. The obtained curves can be roughly approximated by three lengths of straight lines which divide the stand into three layers. The stand upper layer structure corresponds to the worst penetration. Penetration is best in the middle layer.

A similar situation was observed in 1963 in another stand (experiment No. 6, 01–02.09, tassel flowering stage), in which measurements were made with the Niilisk rod at a 10-cm height spacing.

Graphs similar to those of Figure 72 were plotted and from them $\tilde{C}_S$ factors were determined separately for the upper, middle and lower layers of the stand for different solar elevations $h_\odot$.

Figure 73 shows $\tilde{C}_S$ as a function of $h_\odot$ (experiment No. 19, Jõgeva, 26.07.66, emergence of the 13th leaf; 02.08, tasseling). The curves of $\tilde{C}_S(h_\odot)$ for the upper layer are not very accurate as the number of levels chosen was small. It is seen in the drawing that the $\tilde{C}_S(h_\odot)$ variation has the same character as in the case of a

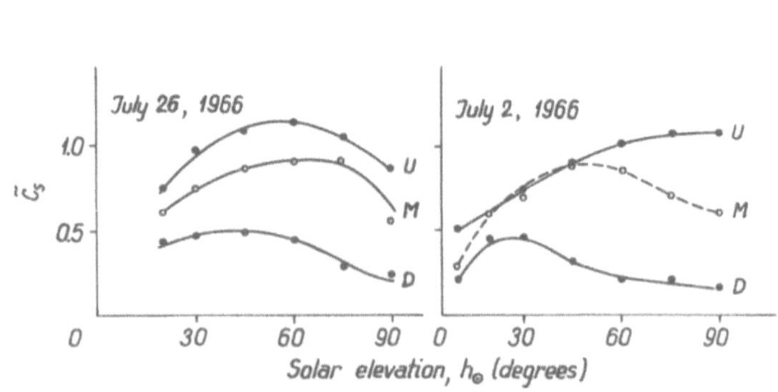

Fig. 73. Coefficient $\tilde{C}_S$ as a function of solar elevation, in the upper (*U*), middle (*M*), and deeper (*D*) layers of a dense maize stand. Experiment No. 19, Estonia, as measured by Laisk using the 'Mouse' device.

Table II.4.1 *Height and depth of maize stand layers having different penetration*

Data	Experi-ment No.	Layer									
		upper				middle			lower		
		L_0	z_0	Δz	ΔL	z_0	Δz	ΔL	z_0	Δz	ΔL
Sept. 1–2 1963	6	4.9	220	40	0.4	180	60	3.0	120	110	1.5
July 26 1966	19	5.0	145	35	1.1	110	40	2.9	70	60	1.0
Aug. 2 1966	19	7.3	175	45	1.2	130	40	3.9	90	70	2.2

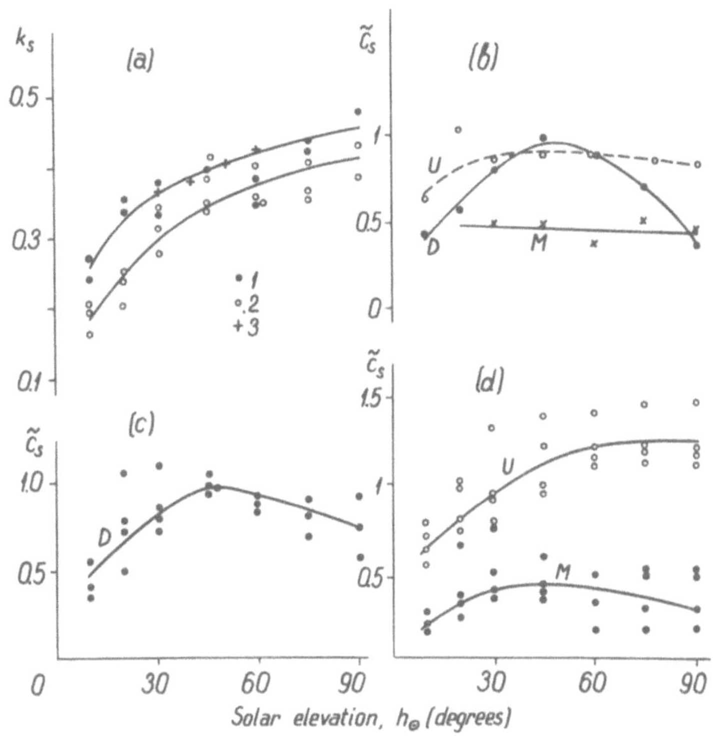

Fig. 74. Dependence of empirical attenuation coefficient k_S (and) coefficient $\tilde{C}_S$ (b), (c), (d) on solar elevation. Sorghum crop, experiment No. 15, Tadjikistan, 1965.
(a) – July 11–12, tasseling stage, $L_0 = 4.0$: 1 – instrument is perpendicular to rows, 2 – instrument is parallel to rows; 3 – Lopukhin rod data; (b) – July 30, August 1, tassel flowering stage, $L_0 = 6.9$; (c), (d) – July 11–16, tasseling stage, $L_0 \approx 4.0$; U, M, D – upper, middle and deeper layers of foliage, respectively. All data, except curve 3, are obtained using the 'Mouse' device.

sparse stand. The shadow effect is greatest in the solar elevation range 30–60°; leaf distribution is nearly random in the upper layer; the middle layer foliage has a noticeable clumped distribution and good penetration; the clumped distribution is not so pronounced in the lower layer.

Presented in Table II.4.1 are data on the height z and depth ΔL for each layer. As seen from the table the upper and middle layers have a comparatively stable height of 35–40 and 40–60 cm, respectively. Visual observations indicate that approximately at the boundary between the upper and lower layers individual plant volumes in neighbouring rows start to overlap each other.

Gulyayev et al. in their works (1967, 1969) presented data on the penetration of direct radiation in maize stands in the Ukraine, obtained with the Lopukhin rod. They found that the experimental data are readily approximated by formula (II.4.60). Assuming that for maize $G=\frac{1}{2}$ Gulyayev's data give $\tilde{C}_S=0.7$ at $h_\odot=15$–$60°$ for a crop on highly fertilized soil and $\tilde{C}_S=1$ for a crop on poorly fertilized soil. These data agree with ours in Figure 71.

Sorghum. Experiment No. 15, Tadjikistan. Penetration was determined with the 'Mouse' device and the Lopukhin rod. Initially the experimental data were approximated by semi-empirical formula (II.4.60). Figure 74A gives the empirical attenuation coefficient k_S as a function of solar elevation, as determined from the graph $-\ln a_S$, L. Analysis of this figure shows a clumped distribution of leaves to be characteristic of a crop of sorghum. With solar elevation increasing from 15 to 90° the value of k_S increases nearly twofold, while the tendency to clumping decreases and the distribution approaches a random one. The data measured with the Lopukhin rod are in good agreement with those obtained with the 'Mouse' device in the position $\perp$.

However, a more detailed analysis of the experimental data reveals the layering of foliage in a closed sorghum stand, just as in a maize stand, therefore curve of $\ln a_S(L)$ assumes an S-shape instead of being linear as was the case when the problem was first treated in approximation.

Given, by way of example, in Figure 75 is the dependence of a_S on the depth L for different solar elevations. When this dependence was approximated by lengths of straight lines the factors $\tilde{C}_S$ were found separately for the upper, middle and lower layers of the stand for different solar elevations (see Figure 74B). The data of 11.07–16.07, the tasseling stage, were given a similar treatment (see Figure 74C and D). Consideration of Figure 74 and its comparison with Figure 73 for a maize stand show that the distribution of leaves in sorghum and maize stands is of the same general character. Leaves in the upper layer tend to clump at low solar elevations and as the latter increases, their distribution approaches a random one (at tassel flowering) or even semiregular at high values of $h_\odot$ (see Figure 74D, tasseling). In the middle layer, leaves with a relative variance $V'=3$–9 tend to clump most of all. The dependence of $\tilde{C}_S$ on $h_\odot$ is weak. The lower layer occupies an intermediate position. Leaf distribution tends from random to clumped. The latter distribution is least pronounced at solar elevations $h_\odot=40$–$60°$.

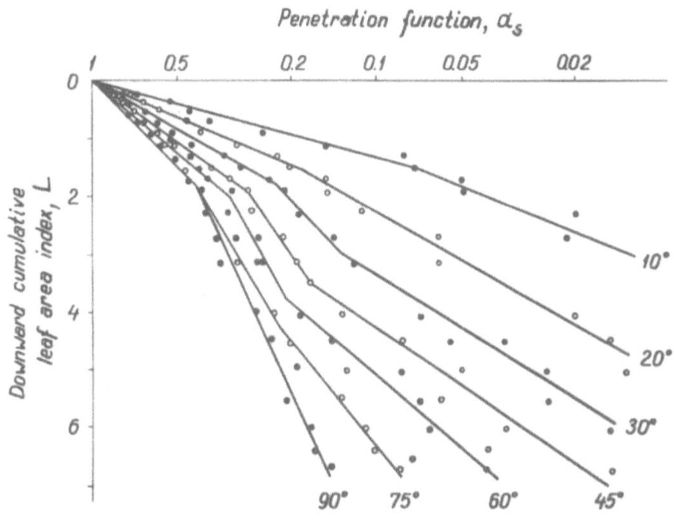

Fig. 75. Penetration function a_S (plotted logarithmically) in a sorghum stand as a function of depth L at different solar elevations. Experiment No. 15, Tadjikistan, July 30, August 1, 1965, tassel flowering stage, $L_0 = 6.9$. Measured by Laisk using the 'Mouse' device.

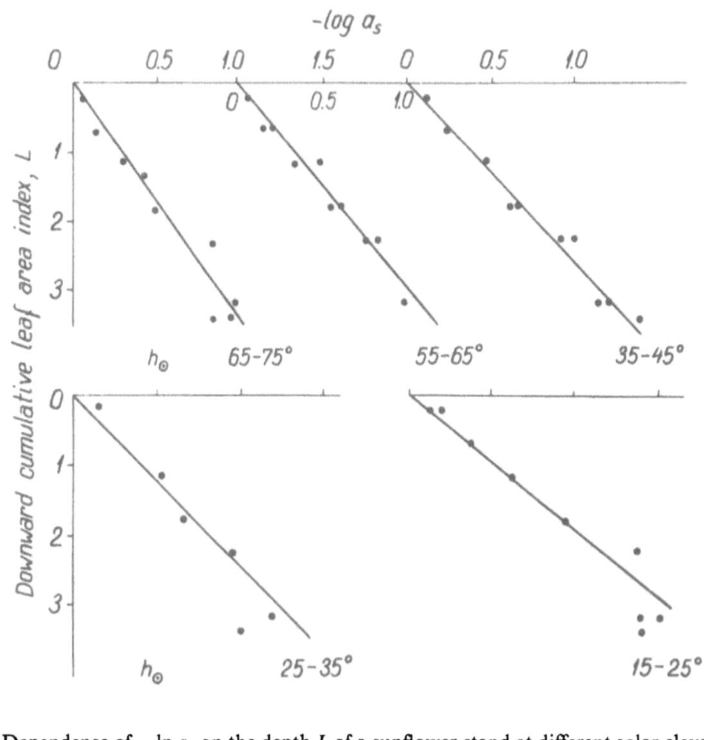

Fig. 76. Dependence of $-\ln a_S$ on the depth L of a sunflower stand at different solar elevations $h_\odot$. Experiment No. 17, Tadjikistan, July 12, 1965, formation of inflorescence.
Measured with Lopukhin rod.

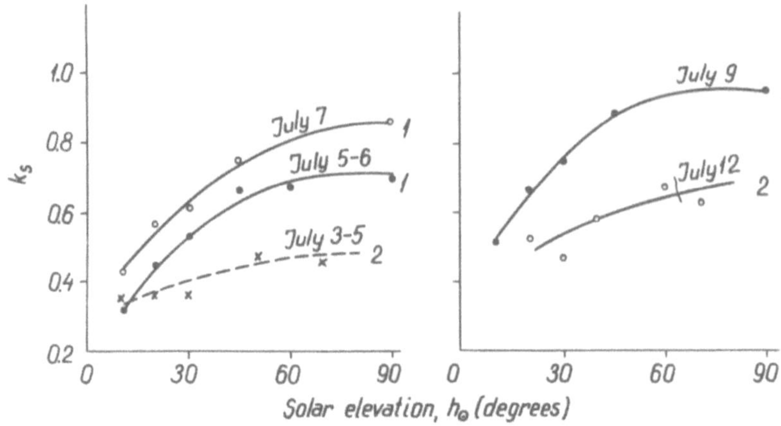

Fig. 77. Attenuation coefficient k_S as a function of solar elevation. Sunflower, experiment No. 17, Tadjikistan, 1965.
1 – as measured by Laisk using the 'Mouse' device, 2 – Lopukhin rod data.

Sunflower. Experiment No. 17, Tadjikistan. Penetration was measured with the 'Mouse' device and Lopukhin rod. The data produced by the Lopukhin rod revealed a notable azimuthal effect in before- and afternoon data on 03–05.07. The effect deminished towards 12.07, just as in the case of maize (see Figure 67B). The dependence $-\ln a_S, L$ appeared to be linear (Figure 76). The same dependence, according to the 'Mouse' device measurements, is slightly S-shaped. The data produced by both the instruments were approximated by empirical formula (II.4.60). As seen from Figure 77, the empirical attenuation coefficient k_S increases with solar elevation, which is indicative of a horizontal orientation of leaves. This is also corroborated by the data of Table I.6.2. The Lopukhin rod overestimates penetration values in comparison with the 'Mouse' device.

The 'Mouse' device overestimates data on penetration (data gathered on 06 and 07.07) when it is placed parallel to rows, as compared with the perpendicular position. Penetration decreases between 03–12.07, which is the inflorescence formation period. The 'Mouse' device measurements show a stronger dependence of k_S on solar elevation: with $h_{\odot}$ increasing from 10 to 90° k_S grows 1.5–2 times. This means that the lower layers of a sunflower stand are sunlit predominantly during the morning and evening hours.

Winter wheat. Experiment No. 7, Estonia. Budagovsky's measurements were performed with the Lopukhin rod on 17.07.67 at the milk ripeness stage. Figure 78A presents the initial data, viz. dependence of the penetration function a_S on solar elevation $h_{\odot}$ for different heights. Smoothed curves were used to determine the mean values of $a_S(h_{\odot}, z)$. The approximation of experimental data was accomplished with formula (II.4.58) and indicated that in the interval of $h_{\odot} = 30$–60° the dependence $-\ln a_S$ on $\sum_j k_j$ presented in Figure 78B was the same. At a low solar elevation $h_{\odot} = 15°$ this dependence is different. The factor $\tilde{C}_S$ has different values in the upper part of the stand, where no leaves grow and spikelets

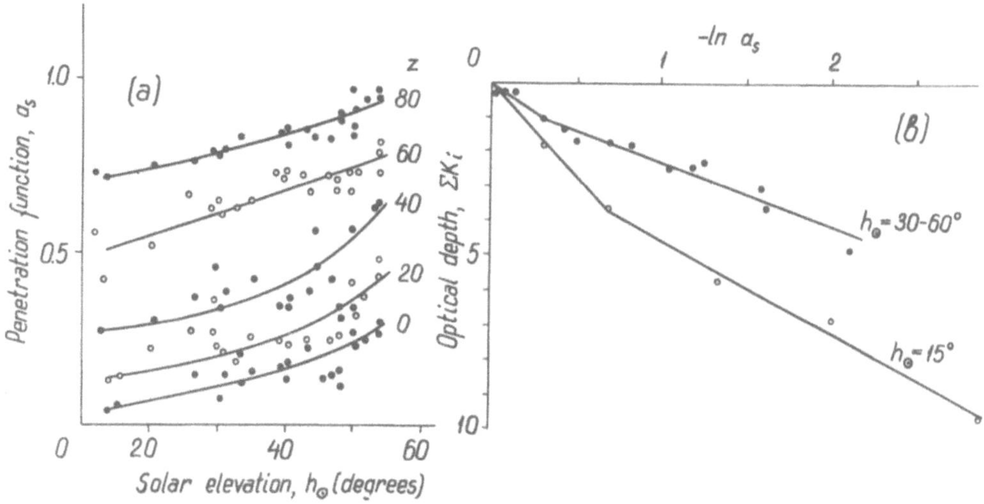

Fig. 78. Penetration of direct solar radiation in a stand of winter wheat. Experiment No. 7, Estonia, 1963.
(a) – dependence of penetration function a_S on solar elevation $h_\odot$ at different heights above ground z (cm), (b) – dependence of $-\ln a_S$ on the optical depth $\sum_i k_i$ at different solar elevations $h_\odot$.

$h_\odot$	15°	30–60°
$\tilde{C}_S$ for upper layer	0.18	0.30
$\tilde{C}_S$ for lower layers	0.37	0.55

are the main shading elements, and in the lower part, where leaves are the main source of shading. It was assumed that for leaves $G=\frac{1}{2}$. The different factors $\tilde{C}_S$ indicate that penetration is greater in the stand upper part than in its lower part. Contradictory to it is the fact that $\tilde{C}_S$ in the upper part of the stand is nearly two times less at low values of $h_\odot$. This gives reasons to believe that the Lopukhin rod systematically underestimates data in cereal crops at low solar elevations, where it 'ignores' fairly distant thin phytoelements, being affected by the penumbra effect. Therefore the data obtained at $h_\odot=15°$ should be regarded with some reserve. Our data, of penetration are 2–3 times as high as those of Gulyayev et al. (1969) obtained also with the Lopukhin rod. The reason for such a great discrepancy is not clear.

Cotton. Experiment No. 18. The measurements were taken on 07.07.65 with the Lopukhin rod. An azimuthal effect and a difference in before- and afternoon measurements were observed. The dependence of k_S on $h_\odot$ is given in Table II.4.2.

Table II.4.2 *Dependence of coefficient k_S on solar elevation $h_\odot$ in a cotton stand*

$h_\odot$........	20	30	40	50	60
k_S before noon	0.72	–	0.68	0.44	0.61
k_S after noon	0.44	0.44	0.49	0.51	0.50

Analysis of experimental data has shown that the exponential law of the attenuation of direct solar radiation in the stand is but a first, comparatively rough, approximation. Actually the situation is more complex. The distribution of phytoelements does not always follow the Poisson distribution; it is different in different layers of the stand; it changes with time and is dependent on the direction of sighting. A feature in common is a tendency from a random to a clumped distribution of phytoelements, though in the upper layer of maize and sorghum, the tendency is towards a regular distribution. The methods based on sunfleck area measurements are subject to various systematic errors resulting from the penumbra effect. Averaging and correlating actinometric and phytometric data present difficulties. Therefore the results of this chapter give us reasons to state that the problem of determining the penetration of direct solar radiation in a plant stand has not been solved, despite the recent achievements. The solution of this problem requires, first of all, more detailed experimental data on the distribution of phytoelements in a stand, due regard for the penumbra effect and a further development of the theory of penetration in a nonhomogeneous plant stand.

II.5 Penetration of diffuse sky radiation into plant stand

II.5.1 GENERAL

The importance of diffuse sky radiation penetrating the stand is not great in the presence of direct solar radiation. When the Sun's disk, however, is covered with clouds, diffuse sky radiation becomes a decisive factor.

The theory of penetration of diffuse sky radiation into a plant stand is based on the theory of penetration of direct solar radiation, and the transfer equation for diffuse sky radiation intensity (II.1.64) coincides formally with the corresponding equation (II.1.61) for direct solar radiation. All theoretical considerations as regards the penetration of direct solar radiation given in Chapter II.4 are, therefore, quite applicable to diffuse sky radiation. Specific problems only arise when converting intensity to irradiance in accordance with formula (II.1.67) where integration of intensity over all points of the sky is required.

The fundamental theory and calculations of the penetration of diffuse sky radiation are published in the works of Ross and Nilson (1963, 1968a, b), Tooming and Ross (1965), Anderson (1964, 1966a), Nilson (1968a), Cowan (1968), Isobe (1969), Anderson and Denmead (1969). The most interesting experimental data are contained in the works of Evans et al. (1960), Evans (1956), Anderson (1964, 1966b), Alekseyev (1963a) and Zelniker (1969). Mathematical models of the radiation regime in plant stands proposed by Idso and de Wit (1970) and Duncan et al. (1967) also include diffuse sky radiation, but regrettably, it is not given a separate treatment.

II.5.2 PENETRATION FORMULAE

Solution of the transfer equation (II.1.64) for the intensities of diffuse sky radiation averaged over the horizontal surface $d(z, r)$, with the boundary condition (II.1.65) taken into account, gives the following formula for the penetration function $a_d(z, r)$ (II.1.66)

$$a_d(z,r) = \prod_k \exp\left[-\sec\theta \int_z^{z_0} u_k(z')G_k(z',r)\mathrm{d}z'\right], \tag{II.5.1}$$

which corresponds to expression (II.4.2) for direct solar radiation. In accordance with (II.1.67) and (II.2.7) the irradiance of diffuse sky radiation penetrating the stand at the height z is in the form

$$D(z) = D_0 \int_0^{\pi/2} D'_{0Z} \exp\left(-\frac{GL}{\cos\theta}\right)\mathrm{d}\theta =$$

$$= \int_0^{2\pi} \mathrm{d}\varphi \int_0^{\pi/2} d_0(r)a_d(z,r)\cos\theta\sin\theta\mathrm{d}\theta, \tag{II.5.2}$$

where integration is taken over all points of the sky. The average irradiance of the leaf upper side at the height z due to the sky irradiance $d_0(r)$ is determined by the following expression in accordance with (II.1.78)

$$D_1(z,r_L) = \int_{\Omega_{11}^L} d_0(r)a_d(z,r)\,|\cos\widehat{r_L r}|\,\mathrm{d}\Omega. \tag{II.5.3}$$

The average irradiance on the leaf's lower side at the height z due to the sky irradiance

$$D_2(z,r_L) = \int_{\Omega_{12}^L} d_0(r)a_d(z,r)\,|\cos\widehat{r r_L}|\,\mathrm{d}\Omega. \tag{II.5.4}$$

The total of the downward and upward irradiance is

$$D_1 + D_2 = \int_{\Omega_1} d_0(r)a_d(z,r)\,|\cos\widehat{r_L r}|\,\mathrm{d}\Omega. \tag{II.5.5}$$

The penetration function for zonal sky radiation, in accordance with (II.2.7) and (II.5.2), has the form

$$a_{Dz}(L,\theta) = \cos\theta\sin\theta \int_0^{2\pi} d'_0(r)a_d(L,r)\mathrm{d}\varphi,. \tag{II.5.6}$$

the penetration function for irradiance on a horizontal surface is

$$a_D(L) = \int_0^{\pi/2} a_{Dz}(L,\theta)\mathrm{d}\theta = \int_{\Omega_1} d'_0(r)a_d(L,r)\cos\theta\mathrm{d}\Omega. \tag{II.5.7}$$

In the case of a uniformly bright sky $d'_0 = 1/\pi$ and

$$a_D(L) = 2\int_0^{\pi/2} \exp\left(-\frac{LG(\theta)}{\cos\theta}\right)\cos\theta\sin\theta\mathrm{d}\theta. \tag{II.5.7a}$$

Expressions (II.5.1)–(II.5.7) are the basic formulae for the diffuse sky radiation penetrating a plant stand. It is seen from formula (II.5.3) that irradiance in the direction r_L is determined by three factors: the spatial distribution of the intensity $d_0(r)$ of the diffuse sky radiation arriving at the plant stand, the penetration function $a_d(z,r)$ and the orientation of the normal r_L expressed by means of the solid angle of integration Ω_{11}^L or Ω_{12}^L and $|\cos\widehat{r_L r}|$.

The most important of these is the penetration function a_d given by formula (II.5.1) which corresponds to the exponential penetration model.

In accordance with (II.4.13) the penetration function for the binomial model has the form

$$a_d(L, r) = \left[1 - \Delta L \frac{G(r)}{\cos\theta}\right]^{L/\Delta L}.$$

The exponential formula (II.5.1) has been used in most of the theoretical works which have been published. Calculations in the work of Idso and de Wit (1970) were performed in accordance with the binomial model.

Integration over the upper hemisphere is the major difficulty in the calculation of irradiance. Even if $d_0(r)$ is given by an empirical formula, integration in quadratures is only practicable in special cases. Some of such cases will be discussed below.

Calculations of irradiance consider three cases of diffuse sky radiation.

1. *Isotropic sky*: $d'_0(r) = 1/\pi$.
2. *Overcast sky*. In our calculations (Tooming & Ross 1965) for $D'_{0Z}(\theta)$ we used experimental data of Põldmaa (1963) (curve 3 in Figure 41). Some authors (Anderson 1966a, Cowan 1968) have used the so called standard overcast sky (curve 2 in Figure 41).
3. *Clear sky*. In our calculations for $d'_0(r)$ we used calculated data of Hess and Linke borrowed from the book of Kondratyev (1954).

II.5.3 CALCULATION OF INTENSITIES AND ZONAL RADIATION

Ross and Nilson (1968b) performed a series of calculations to elucidate the basic regularities of the penetration of intensities and fluxes of diffuse sky radiation on the assumption that the plant stand consists of leaves only and their orientation is the same in all layers of the stand, i.e. $G(L, r) = G(r)$. In this case formula (II.5.1) has the form

$$a_d(L, r) = \exp\left[-\frac{LG(r)}{\cos\theta}\right], \quad \text{(II.5.1a)}$$

whence it follows that radiation intensities decrease exponentially with depth and are dependent on the direction and the foliage orientation.

It follows from the analysis of formula (II.5.1a) that in the case of horizontal leaves $a_d(L) = \exp(-L)$, i.e. the penetration is isotropic.

Presented in polar coordinates in Figure 79 is the penetration $a_d(L, \theta)$ as a function of the zenith angle θ for different depths L in two model stands and two real ones.

It will be seen from Figure 79 that even at small values of depth L the part of the sky near the horizon is almost fully shaded by plant parts, notably when they are oriented vertically. The role of near-zenith directions increases with the increasing depth. The shading effect of nearly horizontal and medium directions also

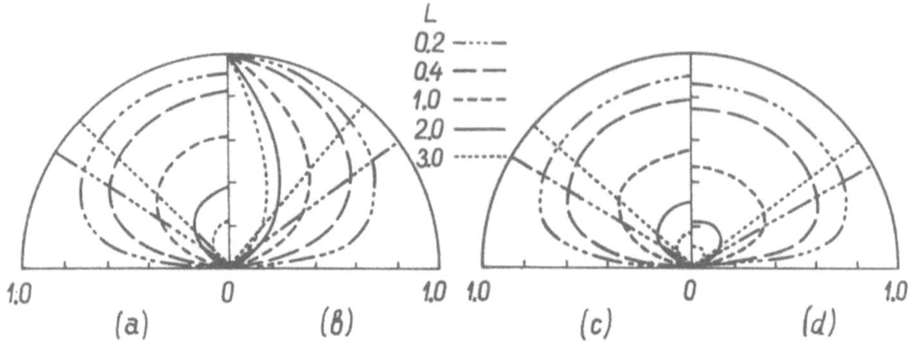

Fig. 79. Penetration function a_d as a function of zenith angle θ at various depths L.
(a) – uniformly oriented leaves, (b) – vertical leaves, (c) – maize foliage, (d) – sunflower foliage. Radial lines correspond to effective angles θ for $L=0.2$ and 3.0 at $d_0=\text{const}$.

increases. It is also seen that the dependence of a_d in maize and sunflower stands is closer to the dependence of a_d for a uniform orientation of leaves than for a vertical orientation.

The largest penetration from the near-zenith part of the sky ($\theta=0$–$50°$) is obtained with a vertical leaf orientation; that in the range of zenith angles 50° to 60°, with a uniform orientation; and that from near-horizontal directions ($\theta=60$–$90°$), with a horizontal orientation.

In the case of the real sky the angular dependence of intensities $d(L, r) = d_0(r)a_d(L, r)$ differs from that of the penetration function $a_d(L, r)$ because of the term $d_0(r)$.

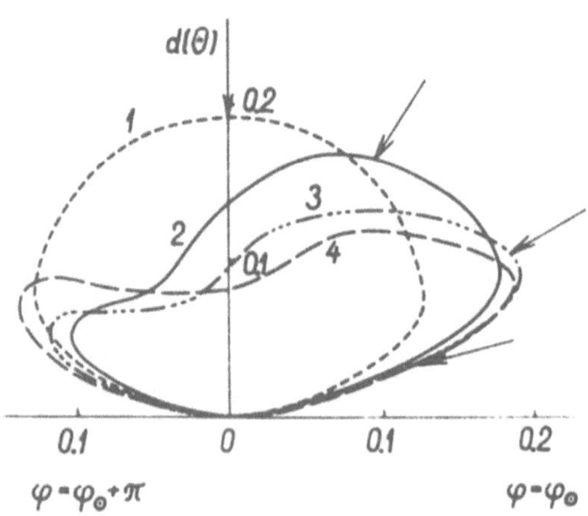

Fig. 80. Angular distribution of sky brightness $d(\theta)$ within a plant stand at the depth $L=0.6$ in the solar vertical plane at different solar elevations $h_\odot$. Uniform orientation of leaves.
1 – $h_\odot=90°$, 2 – $h_\odot=60°$, 3 – $h_\odot=30°$, 4 – $h_\odot=15°$.

Shown in Figure 80 is the hodograph of the intensities of diffuse radiation from the clear sky at the depth $L=0.6$ for four solar elevations, with the leaves uniformly oriented. Figure 81 shows an analogous dependence but at $h_{\odot}=60°$ at different depths L of the stand, from 0 to 3. These illustrations show that the spatial asymmetry of the intensities of the diffuse sky radiation incident on the stand which is due to the solar aureole is extended to deeper layers, but in this case the near-horizontal intensities disappear.

The penetration function for zonal radiation $a_{Dz}(\theta)$ has quite a different angular dependence under an isotropic sky at the depth $L=0.6$ (Figure 82). The zonal radiation coming from the region of the sky near the zenith is insignificant

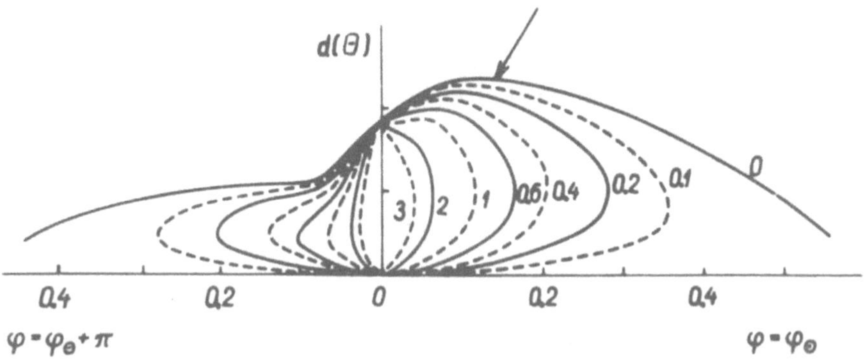

Fig. 81. Angular distribution of sky birghtness $d(\theta)$ at different depths L within a stand in the solar vertical plane. Vertical leaves, $h_{\odot}=60°$.

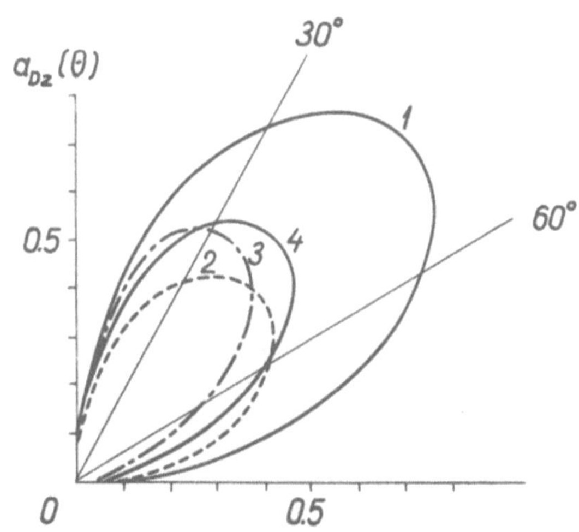

Fig. 82. Angular distribution of the penetration function for zonal sky radiation $a_{Dz}(\theta)$ in polar coordinates at the depth $L=0.6$.

1 – relative zonal brightness of a uniformly bright sky above the stand, 2 – penetration function A_{Dz} for horizontal leaves, 3 – same for vertical leaves, 4 – same for uniform orientation of leaves.

because of the small area of the zone (see Figure 41). The maximum amount of zonal radiation arrives at the plant stand from the zone at an angle of 45°. As the diffuse radiation penetrates inside the stand, this direction is shifted somewhat depending on the orientation of leaves. In the case of a vertical and uniform orientation the maximum penetration is displaced towards a higher zone at the angles of 30° and 40°, respectively; in the case of a horizontal orientation the angular dependence of zonal radiation is unchanged. In the case of a uniform orientation the maximal penetration of radiation is from the zone at an angle of 40° and is equal to 0.66 at $L=0.6$.

II.5.4 CALCULATION OF DOWNWARD FLUXES

Calculation of the downward flux $D(L)$ is performed using formula (II.5.2), and that of the penetration function $a_D(L)$, by formula (II.5.7). Let us discuss some special cases of formula (II.5.7).

Horizontal leaves.

$$a_d(L,r)=\exp(-L), \quad D(L)=D_0\exp(-L).$$
$$a_D(L)=\exp(-L), \tag{II.5.8}$$

In the case of horizontal leaves the diffuse sky radiation flux $D(L)$ inside the stand does not depend on the spatial structure of the incoming diffuse sky radiation, which is characterized by $d_0(r)$.

Vertical leaves.

$$a_d(L,r)=\exp\left(-\frac{2}{\pi}L\tan\theta\right)$$

and

$$a_D=\int_0^{2\pi}\mathrm{d}\varphi\int_0^{\pi/2}d_0'(\theta,\varphi)\times$$
$$\times\exp\left(-\frac{2L}{\pi}\tan\theta\right)\cos\theta\sin\theta\mathrm{d}\theta. \tag{II.5.9}$$

For an isotropic sky

$$a_D(L)=2\int_0^{\pi/2}\exp\left(-\frac{2L}{\pi}\tan\theta\right)\cos\theta\sin\theta\mathrm{d}\theta. \tag{II.5.10}$$

According to Cowan (1968) formula (II.5.10) may be expressed by means of the integral sine and cosine in the form

$$a_D=1-\frac{2L}{\pi}\left(\sin\frac{2L}{\pi}Ci\frac{2L}{\pi}-\cos\frac{2L}{\pi}Si\frac{2L}{\pi}\right). \tag{II.5.10a}$$

Uniform orientation of leaves.

$$a_d(L,r)=\exp\left(-\frac{L}{2}\sec\theta\right)$$

and

$$a_D(L,r)=\int_0^{2\pi}\mathrm{d}\varphi\int_0^{\pi/2}d_0'(r)\times$$

$$\times\exp\left(-\frac{L}{2}\sec\theta\right)\cos\theta\sin\theta\mathrm{d}\theta. \qquad \text{(II.5.11)}$$

In the case of an isotropic sky

$$a_D(L)=2\int_0^{\pi/2}\exp\left(-\frac{L}{2}\sec\theta\right)\times$$

$$\times\cos\theta\sin\theta\mathrm{d}\theta=2E_3\left(\frac{L}{2}\right), \qquad \text{(II.5.12)}$$

where $E_3(x)$ is an integral exponential 3rd order function.

Ross and Nilson (1968b) and Tooming and Ross (1965) performed calculations of $a_D(L)$ for model stands with a horizontal, vertical and uniform orientation of leaves, and also for maize and sunflower stands. The sky was considered in three states: isotrophic, overcast and clear sky at different solar elevations. The results of calculations are given in Table 2.5.1. The following conclusions can be drawn from this table.

1. The spatial distribution of the diffuse sky radiation $d_0'(r)$ incident on the stand has a certain effect on the penetration, which depends on the foliage area orientation and on the stand depth. In the case of a clear sky the maximum

Table II.5.1 *Penetration function for diffuse sky radiation flux in plant stands*

L	Horizontal orientation	Uniform orientation					
		isotropic sky	overcast sky	clear sky			
				15°	30°	60°	90°
0	1.000	1.000	1.00	1.000	1.000	1.000	1.000
0.1	0.905	0.909	0.91	0.873	0.886	0.905	0.894
0.2	819	833	82	773	793	825	808
0.4	670	704	68	623	648	695	672
0.6	549	600	56	513	538	592	567
1.0	368	443	42	360	381	438	414
1.5	223	308	30	240	254	307	286
2.0	135	219	21	164	174	218	202
3.0	50	113	10	80	85	113	103
5.0	7	33	3	22	23	33	29

Table II.5.1 (*Continued*)

Vertical orientation					
L	Isotrophic sky	Clear sky			
		15°	30°	60°	90°
0	1.000	1.000	1.000	1.000	1.000
0.1	0.911	0.861	0.878	0.906	0.891
0.2	841	764	788	835	812
0.4	725	617	646	718	686
0.6	696	518	546	631	596
1.0	504	384	407	502	467
1.5	393	282	298	393	359
2.0	314	216	227	316	286
3.0	213	119	144	218	193
5.0	117	73	74	123	106
Maize					
0	1.000	1.000	1.000	1.000	1.000
0.1	0.908	0.876	0.888	0.905	0.895
0.2	828	777	795	822	808
0.4	696	625	649	689	688
0.6	590	513	537	583	561
1.0	429	356	376	424	403
1.5	292	232	246	289	272
2.0	201	154	164	199	186
3.0	98	71	76	97	89
5.0	24	17	18	24	22
Sunflower					
0	1.000	1.000	1.000	1.000	1.000
0.1	0.907	0.883	0.893	0.904	0.897
0.2	824	787	801	820	809
0.4	685	632	652	680	664
0.6	573	515	535	567	550
1.0	404	345	367	399	383
1.5	263	220	283	260	248
2.0	173	140	149	170	161
3.0	75	59	63	75	70
5.0	15	11	12	15	14

penetration is observed at $h_{\odot}=60°$. The penetration decreases with decreasing solar elevation, and at $h_{\odot}=15°$ it is equal to 85–90% of its maximum value at middle depths $L\approx 1$ and 70–80% at greater depths. This phenomenon is due to the concurrent action of three factors: position of the maximum of $d_0'(\theta)$, horizontal projection area of the zone, and the penetration function for intensities $a_d(\theta)$.

The penetration function has its maximum in the zenith and decreases with the increasing θ, but the zonal area is small in the zenith. The projected area of the zone reaches its maximum value at $\theta = 45^\circ$ and then decreases with the increasing θ (Figure 82). The maximum intensity $d_0'(\theta)$ depends on the position of the Sun, that is why the maximum of the flux penetration function a_D is not in the zenith but at $h_\odot = 60$–70°.

2. An isotropic sky provides greater penetration than a clear sky, as a rule. The difference is not large at $h_\odot = 40 \div 60^\circ$ and amounts only to 1–2%. An exception exists with vertically oriented leaves, in which case the penetration at great depths below a cloudless sky at $h_\odot = 60^\circ$ is several percent larger than for an isotropic sky. This is due to the zone at an angle of 60° that gives the maximum penetration.

3. We have calculated the penetration below an overcast sky for a uniform orientation. It approximately coincides with the penetration under a clear sky at $h_\odot = 50^\circ$. But Cowan (1968) compared the penetration calculated by Anderson (1966a) for the standard overcast sky with the penetration for a uniformly bright sky and found that in the case of a vertical and uniform orientation of leaves the standard overcast sky systematically overestimated values by 5 to 10%. This is easily understood because in the case of the standard overcast sky the intensities increase 30% from horizon to zenith.

It is interesting to find out the effective zenith angles θ_E at which the penetration function for intensities is equal to the penetration function for the flux

$$\exp\left[-\frac{LG(\theta_E)}{\cos\theta_E}\right] = \int_0^{\pi/2} d_0'(\theta)\exp\left[-\frac{LG(\theta)}{\cos\theta}\right]\cos\theta\sin\theta d\theta.$$

It is clear that θ_E must depend on the depth L and orientation (see Figure 79 in which the effective angles are marked with radial strokes). The upper stroke corresponds to $L = 3.0$, the lower to $L = 0.2$, i.e. the effective angle θ_E increases with the increasing depth of the stand, since the role of the near-zenith sky region in the penetration increases. On an average the effective angles of inclination range from 40 to 60°.

Formulae (II.5.8)–(II.5.12) show that the penetration function $a_D(L)$ is exponential in form only in the case of a horizontal orientation of leaves. It is interesting to investigate the character of deviation of $a_D(L)$ from the exponential relation for other types of orientation. It will be seen from Figure 83 showing the dependence of $\ln a_D$ on the depth L for uniform orientations that the deviation from the exponential relation in the case of different orientations of leaves is towards increasing penetration and that in deep layers the maximum penetration is typical of a vertical orientation and exceeds the penetration for a horizontal orientation at $L = 2$ by a factor of 2.5.

The penetration function a_D calculated by Anderson (1966a, 1970) for different leaf inclination angles confirms the above conclusions. It is also shown by her calculations that the vertical orientation of leaves below a clear sky during the summer solstice in Australia exceeds the penetration in the case of the standard overcast sky.

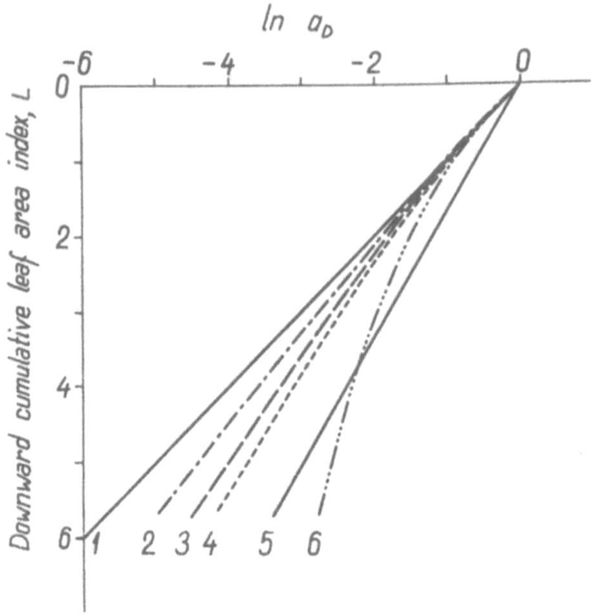

Fig. 83. Variation of ln a_D as a function of the depth L in a stand.
1 – horizontal leaves (or ln a_S at $h_{\odot}=30°$ and uniform orientation of leaves), 2 – sunflower, 3 – maize, 4 – uniform orientation, 5 – ln a_S at $h_{\odot}=60°$ and uniform orientation, 6 – vertical leaves.

The work of Alekseyev (1970) contains calculations of the complementary diffuse radiation arriving inside a forest through gaps between tree crowns. The calculations took account of the actual brightness of the sky $d_0(r)$ at different solar elevations and angular dimensions of gaps. The results of calculations show that the incoming radiation increases with increasing gaps, with a concurrent increase in the difference of the amount of complementary radiation between the northern and southern sides of the gaps. The amount of diffuse radiation arriving through the gaps increases with the increasing solar elevation.

II.5.5 METHOD OF HEMISPHERICAL PHOTOGRAPHS

The method of hemispherical photographs for the determination of the penetration function of direct a_S and diffuse a_D radiation was developed in works of Evans and Coombe (1959), Coombe and Evans (1960) and Anderson (1960). It was further used and developed by Madgwick and Brumfield (1969), Nilson and Ross (1971), Hutchinson (1975), Ducrey (1975a, b) and Nilson (1977b) in the case of forests and by Bonhomme (1970), Bonhomme and Chartier (1972) and Bonhomme (1974) in the case of corn and sweet potato.

Essentially this method consists in the following. A photo camera is fitted with a special lens, the so called 'fisheye', which images the hemisphere on photosensitive film so that proportionality between parts of the sky is maintained in the

Fig. 84. 'Fisheye' photograph of upper hemisphere in a maize stand.

print, i.e. equal angular distances from the zenith give equal linear distances on the photographic image. That is why the shots of the upper hemisphere taken from a point P inside a plant stand reveal all gaps that are visible against the sky from the point P (see Figure 84). If a special reference grid is available, the photograph can be processed by using one of the following three methods (Nilson & V. Ross 1971): (1) averaging the part of gaps over area in each subregion of the grid; (2) averaging the part of gaps over separate lengths on the reference grid; (3) determining visibility of the sky at the grid nodes only ('yes' or 'no'). To expedite the handling of photographs, Bonhomme and Chartier elaborated a device, and Lakso (1976) used a false colour densitometer for the same purpose. By moving the camera in horizontal directions at a constant height z an adequate number of shots can be taken to ensure a reliable degree of averaging. Hemispherical photographs allow determining the angular dependence of the penetration function for diffuse sky radiation intensities $a_d(r)$. After drawing on the reference grid the path of the sun over the sky traversed on some day the diurnal course of the penetration function for direct solar radiation a_S can be calculated, and the penetration function for diffuse radiation flux a_D can be calculated from formula (II.5.7).

This method is advantageous in that it is graphic and very informative: the photograph clearly displays the degree and character of the shading of the whole sky by phytoelements at a given point. In addition, a set of photographs taken at different points in a plant stand allows determining not only mean values of the penetration functions a_S, a_d, a_D but also their statistical parameters and the correlation of the transmissivity of the plant canopy in different directions of the atmosphere.

The drawbacks of this method: (i) the method is laborious. In order to obtain a reliable statistical set of data on penetration we must have numerous photographs, the number of which depends on the uniformity of stand structure and amounts to about one hundred in the case of a forest (Nilson 1977b) and 25 to 100 in the case of crops (Bonhomme & Chartier 1972). Proceeding from a theoretical conception of the autocorrelation of plant stand transmissivity, Nilson (1977b) proposed the following formula for the minimum distance x^* between adjacent points at which photographs are to be taken, in the case of a tree stand

$$x^* = \frac{2R\left[\left(\frac{H}{2}\right)^2 \tan^2 \vartheta_\odot + R^2\right]^{\frac{1}{2}}}{\left\{\left[\left(\frac{H}{2}\right)^2 \tan^2 \vartheta_\odot + R^2\right] \sin^2 \psi + R^2 \cos^2 \psi\right\}^{\frac{1}{2}}},$$

where H and R are the 'mean' tree height and the crown radius, respectively, and ψ is the azimuth of the direction along which photographs are to be taken.

In a forest x^* varies from a few metres to hundreds of metres.

(ii) Where phytoelements are small and placed at a great distance from the photo camera, the penumbra effect makes it difficult to distinguish the boundaries between the sky and phytoelements. The boundaries are particularly blurred in the case of conifers.

(iii) On a cloudless day the photographic brightness of well illuminated phytoelements in upper layers may be close to the sky brightness, and such phytoelements are hardly detectable.

Notwithstanding the above-mentioned drawbacks the method of hemispherical photographs allows solving some problems of phytometry and phytoactinometry. Bonhomme and Chartier (1972) came to the conclusion that this method can be used, with a 10% error, for determining the leaf area index L_0 in crops if L_0 is determined from formula (II.4.9). Anderson (1964) and Madgwick and Brumfield (1969) used this method for determining mean values of the penetration function a_S and a_D. According to Ducrey (1975b) the total radiation flux determined by the method of hemispherical photographs in a bush stand is in good agreement with pyranometer data if the transmissivity of the canopy ranges from 0.15 to 0.20, and hemispherical photographs underestimate the actual irradiance if the penetration is less than 0.10. Hutchinson (1975) used this method for determining the gap image diameter distribution and for predicting the space distribution of direct beam flux densities at various levels in a forest. Nilson (1977b) determined the covariance matrix of the penetration function for different sky zones with zenith angles ϑ_j and ϑ_k for a stand of fires, averaged over 113 photograps. As solar elevation increases from 15° to 60°, the mean value of the penetration function increases from 0.18 to 0.25, but the relative variance V_{a_D} remains practically constant, equal to 0.26.

II.5.6 STATISTICAL CHARACTER OF THE PENETRATION OF DIFFUSE SKY RADIATION

Attempts of a statistical treatment of the penetration of diffuse sky radiation in a plant stand have been made in works of Isobe (1962a, 1969). According to Isobe the mean sky brightness inside the stand at the point P in the direction r is, analogously to direct solar radiation, a random function X_d which has only two values

$$X_d = \begin{cases} 0 \text{ if the sky is not visible through the stand,} \\ d_0(r), \text{ if the sky is visible.} \end{cases}$$

Therefore what has been said in II.4.2 about the statistical character of the penetration of direct solar radiation is likewise applicable to the intensities of diffuse radiation (sky brightness). The problem, however, is considerably more complicated in the case of fluxes. According to formula (II.1.67) the diffuse radiation flux is obtained by double summation of all intensities over the parts of the sky $\Delta\Omega_i^k$, with the weight function of projection $\cos\theta_i$ taken into account. Thus, the diffuse sky radiation flux penetrating into the stand is the double sum of the random functions of intensities, i.e

$$X_D = \sum_i \sum_k X_d(r_{ik}) \cos\theta_i \Delta\Omega_{ik} \tag{II.5.13}$$

and the calculation of statistical characteristics of X_D from formula (II.5.13) requires the knowledge of the inter-correlations between the functions $X_d(r)$ in addition to the information on these functions. The statistical character of the diffuse radiation penetrating inside the stand has received some qualitative interpretation in the works of Anderson (1964, 1966b).

Proceeding from his silhouette theory Nilson (1977d) derived formulae and calculated the autocorrelation functions of penetration for two directions r and r', for two azimuths at fixed almucantars (ϑ = const.) at two heights z and z' and the covariance matrix $K_a(\vartheta_j, \vartheta_k)$. These calculations show that the mode of the autocorrelation functions decreasing depends strongly on the observation height z and the zenith angle of sight ϑ. The larger the angular size of shading elements (crowns, branches, leaves) the slower the decrease of $K_a(\vartheta_j, \vartheta_k)$. Both the mean value of the penetration function of diffuse radiation $a_D(L)$ and the standard deviation of σ_{a_D} increase with the increasing penetrability of crowns, but the variance V_{a_D} decreases. Both σ_{a_D} and V_{a_D} increase considerably with the increasing.

Among the experimental studies, the method of hemispherical photography is of interest.

Another experimental method for obtaining information on the random functions $X_d(r_{ik})$ uses the 'Mouse' device designed and constructed by Laisk, see II.4.6. If this device is adapted to record data as it is moved along the x-axis, we obtain the realization of one random function $X_d(r_{ik})$ at fixed values of i and k. An adequate number of realizations of the random functions at the depth L could be obtained by using various values of i and k.

The third and easiest method for the experimental investigation of the variability of the diffuse radiation flux is that of direct measurement by means of a photopyranometer shielded from direct sunrays and moving along a horizontal line. The data thus obtained, however, are very integrated ones, they contain little information on the spatial structure of the flux, are hardly compatible with phytometric characteristics and are thus mainly empirical in character.

What can be said about the statistical variability of the diffuse radiation flux? No dependable experimental data are available at present. Since in accordance with formula (II.5.13) X_D consists of the double sum of random functions of more or less the same order of weight and it can be assumed that inter-correlations between X_d are not very strong, it follows from the rule of addition for a large number of weakly dependent random variables that the distribution function of probability density X_D must be close to a normal distribution or, at any rate, must be represented by a curve having a single maximum. It can be assumed that the variance of the function X_D is dependent on the geometrical structure of the stand, namely the taller the stand and the smaller the leaves, the smaller the variance, i.e. the more uniform the diffuse radiation field is.

II.6 Scattering of radiation inside plant stands

II.6.1 GENERAL

As was shown in Chapter II.1, the scattering of direct and diffuse solar radiation by phytoelements in a plant stand sets up complementary radiation fields i_S and i_D given by transfer equations (II.1.69) and (II.1.74). Unlike the penetration of direct and diffuse solar radiation, which depends only on geometrical quantities, scattering also depends on the optical properties of phytoelements and is characterized by the scattering phase function of the medium $(1/\pi)\Gamma_L(L, r', r)$ in the transfer equations. The spectral quality of the penetrated radiation does not change inside the stand and is determined by the spectral composition of the incoming direct and diffuse radiation. Since scattering on phytoelements is selective, the spectral composition of complementary radiation is changed. As shown in Chapter II.3 scattering is largest in the NIR region. The scattering of PAR is largest in the green. The near infrared and green parts can therefore be expected to dominate in the complementary radiation spectrum. Since the solution of transfer equations (II.1.69) and (II.1.74) involves great mathematical difficulties, methods of approximate solution and various simplifications are widely used. Most substantial among them are the following.

1. Neglect of multiple scattering. In this case the term under the integral sign in transfer equations (II.1.69) and (II.1.74) is omitted. This is quite valid in the PAR region where the scattering coefficient of a living leaf $\omega_L \approx 0.15$, and therefore the radiation is attenuated by double scattering by about a factor of 50.

2. Solving the transfer equations by means of the Schwarzschild approximation. In this case assuming isotropic complementary radiation fields i_S and i_D we can change from integral-differential transfer equations to usual differential ones, which are more easily solved.

3. Assumption of mat phytoelements, i.e. that the scattering phase function of the phytoelements $(1/\pi)\gamma_L = \text{const}$. This simplification considerably reduces calculation work when taking double integrals over Ω_1 and Ω_2.

4. Assumption of an infinitely deep layer of vegetation ($L_0 \to \infty$). In this case the effect of reflection from the ground surface below the stand is neglected, and the formulae for intensities and fluxes are considerably simplified.

The general formulation of the problem and the transfer equations for determining the complementary radiation fields i_S and i_D were given by the author (Ross 1962, 1964, 1968) and the author jointly with Nilson (Ross & Nilson 1963, 1967e, 1968a, b) and later independently by Lemeur (1971). The calculation of spectral fluxes was discussed by Nilson (1968a, 1972). A transfer equation has also been derived independently by Isobe (1962a, b, c). A later, valuable work of Isobe (1969) also estimates the effect of multiple scattering.

II.6.2 SCATTERING AND ABSORPTION COEFFICIENTS FOR AN ELEMENTARY VOLUME IN A PLANT STAND. SCATTERING PHASE FUNCTION

The coefficients of scattering ω and absorption κ and the scattering phase function $(1/\pi)\Gamma_L$ are important radiation characteristics of the stand. In accordance with formulae (II.1.13), (II.1.38), (II.1.40) and (II.1.47) they depend both on the optical properties of leaves and on the geometrical structure of the stand.

Let us discuss their angular dependence for some special cases of foliage orientation, assuming mat leaves and similar optical characteristics for the upper and lower leaf sides.

Horizontal leaves. According to (I.2.18c) $g_L(r_L)=\delta(\vartheta_L-0)$ and $G(r)=|\cos\theta|$. In formula (II.1.42) $G_{11}=G$, $G_{12}=0$ and from (II.1.45) and (II.1.47a) we obtain, respectively,

$$\omega(\theta)=\omega_L\,|\cos\theta|, \tag{II.6.1}$$

$$\kappa(\theta)=\kappa_L\,|\cos\theta|. \tag{II.6.2}$$

From (II.1.38), taking into account (II.1.38a), we obtain

$$\Gamma_L(\theta',\theta)=\begin{cases} R_L\,|\cos\theta'\cos\theta|, & \text{if the leaf surface intersects the angle } \widehat{r'r},\\ T_L\,|\cos\theta'\cos\theta|, & \text{if the leaf surface does not intersect the angle } \widehat{r'r}, \end{cases} \tag{II.6.3}$$

here θ' and θ vary from 0 to π.

It is seen from (II.6.1) and (II.6.2) that scattering and absorption in the 'elemental' volume of the stand decrease cosinusoidally as the direction changes from vertical to horizontal.

Shown in polar coordinates in Figure 85 is the dependence of the scattering phase function $(1/\pi)\Gamma_L(\theta',\theta)$ on the zenith angle of the reflected beam θ for different types of leaves at the incident angle $\theta'=60°$. Curves 1 and I in this diagram correspond to a 'mean' living leaf, curves 2 and II to succulents, and curves 3 and III to thin leaves. It will be observed that in the case of horizontal

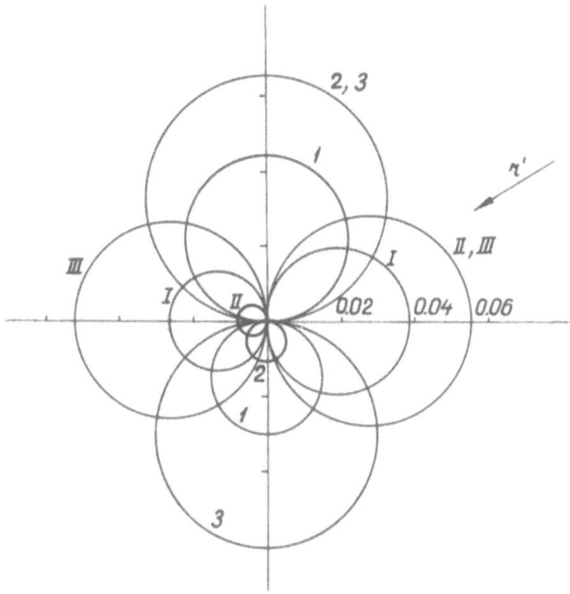

Fig. 85. Generalized phase function Γ_L as a function of zenith angle θ at incident angle $\theta' = 60°$. 1 – horizontal leaves, $T_L = 0.06$, $R_L = 0.09$, 2 – same, $T_L = 0.02$, $R_L = 0.13$, 3 – same, $T_L = 0.12$, $R_L = 0.13$, I – vertical leaves, $T_L = 0.06$, $R_L = 0.09$, II – same, $T_L = 0.02$, $R_L = 0.13$, III – same, $T_L = 0.12$, $R_L = 0.13$.

leaves most radiation is scattered upwards and downwards in the vertical direction. The scattering approaches zero in proximity to the horizon. The scattering phase function is independent of the azimuth angle. The scattering phase function for a plant stand with horizontal leaves is represented geometrically as two spheres placed one over the other, the upper sphere, R_L in diameter, representing the reflection function, and the lower, T_L in diameter, representing the transmission function. It follows from the form of the scattering phase function that, for a stand with horizontal leaves, transmission is small but absorption and scattering are large. In horizontal directions the situation is inverted: transmission through the stand is large, but absorption and scattering are small.

Vertical leaves. According to (I.2.18d) $g_L = \delta(\vartheta_L - \pi/2)$ and $G(\theta) = 2/\pi \sin\theta$. In formula (II.1.42) $G_1 = G_2 = G$ and

$$\omega = \frac{2}{\pi}\omega_L \sin\theta, \tag{II.6.4}$$

$$\kappa = \frac{2}{\pi}\kappa_L \sin\theta. \tag{II.6.5}$$

Upon integrating over φ_L we obtain from (II.1.38)

$$\Gamma_L(\theta', \theta, \beta) = \Gamma_1(\beta)\sin\theta' \sin\theta, \tag{II.6.6}$$

where $\beta=\varphi'-\varphi \qquad 0\leqslant\beta\leqslant\pi$ and

$$\Gamma_1(\beta)=\frac{\omega_L}{2\pi}(\sin\beta-\beta\cos\beta)+\frac{T_L}{2}\cos\beta. \tag{II.6.7}$$

From (II.6.4) and (II.6.5) it is seen that in the case of vertical leaves we have an inverse situation as compared with horizontal leaves. In directions near the vertical, absorption and scattering are close to zero and, correspondingly, transmission is large. As the direction changes towards horizontal, absorption and scattering increase in proportion to the sign of the angle of inclination of both the incident and the scattered beam. It is also seen from Figure 85 that scattering is largest in the horizontal direction and decreases as the direction changes towards the vertical. It is the same in the upper and lower hemispheres. Unlike a stand with horizontal leaves, there is for vertical leaves an azimuthal dependence of scattering, which is characterized by the function $\Gamma_1(\beta)$.

Analysis of formula (II.6.7) shows that the maximum values of $\Gamma_1(\beta)$ are equal to $T_L/2$ and $R_L/2$ and occur at β values equal to 0 and π, respectively. The minimum of $\Gamma_1(\beta)=(\omega_L/2\pi)\sin(\pi T_L/\omega_L)$ corresponds to $\beta=\pi(T_L/\omega_L)$.

The function $\Gamma_1(\beta)$ is shown in Figure 86. It is seen that in succulents (curve 2) scattering is mainly in the backward directions. We have the most uniform scattering in the case of $T_L=R_L=\omega_L/2$ where it is equally large forward and backward and is equal to $\omega_L/2$. In this case minimum scattering is at $\beta=\pi/2$ and is equal to $\omega_L/2\pi$.

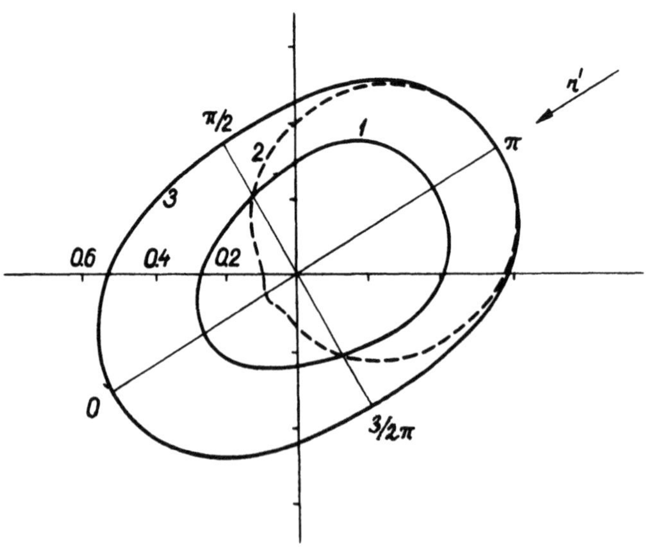

Fig. 86. Generalized phase function $\Gamma_L(\gamma)$ as a function of the angle $\gamma=\widehat{r'r}$ for uniformly oriented leaves at the incident angle $\theta=60°$.
1 – 'mean' leaf, $T_L=0.06$, $R_L=0.09$, 2 – succulent leaf, $T_L=0.02$, $R_L=0.13$, 3 – thin leaf, $T_L=0.12$, $R_L=0.13$. Function $\Gamma_L(\gamma)$ coincides with function $\Gamma_1(\beta)$ which describes the azimuthal dependence of phase function $\Gamma_L(\theta', \theta\beta)$ in the case of vertical leaves.

The function $\Gamma_L(\theta', \theta, \beta)$ is represented geometrically as a distorted tube around the vertical axis, its diameter changing with β and equalling $\Gamma_1(\beta) \sin \theta'$. Vertical sections of the tube are in the shape of a circle (Figure 85). The top view of the shape defined by the function $\Gamma_L(\theta', \theta, \beta)$ is shown in Figure 86.

Uniform orientation. According to (II.2.18b), $(1/2\pi)g_L = 1$ and $G = \frac{1}{2}$. In formula (II.1.42), with (II.1.43), (II.1.44) and (II.1.6) taken into account,

$$G_1 = \frac{1 + \cos \theta'}{4}, \qquad G_2 = \frac{1 - \cos \theta'}{4}$$

and

$$\omega = \frac{\omega_L}{2}, \qquad \kappa = \frac{\kappa_L}{2}. \tag{II.6.8}$$

Upon integration we obtain the following from (II.1.38)

$$\Gamma_L(\gamma) = \frac{\omega_L}{3\pi}(\sin \gamma - \gamma \cos \gamma) + \frac{T_L}{3} \cos \gamma, \tag{II.6.9}$$

where $\gamma = \widehat{r'r}$.

Comparison between (II.6.7) and (II.6.9) shows that these formulae differ from each other only by the factors $\frac{1}{2}$ and $\frac{1}{3}$, respectively. Shown in polar coordinates in Figure 86 is the function $\Gamma_L(\gamma)$ at $\theta' = 60°$ for three types of leaves. The scattering phase function $(1/\pi)\Gamma_L(\widehat{r'r})$ for a uniform leaf orientation is represented geometrically as a body whose axis of revolution coincides with the direction of the incident beam r' (Figure 86).

It will be observed from comparison of the scattering phase function $(1/\pi)\Gamma_L(r', r)$ for plant stands with different model orientations, that these functions differ considerably. With horizontal leaves the forward and backward scattering is mainly in directions close to the vertical, and with vertical leaves in nearly horizontal directions. In both cases scattering depends noticeably on the incident beam direction r', but the symmetry axis of the scattering phase function does not change with the angle of incidence θ'. The situation is different in the case of a uniform orientation. The total scattering and the shape of the phase function do not change with the direction of the beam r' and only the turning of the symmetry axis of the phase function together with r' takes place.

Figure 87 shows the angular dependence of the scattering phase function $\Gamma_L(\gamma)$ of NIR (a) and PAR (b) in a stand with a uniform orientation of leaves. The arrows indicate the direction of the incident beam r'. Curve 1 corresponds to the real scattering phase function of the birch leaf, as measured by Moldau, curve 2 corresponds to mat leaves at $T_L = R_L = \omega/2$, and curve 3 to mat leaves at real values of T_L and R_L.

It is seen from Figure 87 that in the case of birch leaves the scattering phase function $(1/2\pi)\Gamma_L$ does not run smooth because of the specular reflection component, and even with a uniform orientation of leaves in a plant stand the maximum scattering is observed at the angle $\widehat{r'r} = 60°$. This elongation is par-

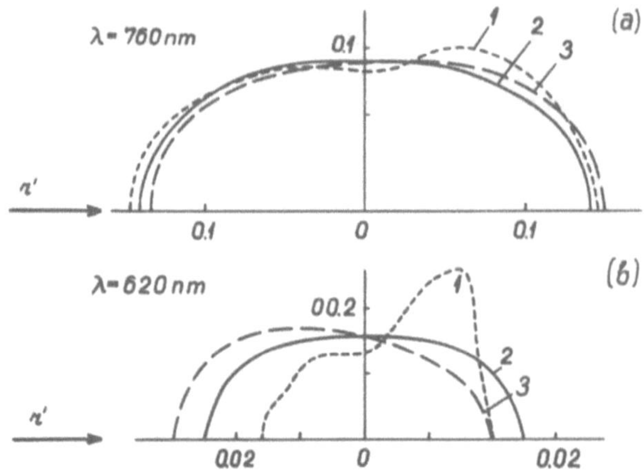

Fig. 87. Generalized phase function $\Gamma_L(\gamma)$ as a function of angle $\gamma=\widehat{r'r}$ in the case of uniformly oriented leaves at $\lambda=760$(a) and $\lambda=620$ nm(b).
1 – for real phase function of birch leaves (data of Moldau), 2 – mat leaves, $T_L=R_L=\omega_L/2$, 3 – mat leaves, real values of T_L and R_L.

ticularly noticeable in the PAR region and is less prominent in the NIR. In the latter case the scattering is mainly forward and backward. Since for birch leaves at $\lambda=620$ nm $R_L>T_L$, curve 3 shows a larger backward scattering in comparison with curve 2, where $R_L=T_L$. At $\lambda=760$ nm the situation is reversed.

II.6.3 SOLUTION OF RADIATION TRANSFER EQUATION FOR HORIZONTAL LEAVES

This case has been discussed in greater detail in our work (Ross & Nilson 1967c), with the following simplifying assumptions:

1. Mat ground surface below the stand. Then from (II.1.22) we obtain

$$\pi\gamma_B(r', r)=A_B=\text{const.},$$

where A_B is the reflection coefficient of the ground surface below the stand.

2. Mat leaves, the transmission and reflection coefficients differ but are equal for the upper and lower sides. In this case the scattering phase function $(1/\pi)\Gamma_L(\theta', \theta)$ is given by formula (II.6.3).

Add up the complementary radiation fields $i_S(z, r)$ and $i_D(z, r)$. Taking into account (II.1.76) and (II.1.77) we obtain

$$\begin{aligned} i_{Q1}(z,r)&=i_{S1}(z,r)+i_{D1}(z,r)\\ i_{Q2}(z,r)&=i_{S2}(z,r)+i_{D2}(z,r). \end{aligned} \qquad \text{(II.6.10)}$$

Taking into account (II.1.69a), (II.1.70a), (II.1.74a) and (II.6.3), converting them from the coordinate z to the depth L and dividing both parts of the equation by $\cos\theta$ we obtain transfer equations for the complementary radiation field in a plant stand with horizontal leaves in the form

$$\left.\begin{aligned} \frac{\partial i_{Q1}}{\partial L} &= -i_{Q1}(L,r) + \frac{T_L}{\pi}\int_{\Omega_1} i_{Q1}(L,r')\cos\theta' \mathrm{d}\Omega' + \\ &+ \frac{R_L}{\pi}\int_{\Omega_2} i_{Q2}(L,r')\cos\theta' \mathrm{d}\Omega' + \frac{T_L}{\pi} Q(L), \\ \frac{\partial i_{Q2}}{\partial L} &= i_{Q2}(L,r) - \frac{T_L}{\pi}\int_{\Omega_2} i_{Q2}(L,r')\cos\theta' \mathrm{d}\Omega' - \\ &- \frac{R_L}{\pi}\int_{\Omega_1} i_{Q1}(L,r')\cos\theta' \mathrm{d}\Omega' - \frac{R_L}{\pi} Q(L) \end{aligned}\right\} \tag{II.6.11}$$

with the boundary consitions

$$\left.\begin{aligned} & i_{Q1}(0,r) = 0, \\ & i_{Q2}(L_0,r) = \frac{A_B}{\pi}[Q^*(L_0) + \pi i_{Q1}(L_0)], \end{aligned}\right\} \tag{II.6.12}$$

where

$$Q^*(L) = Q_0(r_S)\exp(-L), \tag{II.6.13}$$

is the irradiance of that part of the total solar and diffuse sky radiation which reaches the depth L without being intercepted by leaves. It is generally assumed in the set of equations (II.6.11) that the intensities i_{Q1} and i_{Q2} depend on the direction r. But as examination of the set (II.6.11) shows, neither its coefficients nor the boundary conditions (II.6.12) include the direction r. Consequently, the solutions of the set (II.6.11) must not depend on r, i.e.

$$\left.\begin{aligned} i_{Q1}(L,r) &= i_{Q1}(L) \\ i_{Q2}(L,r) &= i_{Q2}(L) \end{aligned}\right\}. \tag{II.6.14}$$

Thus, in the case of spherical reflection and transmission functions of leaves the downward and upward intensities i_{Q1} and i_{Q2} in a plant stand with horizontal leaves are independent of direction, i.e. the radiation field set up in the stand by the interaction of radiation and vegetation is isotropic both in the upper and in the lower hemisphere.

The horizontal plane corresponding to $\theta = \pi/2$ is a discontinuity plane which has one-sided limits for i_{Q1} and i_{Q2}.

Taking into account (II.6.14), i_{Q1} and i_{Q2} in (II.6.11) can be taken outside the integral sign and a set of equations for determining $i_{Q1}(L)$ and $i_{Q2}(L)$ found:

$$\left.\begin{aligned} \frac{\mathrm{d}i_{Q1}}{\mathrm{d}L} &= -(1-T_L)i_{Q1}(L) + R_L i_{Q2}(L) + \frac{T_L}{\pi}Q^*(L), \\ \frac{\mathrm{d}i_{Q2}}{\mathrm{d}L} &= -R_L i_{Q1}(L) + (1-T_L)i_{Q2}(L) - \frac{R_L}{\pi}Q^*(L), \end{aligned}\right\} \tag{II.6.11a}$$

with the boundary conditions (II.6.12).

Since in view of (II.6.14) the flux is

$$H_l(L)=\int_{\Omega_l} i_{Ql}(L)\,|\cos\vartheta|\,d\Omega=\pi i_{Ql}(L), \quad l=1,2,$$

we can easily convert from the equations for intensities (II.6.11a) to equations for fluxes by substituting $\pi i_{Q1}(L)=H_1(L)$ and $\pi i_{Q2}(L)=H_2(L)$; the set (II.6.11a) is a set of ordinary differential equations, which allows an accurate solution.

From the abovesaid it follows that the complementary radiation field $i_Q(L)$ in a plant stand with horizontal leaves is isotropic and can be determined by an accurate solution of the corresponding transfer equation.

The solutions of equations (II.6.11a) with the boundary conditions (II.6.12) have the following form:

$$\left.\begin{aligned} i_{Q1}(L)&=\frac{Q_0}{\pi}\,\frac{a_1(e^{-aL}-e^{-L})-a_2\,e^{-2aL_0}(e^{aL}-e^{-L})}{a_1-a_2\,e^{-2aL_0}},\\ i_{Q2}(L)&=\frac{Q_0}{\pi}\,\frac{b_1\,e^{-a(2L_0-L)}-b_2\,e^{-aL}}{a_1-a_2\,e^{-2aL_0}}, \end{aligned}\right\} \quad \text{(II.6.15)}$$

where

$$\left.\begin{aligned} a&=[(1-T_L)^2-R_L^2]^{\frac{1}{2}},\\ a_1&=A_B-\frac{1}{A'},\\ a_2&=A_B-A',\\ b_1&=1-\frac{A_B}{A'},\\ b_2&=1-A_BA',\\ A'&=\frac{1-T_L-a}{R_L}=\frac{R_L}{1-T_L+a}. \end{aligned}\right\} \quad \text{(II.6.16)}$$

Formulae (II.6.15) will be analyzed in Chapter 11.7 when discussing the total radiation field.

The radiation regime in a stand with horizontal leaves has been discussed independently of us in the work of Cowan (1968). But the final formulae proposed by Cowan are erroneous, since the boundary condition $i_{Q1}(0,r)=0$ is not satisfied. Cowan's error was revealed by Verhagen and Wilson (1969). They also discussed the radiation regime in a plant stand with horizontal leaves, but their final formulae are also erroneous, since they have the plus sign before R_L^2 in the expression for $a=[(1-T_L)^2-R_L^2]^{\frac{1}{2}}$ (II.6.15), which generally somewhat underestimates values for fluxes.

II.6.4 THE SCHWARZSCHILD APPROXIMATION FOR MODIFIED RADIATIVE TRANSFER EQUATION

The Schwarzschild approximation for the modified radiative transfer equation is more complicated than the usual transfer equation.

Let us consider the case of mat leaves with different reflection and transmission coefficients for the upper and lower sides.

Re-write (II.1.69a) and (II.1.74a) in the form

$$\frac{\partial i_{k1}}{\partial L}\cos\theta = -Gi_{k1} + \Gamma_{k0} + \frac{1}{\pi}\times$$
$$\times\int_{\Omega'_1}\Gamma_L i_{k1}\,\mathrm{d}\Omega' + \frac{1}{\pi}\int_{\Omega'_2}\Gamma_1 i_{k2}\,\mathrm{d}\Omega' + \eta,$$
$$\frac{\partial i_{k2}}{\partial L}\cos\vartheta = Gi_{k2} - \Gamma_{k0} - \frac{1}{\pi}\times$$
$$\times\int_{\Omega'_1}\Gamma_L i_{k1}\,\mathrm{d}\Omega' - \frac{1}{\pi}\int_{\Omega'_2}\Gamma_L i_{k2}\,\mathrm{d}\Omega - \eta, \tag{II.6.17}$$

where $k = S, D$,

and

$$\left.\begin{aligned}\Gamma_{S0} &= \frac{S(r_S)}{\pi}\Gamma_L(L, r_S, r),\\ \Gamma_{D0} &= \frac{1}{\pi}\int_{\Omega'_1}\mathrm{d}(r')\Gamma_L(L, r', r)\,\mathrm{d}\Omega'\end{aligned}\right\} \tag{II.6.18}$$

Multiplying the upper equation by $\mathrm{d}\Omega_1$ and the lower one by $\mathrm{d}\Omega_2$ and then integrating over the lower and the upper hemisphere, respectively, and also taking into account (II.1.72) gives the following instead of (II.6.17)

$$\left.\begin{aligned}\frac{\mathrm{d}H_{k1}}{\mathrm{d}L} &= -K_{01}^K + \Gamma_{11}^K + \Gamma_{21}^K + K_{k1} + K_{41}\\ \frac{\mathrm{d}H_{k2}}{\mathrm{d}L} &= K_{02}^K + \Gamma_{12}^K - \Gamma_{22}^K - K_{k2} - K_{42}\end{aligned}\right\} \tag{II.6.19}$$

where

$$K_{0m}^K = \int_{\Omega_m} Gi_{Km}\,\mathrm{d}\Omega = \int_{\Omega_m}\mathrm{d}\Omega\int_{\Omega_1}\frac{g_L}{2\pi}i_{Km}\times$$
$$\times|\cos\widehat{r_L r}|\,\mathrm{d}\Omega_L, \tag{II.6.20}$$

$$\Gamma_{ml}^K = \frac{1}{\pi}\int_{\Omega_l}\mathrm{d}\Omega\int_{\Omega'_m} i_{Km}(r')\,\mathrm{d}\Omega'\int_{\Omega_1}\frac{g_L}{2\pi}\gamma_L\times$$
$$\times|\cos\widehat{r_L r'}\cos\widehat{r_L r}|\,\mathrm{d}\Omega_L, \tag{II.6.21}$$

$$K_{Sl}=\int_{\Omega_l}\Gamma_{S0}\mathrm{d}\Omega,\qquad K_{Dl}=\int_{\Omega_l}\Gamma_{D0}\mathrm{d}\Omega \tag{II.6.22}$$

and

$$K_{4l}=\int_{\Omega_L}\eta\mathrm{d}\Omega=\int_{\Omega_L}\mathrm{d}\Omega\int_{\Omega_1}\frac{g_L}{2\pi}\eta_L|\cos\widehat{r_L r}|\mathrm{d}\Omega_L \tag{II.6.23}$$

when $l, m=1, 2$.

Consider the case of mat leaves in which γ_L and η_L are independent of direction, but are different for the upper and lower sides.

Changing the order of integration in (II.6.21), determining the values of γ_L for the cases of reflection or transmission in accordance with Table II.1.1 and taking into account (II.1.6) we obtain

$$\begin{aligned}\Gamma^K_{ml}=&\frac{T_1+R_1}{2}K^K_{m1}+\frac{T_2+R_2}{2}K^K_{m2}+\\&+(-1)^l\left(\frac{R_1-T_1}{2}\bar{K}^K_{m1}-\frac{R_2-T_2}{2}\bar{K}^K_{m2}\right),\end{aligned} \tag{II.6.21a}$$

where

$$K^K_{ml}=\int_{\Omega'_m}i_{Km}(r')G_{1l}(r')\mathrm{d}\Omega', \tag{II.6.24}$$

$$\bar{K}^K_{ml}=\int_{\Omega'_m}i_{Km}(r')\bar{G}_{1l}(r')\mathrm{d}\Omega' \tag{II.6.24a}$$

and

$$\bar{G}_{1l}(r')=\int_{\Omega_{1l}}\frac{g_L(r_L)}{2\pi}|\cos\widehat{r'r_L}\cos\vartheta_L|\mathrm{d}\Omega_L, \tag{II.6.25}$$

and $G_{1l}(r')$ is given by formulae (II.1.43) and (II.1.44). In the special case when $T_1=T_2$ and $R_1=R_2$ we obtain for

$$\Gamma^K_{ml}=\frac{T+R}{2}K^K_{0m}+(-1)^l\frac{R-T}{2}(\bar{K}^K_{m1}-\bar{K}^K_{m2}). \tag{II.6.21b}$$

Taking into account (II.6.22) and (II.6.18) and changing the order of integration we obtain the following for K_{Sl}

$$\begin{aligned}K_{Sl}=S(r_S)\Bigg\{&\frac{T_1+R_1}{2}G_{11}(r_S)+\frac{T_2+R_2}{2}G_{12}(r_S)+\\&+(-1)^l\left[\frac{R_1-T_1}{2}\bar{G}_{11}(r_S)-\frac{R_2-T_2}{2}\bar{G}_{12}(r_S)\right]\Bigg\}\end{aligned} \tag{II.6.26}$$

In analogy to (II.6.24)

$$K_{Dl}=\frac{T_1+R_1}{2}C_{D1}+\frac{T_2+R_2}{2}D_{D2}+$$

$$+(-1)^l\left[\frac{R_1-T_1}{2}\bar{C}_{D1}-\frac{R_2-T_2}{2}\bar{C}_{D2}\right], \tag{II.6.27}$$

where

$$C_{Dm}=\int_{\Omega_1} d(r')G_{1m}(r')d\Omega' \tag{II.6.28}$$

and

$$\bar{C}_{Dm}=\int_{\Omega_1} d(r')\bar{G}_{1m}(r')d\Omega'. \tag{II.6.29}$$

From (II.6.23) we obtain the following for K_{4l}

$$K_{4l}=\pi\frac{\eta_1+\eta_2}{2}+(-1)^l\pi\frac{\eta_1-\eta_2}{2}G_2, \tag{II.6.30}$$

where G_2 is given by formula (I.2.30b).

In the case of short-wave radiation $K_{4l}=0$, and in the case of long-wave radiation $K_{ml}=K_{kl}=0$.

Denoting

$$U_{0l}^K=\frac{K_{0l}^K}{H_{Kl}}=\frac{\int_{\Omega_l} i_{Kl}G d\Omega}{\int_{\Omega_l} i_{Kl}|\cos\vartheta| d\Omega} \tag{II.6.31}$$

and

$$U_{ml}^K=\frac{\Gamma_{ml}^K}{H_{Km}} \tag{II.6.32}$$

we obtain a set of equations (II.6.19) at $=0$ in the form

$$\left.\begin{aligned}\frac{dH_{K1}}{dL}&=-U_{01}^K H_{K1}+U_{11}^K H_{K1}+U_{21}^K H_{K2}+K_{K1},\\ \frac{dH_{K2}}{dL}&=U_{02}^K H_{K2}-U_{12}^K H_{K1}-U_{22}^K H_{K2}-K_{K2}.\end{aligned}\right\} \tag{II.6.33}$$

The Schwarzschild approximation given by the set (II.6.33) is realized when the quantities U_{0l}^K and U_{ml}^K are independent of the depth L in all forms of the functions i_{K1} and i_{K2} which are solutions of the set of equations (II.6.17). In this case (II.6.33) is a set of two linear differential equations with constant coefficients. Thus, the weaker the dependence of U_{ml}^K on the depth L, the closer the Schwarzschild approximation to the real situation.

The requirement on the isotropy of the complementary radiation field $i(L, r)$ is sufficient to fulfil the Schwarzschild approximation. But it is open for discussion whether it is necessary.

In the special case when $T_1 = T_2 = R_1 = R_2 = \omega_L/2$ (II.6.21a) gives $\Gamma^K_{ml} = (\omega_L/2)K^K_{0l}$, (II.6.32) gives $U^K_{ml} = (\omega_L/2)U^K_{0l}$, and (II.6.33) gives a set of equations for the complementary radiation fluxes i_S and i_D. In the case of $U^K_{0l} = 1$:

$$\left.\begin{aligned} \frac{dH_{S1}}{dL} &= -\left(1-\frac{\omega_L}{2}\right)H_{S1} + \frac{\omega_L}{2}H_{S2} + \frac{\omega_L}{2}G(r_S)S(L, r_S), \\ \frac{dH_{S2}}{dL} &= -\frac{\omega_L}{2}H_{S1} + \left(1-\frac{\omega_L}{2}\right)H_{S2} - \frac{\omega_L}{2}G(r_L)S(L, r_S), \end{aligned}\right\} \tag{II.6.34}$$

because

$$K_{S1} = K_{S2} = \frac{\omega_L}{2}G(r_S)S(L, r_S)$$

$$\left.\begin{aligned} \frac{dH_{D1}}{dL} &= -\left(1-\frac{\omega_L}{2}\right)H_{D1} + \frac{\omega_L}{2}H_{D2} + \frac{\omega_L}{2}\int_{\Omega_1} G(r)d(L, r)d\Omega, \\ \frac{dH_{D2}}{dL} &= -\frac{\omega}{2}H_{D1} + \left(1-\frac{\omega_L}{2}\right)H_{D2} - \frac{\omega_L}{2}\int_{\Omega_1} G(r)d(L, r)d\Omega, \end{aligned}\right\} \tag{II.6.35}$$

because

$$K_{D1} = K_{D2} = \frac{\omega_L}{2}\int_{\Omega_1} G(r)d(L, r)d\Omega. \tag{II.6.36}$$

The respective boundary conditions for (II.6.35) and (II.6.36) are

$$\left.\begin{aligned} &H_{S1}(0) = 0 \\ &H_{S2}(L_0) = A_B[S'(L_0, r_S) + H_{S1}(L_0)] \end{aligned}\right\} \tag{II.6.37}$$

$$\left.\begin{aligned} &H_{D1}(0) = 0 \\ &H_{D2}(L_0) = A_B[D(L_0) + H_{D1}(L_0)] \end{aligned}\right\} \tag{II.6.38}$$

The sets (II.6.34) and (II.6.35) have been solved in our works (Ross 1962; Ross & Nilson 1963) for a uniform orientation $G(r) = \frac{1}{2}$ and $\tau = \frac{1}{2}$ and in the work of Nilson (1968a) for any orientation, but at $L_0 \to \infty$.

In the case of a uniform orientation ($G = \frac{1}{2}$)

$$\left.\begin{aligned} H_{S1}(L) &= S'_0\frac{\omega_L}{2}\left[u_1(B_1e^{-\alpha L} - e^{-L/2\mu_S}) + u_2B_2e^{-L_0/2\mu_S}\right], \\ H_{S2}(L) &= S'_0\frac{\omega_L}{2}\left[\frac{u_1}{C_1}B_3e^{-\alpha L} + u_3e^{-L/2\mu_S} + \frac{u_2}{C_1}B_4e^{-L_0/2\mu_S}\right], \end{aligned}\right\} \tag{II.6.39}$$

$$\left.\begin{aligned} H_{D1}(L) &= D_0\frac{\omega_L}{4(1-\omega_L)}[u_4 + u_5B_2 - u_6B_1e^{-\alpha L}], \\ H_{D2}(L) &= D_0\frac{\omega_L}{4(1-\omega_L)}\left[u_7 + \frac{u_5}{C_1}B_4 - \frac{u_6}{C_1}B_3e^{-\alpha L}\right], \end{aligned}\right\} \tag{II.6.40}$$

where

$$\mu_S = \cos\theta_{\odot}, \qquad \alpha = \sqrt{1-\omega_L},$$

$$B_1(L, L_0) = \frac{1 - C_2 e^{-2\alpha(L_0 - L)}}{1 - C_2 e^{-2\alpha L_0}},$$

$$B_2(L, L_0) = \frac{e^{-\alpha(L_0 - L)} - e^{-\alpha(L_0 + L)}}{1 - C_2 e^{-2\alpha L_0}}$$

$$B_3(L, L_0) = \frac{C_1^2 - C_2 e^{-2\alpha(L_0 - L)}}{1 - C_2 e^{-2\alpha L_0}},$$

$$B_4(L, L_0) = \frac{e^{-\alpha(L_0 - L)} - C_1^2 e^{-\alpha(L_0 + L)}}{1 - C_2 e^{-2\alpha L_0}}$$

$$C_1 - \frac{1 - \sqrt{1-\omega_L}}{1 + \sqrt{1-\omega_L}}, \qquad C_2 = \frac{C_1(C_1 - A_B)}{1 - C_1 A_B}$$

$$u_1(\mu_S) = \frac{1 + 2\mu_S}{1 - 4(1-\omega_L)\mu_S^2},$$

$$u_2(\mu_S) = \frac{C_1}{1 - A_B C_1}\left[\frac{2A_B}{\omega_L} - \frac{1 + A_B - 2(1 - A_B)\mu_S}{1 - 4(1-\omega_L)\mu_S^2}\right],$$

$$u_3(\mu_S) = \frac{1 - 2\mu_S}{1 - 4(1-\omega_L)\mu_S^2},$$

$$u_4(L) = E_1\left(\frac{L}{2}\right) + 2E_2\left(\frac{L}{2}\right) - \frac{1-\alpha}{\alpha}F_1\left(\frac{L}{2}\right) + \\ + \frac{1+\alpha}{\alpha}E_1\left((1+2\alpha)\frac{L}{2}\right)e^{\alpha L},$$

$$u_5(L_0 - L) = \frac{C_1}{1 - A_B C_1}\left\{(1 + A_B)E_1\left(\frac{L_0 - L}{2}\right) - 2(1 - A_B)E_2\left(\frac{L_0 - L}{2}\right) + \right. \\ + 8A_B\frac{1-\omega_L}{\omega_L}E_3\left(\frac{L_0 - L}{2}\right) + \frac{1+\alpha}{\alpha}\left[(C_1 - A_B)F_1\left(\frac{L_0 - L}{2}\right) + \right. \\ \left.\left. + (1 - A_B C_1)E_1\left((1+2\alpha)\frac{L_0 - L}{2}\right)e^{-\alpha(L_0 - L)}\right]\right\},$$

$$u_6 = 2 + \frac{1+\alpha}{\alpha}\ln|1 - 2\alpha| - \frac{1-\alpha}{\alpha}\ln(1 + 2\alpha)$$

$$u_7 = -E_1\left(\frac{L}{2}\right) + 2E_2\left(\frac{L}{2}\right) - \frac{1-\alpha}{\alpha}F_1\left(\frac{L}{2}\right) + \\ + \frac{1+\alpha}{\alpha}E_1\left((1+2\alpha)\frac{L}{2}\right)e^{\alpha L},$$

$$\left.\begin{aligned} E_n(L) &= \int_1^{\infty} X^{-n} \mathrm{e}^{-Lx}\,\mathrm{d}x, \\ F_1(L) &= \int_1^{\infty} \frac{\mathrm{e}^{-Lx}}{x-2\alpha}\,\mathrm{d}x, \end{aligned}\right\} \tag{II.6.40a}$$

In the case of $L_0 \to \infty$ and an arbitrary orientation

$$\left.\begin{aligned} H_{S1}(L) &= S'_0 \frac{\omega_L}{2} u'_1 (\mathrm{e}^{-G_S L/\mu_S} - \mathrm{e}^{-\alpha L}), \\ H_{S2}(L) &= S'_0 \frac{\omega_L}{2} (u'_3\, \mathrm{e}^{-G_S L/\mu_S} - u'_1 C_1 \mathrm{e}^{-\alpha L}), \end{aligned}\right\} \tag{II.6.41}$$

where

$$G_S = G(\mu_S), \quad u'_1 = G_S \frac{\mu_S + G_S}{\alpha^2 \mu_S^2 - G_S^2}$$

and

$$u'_3 = G_S \frac{\mu_S - G_S}{\alpha^2 \mu_S^2 - G_S^2}.$$

Analogous solutions can also be obtained for H_{D1} and H_{D2}, but they contain integrals which are not integrable as elementary functions in the general case.

In a special case when the orientation is uniform ($G = \frac{1}{2}$) and $L_0 \to \infty$ (II.6.41) gives:

$$\left.\begin{aligned} H_{S1}(L) &= S'_0 \frac{\omega_L}{2} u_1 (\mathrm{e}^{-\alpha L} - \mathrm{e}^{-L/2\mu_S}) \\ H_{S2}(L) &= S'_0 \frac{\omega_L}{2} (u_1 C_1\, \mathrm{e}^{-\alpha L} + u_3\, \mathrm{e}^{-L/2\mu_S}) \end{aligned}\right\} \tag{II.6.42}$$

$$\left.\begin{aligned} H_{D1}(L) &= D_0 \frac{\omega_L}{4(1-\omega_L)} (u_4 - u_6\, \mathrm{e}^{-\alpha L}), \\ H_{D2}(L) &= D_0 \frac{\omega_L}{4(1-\omega_L)} (u_7 - u_6 C_1\, \mathrm{e}^{-\alpha L}). \end{aligned}\right\} \tag{II.6.43}$$

Special cases of transfer equation (II.1.61), (II.1.64) and (II.1.69a) have been studied by different authors. Isobe (1962a, b, c, 1969) studied a stand consisting of stems and leaves with a constant inclination angle $\vartheta_L = \text{const.}$ and a uniform distribution in azimuth. In this case the function is given by formula (I.2.22). The transfer equations for direct and diffuse radiation will coincide with equations (II.1.61) and (II.1.64), respectively, but are more complicated in form. In Isobe's first work the transfer equation lacked the term $\cos\theta$, but this error was corrected in his later works.

Formulae for the complementary radiation field were derived for single scattering at $T = R = \omega_L/2$, but in a simpler form than (II.6.44) because Γ_L were replaced with G. Some special cases of transfer equation (II.6.33) have been discussed using the Schwarzschild approximation in works of Allen. In Allen's first works on plant stands (Allen & Brown 1965, Allen & Richardson 1968) he uses the Kubelka-Munk approach and assumes $U_{11}=U_{22}=K_{K1}=K_{K2}=0$, $U_{01}=U_{02}=a+r$, $U_{21}=U_{12}=r$, where a and r are the absorption and reflection coefficients of a unit canopy layer, respectively. In Allen's subsequent work (Allen et al. 1970) the Duntley approach is used. The attenuation of incident direct solar radiation inside the stand is characterized by the coefficients of absorption a, back-scattering ω'_R and forward scattering ω'_T. In the case of a complementary radiation field the absorption a and back-scattering ω'_R are only discussed. Then instead of equations (II.1.61) and (II.6.33) we obtain the set

$$\frac{\mathrm{d}S}{\mathrm{d}L}=-(a'+\omega'_R-\omega'_T)S'$$

$$\frac{\mathrm{d}H_{S1}}{\mathrm{d}L}=-aH_{S1}-\omega_R H_{S1}+\omega_R H_{S2}+\omega_T S'$$

$$\frac{\mathrm{d}H_{S2}}{\mathrm{d}L}=aH_{S2}+\omega_R H_{S2}-\omega_R H_{S1}-\omega_R S'$$

II.6.5 APPROXIMATION FOR SINGLE SCATTERING

When single scattering is assumed, the integrands in the transfer equations for short-wave radiation ($\eta_L=0$) are eliminated, and from (II.1.69a) we obtain the following for i_{S1} and i_{S2}, respectively:

$$\left.\begin{aligned}\cos\theta\frac{\partial i_{S1}(L,r)}{\partial L}&=-G(L,r)i_{S1}(L,r)+\\ &+\frac{1}{\pi}\Gamma_L(L,r_S,r)S(L,r_S)\\ \cos\vartheta\frac{\partial i_{S2}(L,r)}{\partial L}&=G(L,r)i_{S2}(L,r)-\\ &-\frac{1}{\pi}\Gamma_L(L,r_S,r)S(L,r_S)\end{aligned}\right\}\qquad\text{(II.6.44)}$$

with the boundary conditions (II.1.70a).

The second term $(\Gamma_L/\pi)S$ in the set (II.6.44) defines an increment of radiation in the plant stand layer due to the scattering of the direct solar radiation penetrating the stand down to the given layer. This increment of radiation is concurrent with a decrease of radiation due to attenuation in the stand. This is described by the first term Gi_{SK} in the set (II.6.44). In the upper layers of the stand $(\Gamma_L/\pi)S$ is comparatively large and i_{S1} is small, therefore i_{S1} increases. In the lower layers of the stand the attenuation of radiation exceeds its increase, and as a result i_{S1} decreases.

Assuming $g_L(L, r)=g_L(r)$ and a mat ground surface, the solution of (II.6.44) gives

$$i_{S1}(L, r)=\frac{S'_0}{\pi}B_0^-(r_S, r)[a(L, r_S)-a(L, r)], \tag{II.6.45}$$

$$i_{S2}(L, r)=\frac{S'_0}{\pi}b'_S(r_S, r)[a(L, r_S)-a(L_0, r_S)a(L_0-L, r)]+$$

$$+\frac{A_B}{\pi}F_{1S}(L_0)a(L_0-L, r), \tag{II.6.46}$$

where

$$a(L, r)=\exp\left[-\frac{LG(r)}{\cos\theta}\right],$$

$$F_{1S}(L_0)=S'(L_0, r_S)+\int_{\Omega_1} i_{S1}(L_0, r)\cos\theta d\Omega,$$

$$B_0^-(r_S, r)=\frac{\Gamma_L(r_S, r)}{G(r)\cos\theta_\odot-G(r_S)\cos\theta} \tag{II.6.47}$$

and

$$b'_S(r_S, r)=\frac{\Gamma_L(r_S, r)}{G(r)\cos\theta_\odot+G(r_S)\cos\theta}. \tag{II.6.48}$$

In the special case of $r_S=r$

$$i_{S1}(L, r_S)=\frac{S'_0}{\pi}L\frac{\Gamma_L(r_S, r_S)}{\cos^2\theta_\odot}a(L, r_S), \tag{II.6.45a}$$

$$i_{S2}(L, r_S)=\frac{S'_0}{\pi}(L_0-L)\frac{\Gamma_L(r_S, r_S)}{\cos^2\theta_\odot}a(L, r_S)+$$

$$+\frac{A_B}{\pi}F_{1S}(L_0)a(L_0-L, r_S). \tag{II.6.46a}$$

When reflection from the ground surface below the stand is interpreted as an act of scattering, the single scattering approximation will treat the reflection of the complementary radiation intensitites i_S and i_D from the ground surface as a second act of scattering. It can therefore be assumed, for the sake of simplification, that

$$\frac{A_B}{\pi}F_{S1}(L_0)\approx\frac{A_B}{\pi}S'_0(L_0, r_S)=\frac{A_B}{\pi}S'_0 a(L_0, r_S)$$

and consequently,

$$i_{S2}(L, r)=\frac{S'_0}{\pi}\{b'_S(r_S, r)[a(L, r_S)-a(L_0, r_S)a(L_0-L, r)]+$$

$$+A_B a(L_0, r_S)a(L_0-L, r)\}. \tag{II.6.46b}$$

This simplification is valid, since in the PAR region the scattering coefficient of a living leaf and the reflection coefficient of the ground surface are of the same order of magnitude (unless the ground is covered with snow). The first term in expression (II.6.46b) for i_{S2} characterizes reflection from plant stand elements, the second term characterizes reflection from the ground surface. The multiplier $a(L_0-L, r)$ in the second term defines the screening effect of the layer L_0-L on the radiation reflected by the ground.

An analogous set of equations is derived for the complementary diffuse radiation field. The intensities $i_{D1}(L, r)$ and $i_{D2}(L, r)$ are found by substituting S_0' by $d_0 \cos\theta$ and r_S by r' in expressions (II.6.45) and (II.6.46), multiplying the expressions by $d\Omega'$ and integrating them over Ω, and Ω_2, respectively.

In the case of an isotropic and mat ground surface we obtain

$$i_{D1}(L, r) = \frac{D_0}{\pi^2} \cos\theta \int_{\Omega_1} B_0^-(r', r)[a(L, r') - a(L, r)] d\Omega' \qquad \text{(II.6.49)}$$

and

$$i_{D2}(L, r) = \frac{D_0}{\pi^2} \cos\theta \int_{\Omega_2} \{b_S'(r', r) a(L, r') +$$

$$+ [A_B - b_S'(r', r)] a(L_0, r') a(L_0 - L, r)\} d\Omega'. \qquad \text{(II.6.50)}$$

Because of the complexity of the functions $b_S'(r', r)$ and $B_0^-(r', r)$ expressions (II.6.49) and (II.6.50) are not integrable as elementary functions in the general case.

Equations (II.6.45), (II.6.46), (II.6.49) and (II.6.50), respectively, for i_{S1}, i_{S2}, i_{D1} and i_{D2} are initial expressions for the calculation of complementary fields of PAR in a stand.

Consider the case of a dense and high stand, $L_0 \to \infty$. Then the form of $i_{S1}(L, r)$ is unchanged, but the following expression is obtained for $i_{S2}(L, r)$ from (II.6.46)

$$i_{S2}(L, r) = \frac{S_0'}{\pi} b_S'(r_S, r) a(L, r_S). \qquad \text{(II.6.46b)}$$

At $L=0$

$$i_{S2}(0, r) = \frac{S_0'}{\pi} b_S'(r_S, r),$$

wherefrom it is seen that the function $(1/\pi) b_S'(r_S, r)$ is the brightness coefficient of the upper boundary of a dense and high stand in the case of direct solar radiation.

The downward and upward fluxes of complementary radiation i_S and i_D are obtained by integrating i_S and i_D over the upper and lower hemispheres according to formula (II.1.72). In the general case, however, i_{S1}, i_{S2}, i_{D1} and i_{D2} are not integrable over r as elementary functions. Therefore it is sometimes useful to solve transfer equations (II.6.44) using the Schwarzschild approximation. When single scattering is assumed, (II.6.35) and (II.6.36) give respective equations for determining the downward and upward fluxes H_{S1}, H_{S2}, H_{D1} and H_{D2} in the form

$$\frac{\mathrm{d}H_{Sk}}{\mathrm{d}L}=(-1)^k\left[H_{SK}-\frac{\omega}{2}G(r_S)S(L,r_S)\right], \tag{II.6.51}$$

$$\frac{\mathrm{d}H_{Dk}}{\mathrm{d}L}=(-1)^k\left[H_{DK}-\frac{\omega}{2}\int_{\Omega_1}G(r)d(L,r)\mathrm{d}\Omega\right], \tag{II.6.52}$$

where $k=1, 2$, with the boundary conditions

$$\left.\begin{array}{l}H_{S1}(0)=0\\ H_{S2}(L_0)=A_B S'(L_0,r_S),\end{array}\right\} \tag{II.6.53}$$

$$\begin{array}{l}H_{D1}(0)=0\\ H_{D2}(L_0)=A_B D(L_0)\end{array} \tag{II.6.54}$$

Solutions of these equations for direct solar radiation give

$$H_{S1}(L)=S'_0 B_1(\mathrm{e}^{-LG_S/\mu_S}-\mathrm{e}^{-L}), \tag{II.6.55}$$

$$H_{S2}L)=S'_0[A'_S\mathrm{e}^{-LG_S/\mu_S}+(A_B-A'_S)\mathrm{e}^{-(L_0-L)-L_0G_S/\mu_S}], \tag{II.6.56}$$

where

$$B_1(r_S)=\frac{\omega_L}{2}\frac{G(r_S)}{\mu_S-G(r_S)} \tag{II.6.57}$$

and

$$A'_S(r_S)=\frac{\omega_L}{2}\frac{G(r_S)}{\mu_S+G(r_S)} \tag{II.6.58}$$

and for diffuse radiation

$$H_{D1}(L)=\int_{\Omega_1}d_0(r')B_1(r')[\mathrm{e}^{-LG(r')/\mu'}-\mathrm{e}^{-L}]\mu'\mathrm{d}\Omega', \tag{II.6.59}$$

$$H_{D2}(L)=\int_{\Omega_1}d_0(r')[A'_S(r')\mathrm{e}^{-LG(r')/\mu'}-$$
$$-(A_B-A'_S)\mathrm{e}^{-(L_0-L)-[L_0G(r')/\mu']}]\mu'\mathrm{d}\Omega'. \tag{II.6.60}$$

If $L_0\to\infty$, then

$$H_{S2}(L)=S'_0A'_S(r_S)\mathrm{e}^{-LG_S/\mu_S} \tag{II.6.61}$$

and

$$\mathrm{H}_{D2}(L)=\int_{\Omega_1}d_0(r')A'_S(r')\mathrm{e}^{-LG(r')/\mu'}\mu'\mathrm{d}\Omega'. \tag{II.6.62}$$

Let us explain the physical meaning of the quantity A'_S. Since $H_{S2}(0)/S'_0=A'_S$, A'_S is the reflection coefficient (albedo) of a dense and high stand ($L_0\to\infty$) in the case of direct solar radiation.

Comparison between formulae (II.6.39) and (II.6.55), and between (II.6.40) and (II.6.60), which represent the Schwarzschild approximation in the case of multiple and single scattering, respectively, shows that multiple scattering is accounted for by the quantity $\alpha=\sqrt{1-\omega_L}$. The value $\alpha=1$ corresponds to single scattering, and formulae (II.6.39) and (II.6.40) change to formulae (II.6.55), (II.6.56) and (II.6.59), (II.6.60).

Comparison of expressions (II.6.55) and (II.6.56) for fluxes with expressions (II.6.45) and (II.6.46) for intensities shows that in the Schwarzschild approximation the conversion from intensities to fluxes is accomplished by substituting $B_0^-(r_S, r)$ by $B_1(r_S)$, b'_S by A'_S and $\exp[-LG(r)/\mu]$ by $\exp(-L)$ in expressions (II.6.45) and (II.6.46).

II.6.6 LEAF SCATTERING COEFFICIENT ω_L AND THE COMPLEMENTARY RADIATION FIELD

As was shown in Chapter II.3, the scattering coefficient of living phytoelements ω_L is specifically wavelength dependent in the short-wave spectral region. In the blue spectral region $\omega_L \approx 0.10$, in the green – 0.30, in the red – 0.10 and in the NIR – 0.75–0.95. Theoretically ω_L can range from 0 (full absorption) to 1 (full scattering). The choice of an approximate method for the solution of the transfer equation (II.1.53)

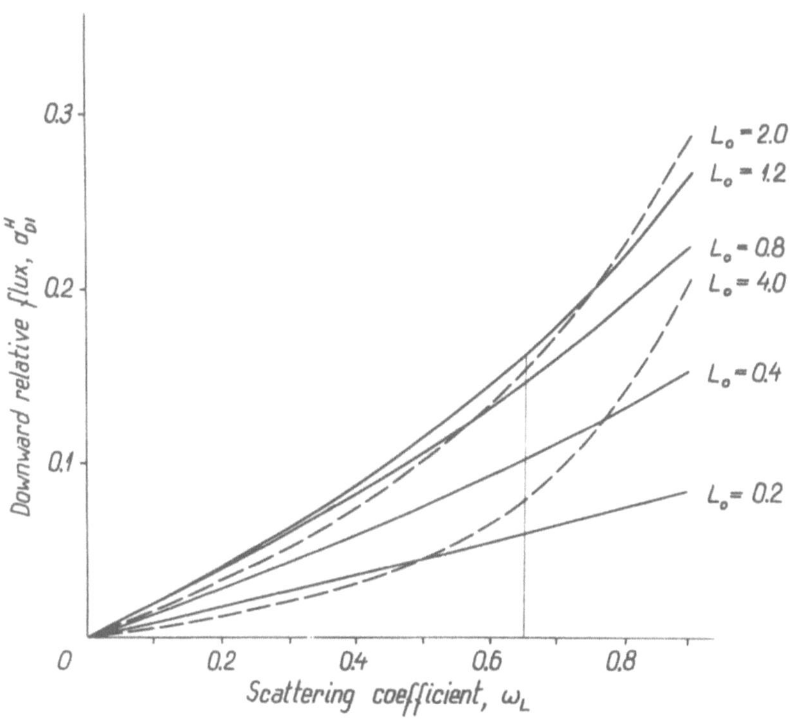

Fig. 88. Dependence of downward relative flux a_{D1}^H on leaf's scattering coefficient ω_L for stands of different leaf area indices L_0. Uniform leaf orientation, reflection coefficient (albedo) of ground surface $A_B=0.15$.

depends to a considerable degree on the value of ω_L. At small values of ω_L the single scattering approximation is sufficient, but it is not permissible at large values of ω_L.

Figures 88 and 89 show the dependence of relative fluxes $a^H_{D1}(L_0)$ and $a^H_{D2}(0) = A_D$ on ω_L for different depths of stand L_0. The calculation was made using the Schwarzschild approximation for a uniform orientation ($G = \frac{1}{2}$) at $A_B = 0.15$. The curves show that this dependence is generally nonlinear, the nonlinearity increasing with L_0 and being more prominent at large values of ω_L, and also in the case of upward fluxes. But in the PAR region, where ω_L varies from 0.10–0.40 the dependence of the complementary radiation field on ω_L can be considered as linear.

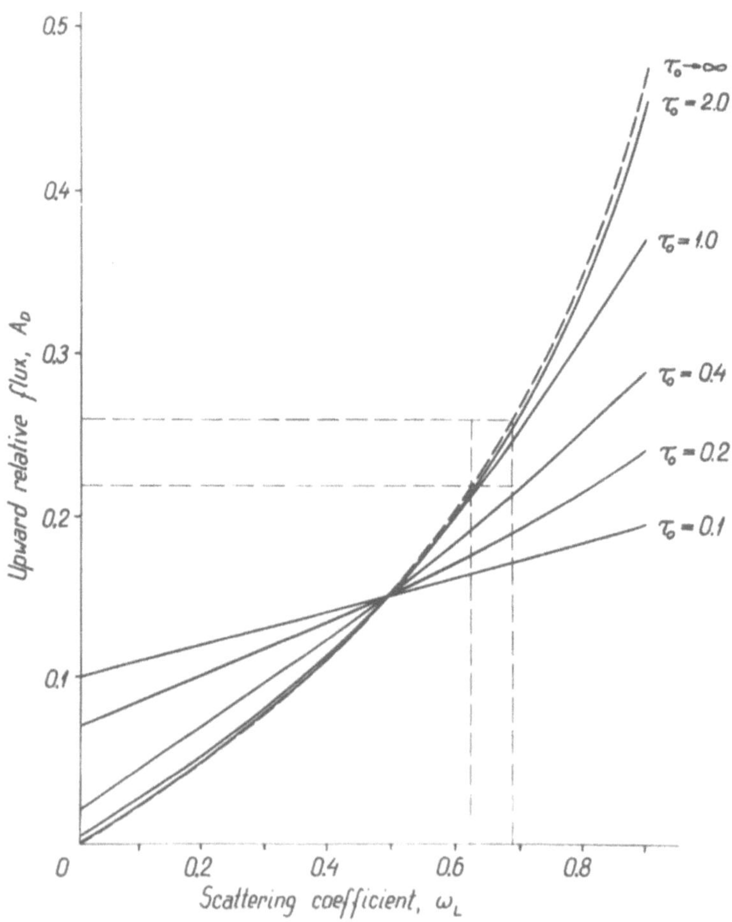

Fig. 89. Dependence of upward relative flux $a^H_{D2}(O) = A_D$ (plant stand reflection coefficient for diffuse radiation) on leaf's scattering coefficient ω_L for stands of different leaf area indices $L_0 = 2\tau_0$. Uniform leaf orientation, reflection coefficient of ground surface $A_B = 0.15$.

Formulae (II.6.45), (II.6.46), (II.6.49) and (II.6.50) allow calculating the intensities of the complementary radiation fields i_S and i_D set up in the stand by the scattering of radiation on phytoelements. The derived formulae for fluxes include (II.6.15) for horizontal leaves, (II.6.39) and (II.6.40) using the Schwarzschild approximation for a uniform orientation, (II.6.41) for an arbitrary orientation at $L_0 \to \infty$ and (II.6.55), (II.6.56), (II.6.59) and (II.6.60) using the single scattering approximation. These formulae define the complementary radiation fields $i_S(L, r)$ and $i_D(L, r)$ as dependent on the leaf's scattering coefficient ω_L, leaf orientation $(1/2\pi)g_L(r_L)$, conditions of incoming radiation $S'_0 = S_0(r_S) \sin h_\odot$ and $d_0(r)$ and on the reflection coefficient (albedo) of the ground surface A_B.

The angular dependence of the intensities $i_S(\vartheta, \varphi)$ and $i_D(\vartheta, \varphi)$ is very complex in nature and is largely determined by the scattering phase function $(1/\pi)\Gamma(r', r)$.

By way of example Figures 90 and 91 show the angular dependence of relative intensities $i_S(\vartheta)/S'_0$ in the solar vertical plane at different depths L. The upper portion of the graphs coresponds to downward intensities, and the lower to upward ones. These graphs, as well as the analysis of the formulae, allow drawing some general conclusions on the angular dependence of the field $i_S(\vartheta)$.

1. The angular dependences of upward intensities $i_{S2}(\vartheta)$ and downward intensities $i_{S1}(\theta)$ differ considerably, particularly in the upper layers of the stand. As the depth L increases these differences decrease.

2. The angular dependence of $i_S(\vartheta)$ is stronger in the upper layers of the stand. The contrasts decrease with the increasing depth.

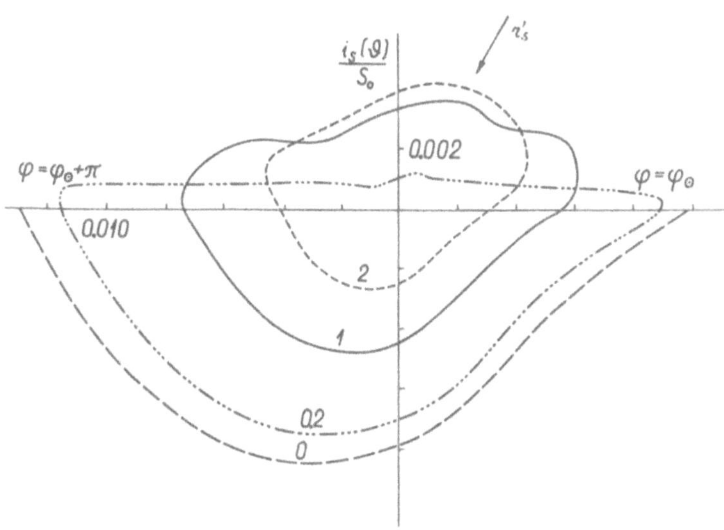

Fig. 90. Angular distribution of relative PAR intensities $i_S(\vartheta)/S_0$ in the solar vertical plane at depths $L=0$; 0.2; 1.0 and 2.0 within a stand. First order scattering, uniformly oriented leaves, $h_\odot = 60°$, $R_L = 0.09$, $T_L = 0.06$.

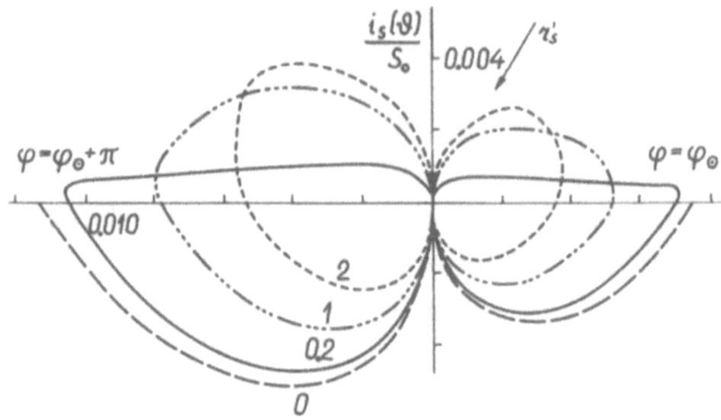

Fig. 91. Angular distribution of relative PAR intensities $i_S(\vartheta)/S_0$ in the solar vertical plane at depths $L=0; 0.2; 1.0$ and 2.0 within a stand. First order scattering, vertical leaves, $h_\odot=60°$, $R_L=0.09$, $T_L=0.06$.

3. The angular dependence is largely due to the orientation of leaves: it is nonexistent for horizontal mat leaves, but is most strong in the case of vertical leaves. Then the intensities are largest in horizontal directions. In the case of a uniform leaf orientation the intensities in horizontal directions are predominant.
4. In the case of a uniform leaf orientation the angular dependence of $i_S(\vartheta)$ at $L=0.2 \div 0.4$ is qualitatively similar to that of the mean sky brightness $d_0(\theta)$ in the solar circle plane. This is due to the fact that in both cases we have planoparallel layers of approximately the same optical depth, which scatter radiation following a law somewhat different from the Rayleigh scattering, and the first order-scattering is predominant in both cases. The elongation of $i_{S1}(\theta)$ in a direction close to $\theta=\pi/2$ is generally characteristic of primary scattering, and the asymmetry of $i_S(\vartheta)$ in Figures 90 and 91 is due to the nonspherical shape of the scattering phase function $(1/\pi)\Gamma(r', r)$.

The situation is more complicated in the case of nonmat leaves with a specular reflection component. Figure 92 shows the angular dependence of the complementary PAR field expressed in relative units of i_S/S_0 for birch leaves. The calculations were made from formulae (II.6.45) and (II.6.46), while the scattering phase function $(1/\pi)\Gamma(\gamma)$ corresponds to curve 1 in Figure 87.

As was shown under II.6.3, the complementary field i_S in the case of horizontal mat leaves must not depend on the direction r. In the case of nonmat leaves the specular reflection component disturbs the isotropy of the field i_S: maximum intensities are concentrated on the sunny side, the maximum downward intensities are located in directions between the zenith and the solar beam direction, and the maximum upward intensities in directions close to the angle of specular reflection. As shown by an analysis of formulae (II.6.45a) and (II.6.46c), this asymmetry of the field i_S does not change with depth. Comparison between Figure 90 and Figure 92 shows that the presence of specular reflection of PAR by birch leaves distorts the field i_S in comparison with mat leaves (uniform orientation of leaves). Maximum intensities in the horizontal direction remain, but they are extended still more towards the sunny side. The complementary ma-

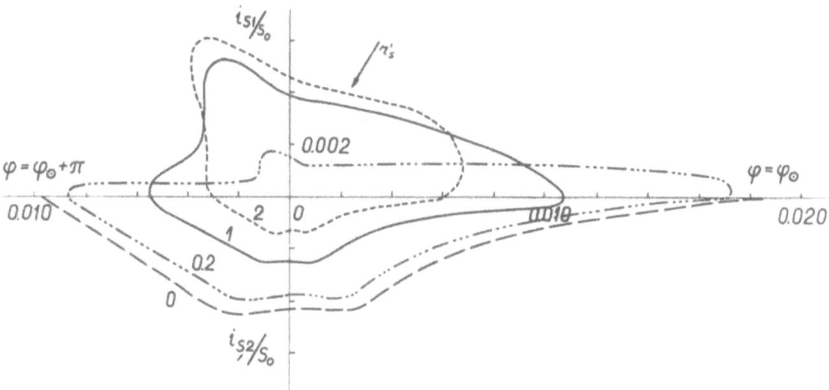

Fig. 92. Angular distribution of relative PAR intensities $i_S(\vartheta)/S_0$ in the solar vertical plane at depths $=0$; 0.2; 1.0 and 2.0 within a stand. First order scattering, uniformly oriented birch leaves, $h_\odot=60°$, $L_0\to\infty$.

ximum in the direction r_S disappears and another maximum appears in the direction of specular reflection. The upward intensities have two complementary maxima: in the direction r_S and in the direction of specular reflection.

The vertical profile of the complementary radiation field is defined by a combination of exponents of the type $e^{-GL/\mu}$, e^{-G_SL/μ_S}, $e^{-\alpha L}$ and e^{-L}. For example, in the case of $L_0\to\infty$ the dependence of H_{S1} on L in formula (II.6.41) is determined by the difference $e^{-G_SL/\mu_S}-e^{-\alpha L}$. According to the boundary condition (II.6.37), $H_{S1}(0)=0$. It is seen from (II.6.41) that $\lim_{L\to\infty} H_{S1}=0$. Consequently, $H_{S1}(L)$ first increases with L, reaches its maximum value at a certain depth L^*, and then decreases with the further increasing L. The value of L^* corresponding to H_{S1} is determined from conditions $\partial H_{S1}/\partial L=0$ and in the case of (II.6.41) will be

$$L^*=\frac{\ln\dfrac{G(h_\odot)}{\alpha\sin h_\odot}}{\dfrac{G(h_\odot)}{\sin h_\odot}-\alpha}. \tag{II.6.63}$$

It is seen from (II.6.63) that the depth corresponding to the maximum downward flux $H_{S1}(L)$ depends both on the geometrical characteristics $h_\odot$ and $G(h_\odot)$ and on the optical properties of leaves characterized by the values of $\alpha=\sqrt{1-\omega_L}$.

Table II.6.1 gives L^* values calculated from formula (II.6.63) for different leaf orientations.

Table II.6.1 allows drawing some conclusions as to the vertical profile of the flux $H_{S1}(L)$.

1. As scattering of radiation by the leaves increases, max H_{S1} moves to deeper layers of the stand, but this displacement is comparatively small.

Table II.6.1 *Location L^* of the maximum of downward flux $H_{S1}(L)$ inside plant stand for different leaf orientations and for solar elevations* 10° *and* 80°

ω_L	$\alpha=\sqrt{1-\omega_L}$	Leaf orientation				
		horizontal	uniform		vertical	
			10°	80°	10°	80°
0.1	0.95	1.0	0.55	1.4	0.50	2.5
0.5	0.7	1.2	0.65	1.6	0.55	3.1
0.75	0.5	1.4	0.75	2.0	0.65	3.9
0.91	0.3	1.7	0.90	2.5	0.75	5.2

2. As solar elevation increases, max H_{S1} also moves to deeper layers in the stand. In this case the dependence of L^* on $h_{\odot}$ is stronger than that of L^* on ω_L.

3. At high solar elevations max H_{S1} is deepest in the stand in the case of vertical leaves and is least deep in the case of horizontal leaves.

4. The situation is reversed at low solar elevations: max H_{S1} is deepest in the case of horizontal leaves.

5. At moderate solar elevations max H_{S1} for PAR is at the depth $L^* \approx 1$; and in the case of NIR, at the depth $L^* \approx 2$. This indicates the presence of the Forbes effect in the stand: the complementary field i_S is increasingly spread in the stand due to the increasing scattering coefficient ω_L. Since ω_L is spectrum-dependent, the spectral composition of the complementary field i_S varies with depth.

As shown further on, these conclusions are proved by concrete calculations.

The dependence of H_{S2} on L is different. When $A_B = A'_S$ in expression (II.6.56), $H_{S2}(L)$ decreases exponentially with depth in accordance with the term e^{-LG_S/μ_S}. When $A_B > A'_S$, i.e. the reflection coefficient of the ground surface is more than that of a dense stand, the second term in (II.6.56) increases exponentially in accordance with e^L. Then $H_{S2}(L)/S'_0$ first decreases, reaches its minimum at

$$L = \frac{\ln \dfrac{\mu_S}{G_S}\left(\dfrac{A_B}{A'_S} - 1\right)}{1 + \dfrac{G_S}{\mu_S}} L_0$$

and then increases to A_B at $L = L_0$. In the case of $A_B < A'_S$ $H_{S2}(L)$ decreases continuously with depth.

The flux H_{S1} of the complementary PAR field is only 1–2% of the direct solar radiation flux over a horizontal surface above the stand. The flux H_{S2} increases due to the decreasing depth L_0 and at $L=1$ reaches 6% of S'_0. The increase in radiation due to the multiple scattering in the PAR is small in its absolute value, not more than 0.5% of S'_0. Its relative contribution, however, is larger and reaches a maximum of 30% of H_S.

The results of calculations for determining the complementary radiation fields i_S and i_D from the single scattering formulae can be improved by increasing the scattering coefficient ω_L. This expedient will somewhat compensate for not taking into account the multiple scattering and the Forbes effect. In the case of PAR, for example, $\omega_L = 0.15$ can be changed to $\omega_L = 0.18 \div 0.20$.

Figures 93 and 94 (Ross & Nilson 1963) show the vertical profiles respectively of the downward a_{S1}^H, a_{D1}^H and upward a_{S2}^H, a_{D2}^H relative fluxes at specified solar elevations in the case of a uniform orientation. The calculations were performed by formulae (II.6.42) and (II.6.43) at $L_0 \to \infty$ and $\omega_L = 0.65$. The value $\omega_L = 0.65$ corresponds to the effective leaf's scattering coefficient for integral short-wave radiation. It is seen from Figure 93 that max $a_{S1}^H(L)$ is displaced from $L = 0.6$ at $h_\odot = 10°$ to deeper layers due to the increasing solar elevation, and at $h_\odot = 50°$ it is at $L = 1.5$. At the same time the maximum value of the relative flux a_{S1}^H decreases from 0.25 to 0.15. It is seen from Figure 94 that the lower the Sun, the more rapid the decrease in $a_{S2}^H(L)$ with depth.

Thus, the complementary radiation field i_S at low solar elevations is concentrated in the uppermost layers. As the solar elevation increases the com-

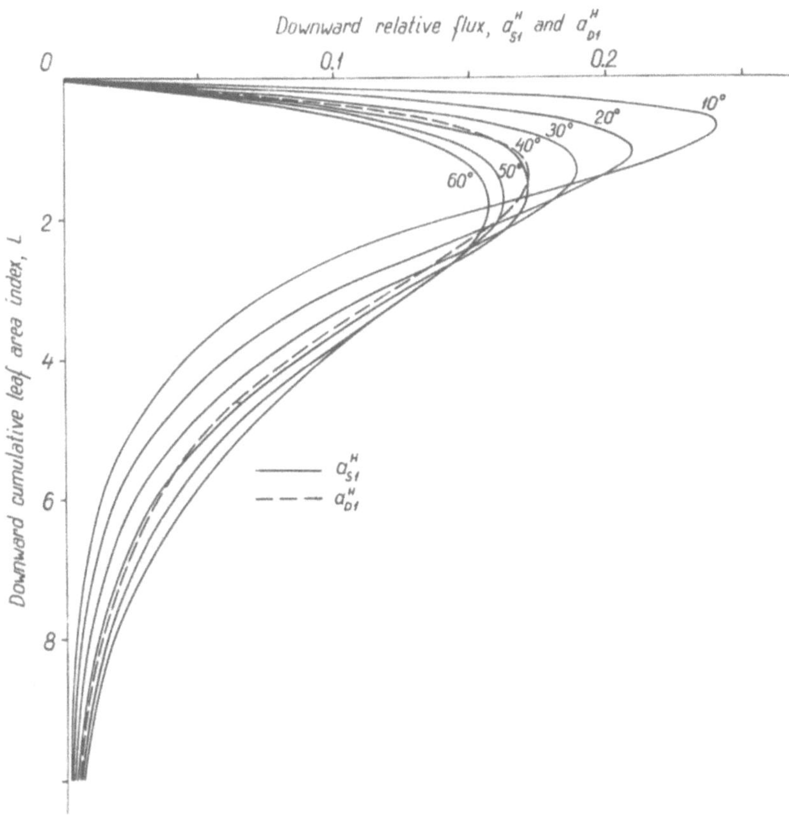

Fig. 93. Vertical profiles of downward relative fluxes $a_{S1}^H(L)$ and $a_{D1}^H(L)$ for different solar elevations. High, dense stand with uniformly oriented leaves, $L_0 \to \infty$, $\omega_L = 0.65$.

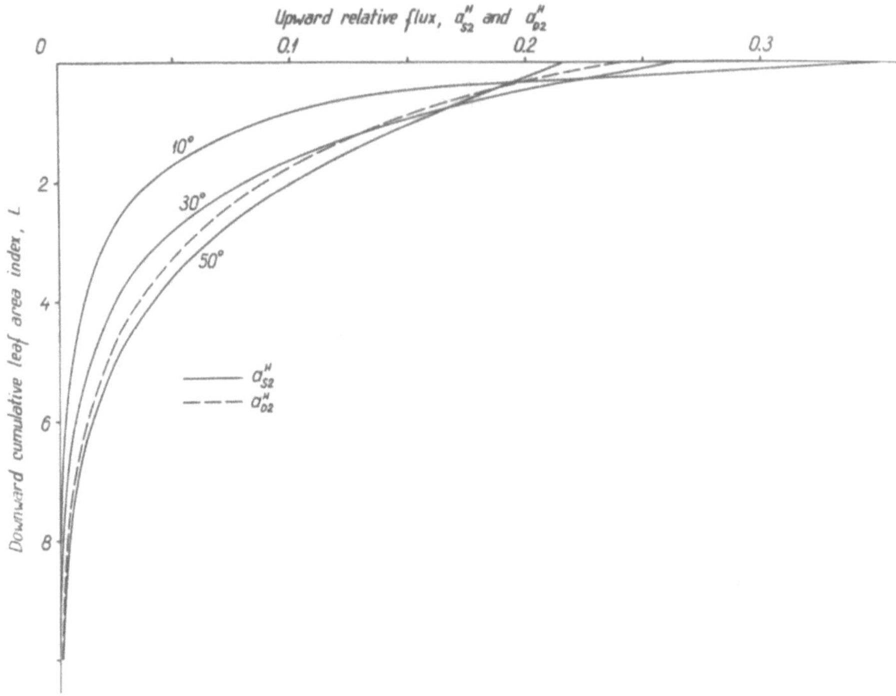

Fig. 94. Vertical profiles of upward relative fluxes $a^H_{S2}(L)$ and $a^H_{D2}(L)$ for solar elevations 10°, 30° and 50°. High, dense stand with uniformly oriented leaves, $L_0 \to \infty$, $\omega_L = 0.65$.

plementary radiation is distributed more uniformly over the individual horizontal layers. The contribution of complementary radiation in the short-wave region exceeds that of PAR by a factor of 5–10 and reaches as high as 20–25% of S'_0.

The fluxes $a^H_{D1}(L)$ and $a^H_{D2}(L)$ are similar in nature to the fluxes a^H_{S1} and a^H_{S2} and correspond to an effective solar elevation of about 40°. The character of the dependence $a^H_{S1}(h_\odot)$ and $a^H_{S2}(h_\odot)$ changes with depth. The upward flux $a^H_{S2}(h_\odot)$ at $L=0$ (reflection coefficient $A_S(h_\odot)$) decreases from 0.09 at $h_\odot = 0°$ to 0.045 at $h_\odot = 90°$. But even at $L=0.2$ $a^H_{S2}=0$ if $h_\odot = 0$. This is due to the screening effect of the overlying layers. It is interesting to note that starting from $h_\odot > 30°$ the dependences $a^H_{S1}(h_\odot)$ and $a^H_{S2}(h_\odot)$ are very weak. This allows a considerable simplification of the calculation of the hourly totals of radiation fluxes inside the stand.

II.6.8 CALCULATION OF THE COMPLEMENTARY NIR FIELD

Calculation of the complementary NIR field is complicated because of the necessity to take into account the multiple scattering on the plant stand elements. The transfer equation (II.1.69a) has an exact solution only in the special case when the leaves are horizontal. The Schwarzschild approximation applies to fluxes and does not show the angular dependence of intensities, and the single scattering approximation is altogether inapplicable in this case.

Generally speaking, solution of the transfer equation for NIR can be performed using common computer calculation methods.

The simple iteration method (Marchuk 1961; Sobolev 1956; Tamme & Sõrmus 1967) is applicable for stands of any geometrical structure. It takes into account all orders of radiation scattering successively starting with the first. In this case the terms under the integral sign in equation (II.1.69a) are substituted by sums by means of cubature formulae, and the integral differential equation is reduced to a set of linear differential equations. The work is laborious for large L_0 because the convergence in the iteration process is weak.

The factorization method developed by Rybicki (1965) provides a stable flow chart and converts an integral differential equation to a matrix equation of Riccati. This method does not take into account the complementary radiation reflected from the ground surface and in the general case is even difficult for computer calculation.

The approximate methods are simplified if calculations are performed for a stand consisting of uniformly oriented leaves, i.e. if $G(r)=\frac{1}{2}$ and $\Gamma(r', r)=\Gamma(\gamma)$. In this case the modified transfer equation reduces to a classical equation which can be solved by a number of approximate methods used in astrophysics and atmospheric physics and based on the expansion of $\Gamma(\gamma)$ as a series in Legendre polynomials.

In the NIR region the scattering phase function $\Gamma(\gamma)$ can be approximated as a Rayleigh type scattering phase function $\Gamma(\cos\gamma)=1+5g_2P_2(\cos\gamma)$, where the second Legendre polynomial $P_2(\cos\gamma)=\frac{1}{2}(3\cos^2\gamma-1)$. For NIR $g_2=0.0646$ according to the data of Figure 87. The transfer equation (II.1.68a) for such a special case has been solved approximately by Nilson (1972) using the method of spherical harmonics, and intensities of the complementary field i_S have been obtained. The approximate calculation of the vertical NIR fluxes was performed by formulae (II.6.39) and (II.6.41) corresponding to the Schwarzschild approximation.

Figure 95 (Nilson 1968a) characterizes the dependence of the geometrical structure of the stand on the vertical fluxes of the complementary field a_{S1}^H and a_{S2}^H in NIR. The calculations are made by formula (II.6.41) for two specified solar elevations: 15° and 60° at $A_B=0.15$, $\omega_L=0.85$ and $L_0\to\infty$.

Maximum downward fluxes H_{S1} at $h_\odot=60°$ occur in the case of horizontal leaves; minimum fluxes, in the case of vertical leaves; at $h_\odot=15°$ the situation is reversed. For upward fluxes there exists some neutral depth L_H, where fluxes are independent of leaf orientation. At $h_\odot=60°$, $L_H=1.3$, at $h_\odot=15°$, $L_H=0.4$.

Figure 96 shows the angular dependence of the relative intensities i_S/S_0' in the solar vertical plane at the depth $L=0$, 1, 2 and 4 in the stand with uniformly oriented leaves at $h_\odot=60°$ and $L_0\to\infty$. A substantial variation of the angular dependence of intensities with depth L is characteristic of NIR. In the upper layers of the stand the intensities are maximum in the horizontal direction, but the near vertical intensities gain in importance with the increasing depth.

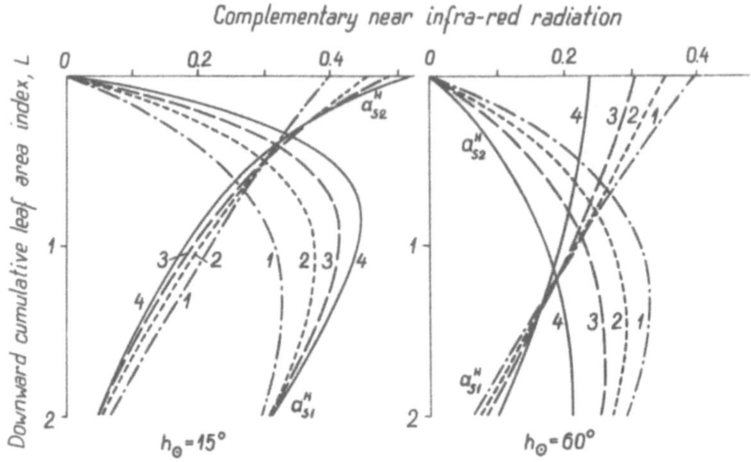

Fig. 95. Vertical profiles of downward $a_{S1}^H(L)$ and upward $a_{S2}^H(L)$ relative complementary fluxes of NIR at solar elevations 15° and 60° for different leaf orientations. $L_0=2$, $\omega_L=0.85$; 1 – horizontal leaves, 2 – sunflower, 3 – uniform orientation of leaves, 4 – vertical leaves.

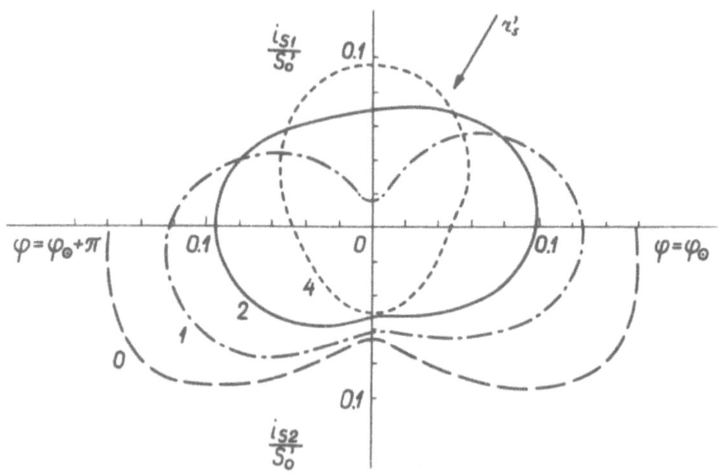

Fig. 96. Angular distribution of relative intensities of NIR $i_S(\vartheta)/S_0'$ in solar vertical plane at depths $L=0$, 1, 2 and 4 within a stand with uniformly oriented leaves. $h_\odot=60°$, $L_0\to\infty$.

Comparison between Figures 96 and 90 shows that the NIR field is more symmetrical about the solar circle than the PAR field, and that the angular dependence of the PAR field is less variable with height. Comparison between Figures 92 and 96 shows that the fluxes of NIR exceed those of PAR in the stand by almost an order of magnitude. Due to the increasing leaf's scattering coefficient ω_L max $a_{S1}^H(L)$ passes to deeper layers, and in the NIR region (ω_L $=0.85$) is at about $L=1$.

The results discussed indicate that the complementary radiation field $i(L, r)$ of NIR differs considerably from the PAR field both in value and in the angular dependence of intensities. This is caused by the different values of the leaf's scattering coefficient ω_L, leaf orientation and the structure of the solar and diffuse sky radiation incident on the stand.

II.7 Total radiation field in plant stands

II.7.1 GENERAL

The foregoing chapters discussed the attenuation with depth and the angular dependence of individual components of the total radiation field, and corresponding calculation formulae were derived. The present chapter deals with the total radiation field and gives calculation formulae for intensities, irradiances and vertical fluxes. Since the individual components of total radiation differ in importance, various approximations can be used depending on the situation, the accuracy of which is estimated at the end of this chapter. The difference in importance of the multiple scattering of PAR and NIR also necessitates the use of different techniques and approximations in calculation.

II.7.2 INTENSITIES OF TOTAL RADIATION FIELD

The intensities of the total downward and upward radiation field inside the stand are defined by formulae (II.1.76) and (II.1.77), respectively. The following general conclusions as to the total radiation field can be drawn from what was said about the quantities $S(L, r_S)$, $d(L, r)$, $i_S(L, r)$ and $i_D(L, r)$ in the previous chapters.

1. The solar beam direction r_S is most important for the total radiation field. The intensity in this direction is several orders of magnitude higher than intensity in other directions, which is expressed by the term $S(L, r_S)\delta(r-r_S)$ in the formulae.

2. Since i_{S1} and i_{S2} values are of the same order, the total downward intensities $i_{Q1}(L, r)$ of PAR exceed the upward intensities $i_{Q2}(L, r)$ by more than an order of magnitude. The difference is less in the NIR region.

3. In the presence of direct solar radiation, the complementary field $i_D(L, r)$ is one or two orders less than the other terms in formulae (II.1.76) and (II.1.77) and can be neglected. On a cloudy day, however, must be taken into account.

4. Since in the case of horizontal mat leaves the complementary fields i_S and i_D as also the penetration function a_S are independent of direction, the spatial

structure of intensities inside the stand is determined by that of the incident radiation.

5. For vertical leaves $d(L, r)$ has its maximum values in near zenith directions (see Figures 79B and 81), but $i_{S1}(L, r)$ is nearly equal to zero (see Figure 90). Therefore, their angular dependences are mutually compensated, and the angular dependence of the total field is less pronounced.

6. In real crops the near zenith intensities of the total field become predominant with the increasing depth L. This is due to the low value of the leaf scattering coefficient and the beam pathway length increasing in proportion to sec θ.

7. The total radiation field has a discontinuity at the interface between atmosphere and vegetation ($L=0, \theta=\pi/2$). In the case of PAR, $i_{Q1}(0, \pi/2) > i_{Q2}(0, \pi/2)$; in the case of NIR $i_{Q1}(0, \pi/2) < i_{Q2}(0, \pi/2)$.

By way of example Figure 97 shows the profile of the total field of PAR intensities at the depth $L=0.2$ (a) and 1.0 (b) and specified solar elevations. The calculations were made from formulae (II.1.76) and (II.1.77), with $i_S(L, r)$ calculated assuming single scattering from formulae (II.6.45) and (II.6.46). As seen from this profile, the total PAR field is characterized by a rapid decrease in intensities when passing to nearly horizontal and upward directions. The asymmetry of the field also extends to deeper layers, it is not existent when the Sun is in the zenith but increases with the decreasing $h_\odot$.

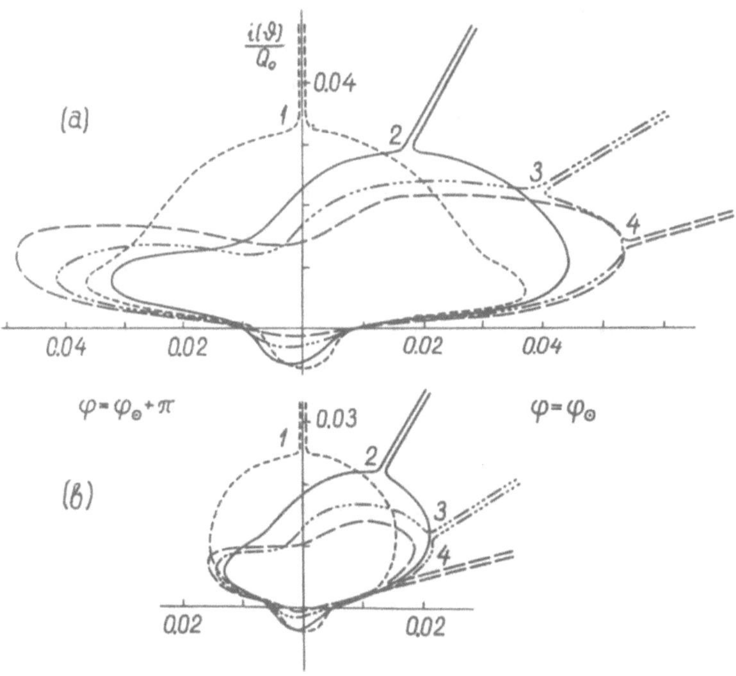

Fig. 97. Angular distribution of total PAR field in units of total incident flux in solar vertical plane at different solar elevations. First order scattering, uniformly oriented leaves, $L_0 \to \infty$, $Q_0=1$, $S_0/D_0=6.5$.

(a) $L=0.2$; (b) $L=1.0$; 1 – $h_\odot=90°$, 2 – $h_\odot=60°$, 3 – $h_\odot=30°$, 4 – $h_\odot=15°$.

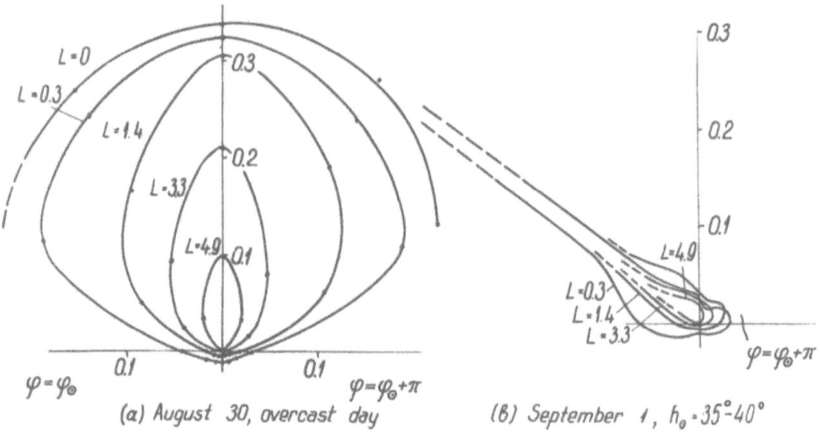

Fig. 98. Angular distribution of total field of monochromatic intensities in a plane perpendicular to rows, at different depths L in a maize stand, in units of incident flux ($Q_0=1$). Experimental data of Laisk (1965b), experiment No. 6, Jõgeva, 1963, $\lambda=558\ \mu$m.

The angular distribution of intensities is different in the NIR region. In upper layers the near horizontal intensities are several times higher than the near vertical ones, the intensities from the lower hemisphere exceed those from the higher hemisphere. In lower layers the intensities are at a minimum in near zenith directions, and the upward intensities are practically isotropic.

The NIR intensities are five-ten times higher than the PAR intensities, the difference being particularly great in the nearly horizontal and upward directions and at great depths in the stand.

The angular distribution of the total field of intensities was determined experimentally by Laisk (1965b) in a maize stand at the tassel flowering stage (Figure 98). The measurements were performed with the Laisk spectrophotometer (Laisk 1964). It is seen from the graph that below a cloudy sky the total field is practically symmetrical about the vertical, with the near zenith intensities predominant. The lateral and upward intensities decrease rapidly with the increasing depth L.

Comparison between the experimental and calculated data shows that in the upper layer of the stand the measured values of intensity i_S in the direction of specular reflection of the solar beam exceed calculated values. This is due to the specular reflection effect of leaves in the underlying layers, which was not taken into account in the calculation model.

II.7.3 TOTAL VERTICAL FLUXES

The total downward and upward irradiances inside a plant stand are defined by formulae (II.1.80) and (II.1.81). In the case of PAR, where single scattering may be assumed for calculating the complementary radiation fields i_S and i_D, formulae

(II.1.80) and (II.1.81) assume the following form when (II.4.3), (II.5.2), (II.6.55), (II.6.56), (II.6.59) and (II.6.60) are taken into account:

(a) below a clear sky

$$Q(L) = S'_0\, \mathrm{e}^{-G_S L/\sin h_\odot} + D_0 \int_0^{\pi/2} D'_{0z}(\theta') e^{-GL/\cos\theta'}\, \mathrm{d}\theta' +$$

$$+ S'_0 \frac{\omega_L}{2} \frac{G_S}{\sin h_\odot - G_S} (\mathrm{e}^{-G_S L/\sin h_\odot} - \mathrm{e}^{-L}), \tag{II.7.1}$$

$$F_2(L) = H_{S2}(L) = S'_0 [A'_S\, \mathrm{e}^{-G_S L/\sin h_\odot} +$$

$$+ (A_B - A'_S) \mathrm{e}^{-(L_0 - L) - (G_S L/\sin h_\odot)}], \tag{II.7.2}$$

(b) below a cloudy sky

$$Q^C(L) = D_0 \int_0^{\pi/2} D'_{0z}(\theta') \Bigg[\mathrm{e}^{-GL/\cos\theta'} +$$

$$+ \frac{\omega_L}{2} \frac{G}{\cos\theta' - G} (\mathrm{e}^{-GL/\cos\theta'} - \mathrm{e}^{-L} \Bigg] \mathrm{d}\theta', \tag{II.7.3}$$

$$F_2(L) = H_{D2}(L) = D_0 \int_0^{\pi/2} D'_{0z}(\theta') [A'_S(\theta')\, \mathrm{e}^{-GL/\cos\theta'} +$$

$$+ (A_B - A'_S(\theta')) \mathrm{e}^{-(L_0 - L)G/\cos\theta'}]\, \mathrm{d}\theta', \tag{II.7.4}$$

where according to (II.6.58), $A'_S(\theta') = (\omega/2) G_S(\theta')/(\cos\theta' + G_S(\theta'))$ is the reflection coefficient (albedo) of a dense and high stand for direct solar radiation.

Taking into account (II.1.80) the penetration function of total solar radiation is expressed

$$a_Q(L) = \frac{Q_L}{Q_0} = \frac{(a_S + a^H_{S1}) \dfrac{S'_0}{D_0} + a_D + a^H_{D1}}{1 + \dfrac{S'_0}{D_0}}. \tag{II.7.5}$$

The penetration function of total solar radiation on a cloudy day,

$$a^C_Q = a_D + a^H_{D1}.$$

From the above formulae we obtain special cases for horizontal ($G_S = \sin h_\odot$) and vertical ($G_S = (2/\pi)\cos h_\odot$) leaves and also for uniformly oriented leaves $G_S = \frac{1}{2}$. Since the special case of horizontal leaves also allows an exact solution of the transfer equation, it is advisable to use formulae (II.6.15) instead of single scattering formulae (II.6.55) and (II.6.56). In this case, taking into account (II.4.6) and (II.5.8) we obtain the following expressions for vertical fluxes below a uniformly bright sky:

$$Q(L)=Q_0\frac{\left(A_B-\dfrac{1}{A'}\right)e^{-aL}-(A_B-A')e^{-a(2L_0-L)}}{A_B-\dfrac{1}{A'}-(A_B-A')e^{-2L_0a}}, \qquad \text{(II.7.6)}$$

$$F_2(L)=Q_0\frac{\left(1-\dfrac{A_B}{A'}\right)e^{-a(2L_0-L)}-(1-A_BA')e^{-aL}}{A_B-\dfrac{1}{A'}-(A_B-A')e^{-2aL_0}}, \qquad \text{(II.7.7)}$$

the reflection coefficient (albedo) is expressed by

$$A_Q=\frac{F_2(L)}{Q(L)}=\frac{\left(1-\dfrac{A_B}{A'}\right)e^{-a(2L_0-L)}-(1-A_BA')e^{-aL}}{(A'-A_B)e^{-a(2L_0-L)}-\left(\dfrac{1}{A'}-A_B\right)e^{-aL}}, \qquad \text{(II.7.8)}$$

where $a=[(1-T_L)^2-R_L^2]^{\frac{1}{2}}$ and $A'=(1-T_L-a)/R_L$.

Formulae for a cloudy day are obtained by changing from Q_0 to D_0. Formulae (II.7.1)–(II.7.7) are basic expressions for the calculation of total vertical PAR fluxes in stands and define the dependence of Q and F_2 on the following parameters: S_0, the irradiance of direct solar radiation above the stand; $h_\odot$, the solar elevation; D_0, the irradiance of diffuse sky radiation above the stand; $D'_{0z}(\theta')$, the distribution of diffuse sky radiation over the sky; A_B, the reflection coefficient (albedo) of the ground surface below the stand; L_0, the depth of stand; $G(\theta')$, the characteristic of foliage orientation; $\omega_L=T_L+R_L$, the leaf scattering coefficient for PAR.

The following simplifying assumptions were made when deriving these formulae:

(1) the stand is horizontally uniform and consists of leaves only;
(2) the leaves are randomly distributed in the stand;
(3) the leaves are mat, with the same optical properties on both sides;
(4) there is no penumbra effect in the stand;
(5) orientation of leaves is the same in all layers.

Let us proceed to discussing the results of calculation to find out the basic regularities of the variation of vertical fluxes.

Figure 99 shows vertical profiles of the relative downward fluxes in the blue (a), green (b), red (c) and near infrared (d) spectral regions. In the blue and red, $\omega_L=0.1$; in the green, 0.35; and in the near infrared, 0.85. The calculations were performed for $h_\odot=60°$ and the leaf orientation corresponding to sunflower (G-function is shown in Figure 32, curve 2). According to Figure 99 the complementary fluxes H_{S1} and H_{D1} in the strong attenuation regions (blue and red) amount but to a few per cent of the total flux, and only the penetrated radiation $S'(L)$ and $D(L)$ need be taken into account for approximate calculation. In the region of moderate scattering (green) the fluxes $D(L)$ and $H_{S1}(L)$ at middle depths

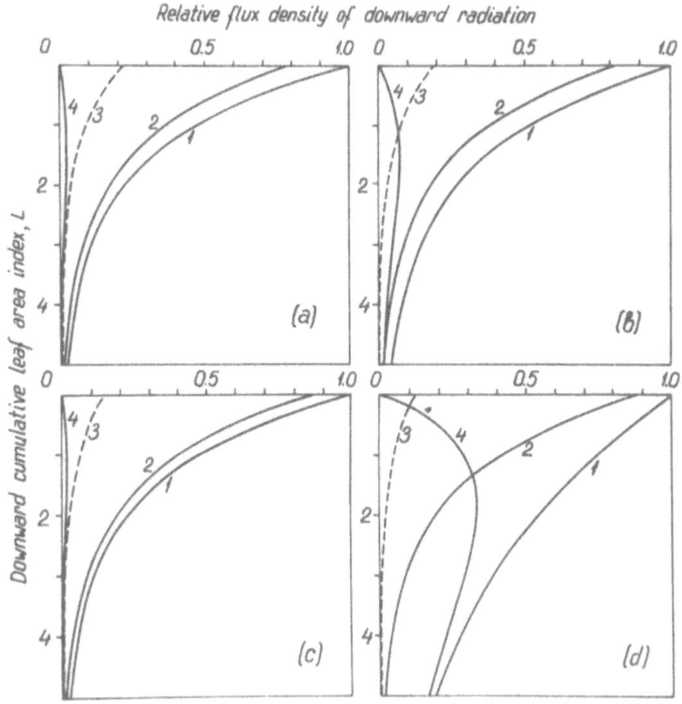

Fig. 99. Calculated vertical profiles of relative total radiation flux and its components in the blue (a), green (b), red (c) and near infrared (d) spectral regions.

1 – Q/Q_0, 2 – S'/Q_0, 3 – D/Q_0, 4 – H_{S1}/Q_0.

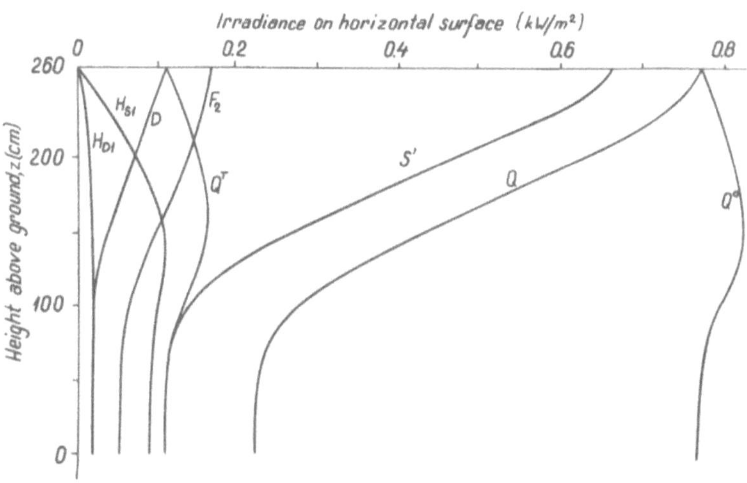

Fig. 100. Calculated vertical profiles of various components of short-wave radiation flux in a maize stand. $S_0 = 835$, $D_0 = 104$, $h_\odot = 50°$, $\omega_L = 0.65$, $L_0 \to \infty$.

are comparable; at great depths $H_{S1}(L)$ exceeds D. In the region of strong scattering (NIR) the role of H_{S1} and H_{D1} is substantial and becomes predominant at depths $L>3$. As scattering increases, the decrease of total downward flux with depth becomes less rapid.

Figure 100 illustrates the variation of vertical fluxes of short-wave radiation with the height z as calculated for a milk ripe maize crop in Moldavia (experiment No. 11). It is seen that variation of the fluxes Q, S', D, H_{S1} and H_{D1} with the height z is practically linear in the upper 150-cm layer where most attenuation occurs. The total downward short-wave radiation flux in gaps Q° and in shaded areas Q^T first increases in the upper layers, reaches its maximum at a 100-cm depth, and then decreases. This is determined by the vertical profile of the complementary flux H_{S1}. In the lower layers the total downward short-wave radiation flux in gaps exceeds the flux in shaded areas by a factor of 9. These data coincide rather well with the experimental data of Bonhomme et al. (1971) obtained in a maize stand in France.

Figure 101 shows the variation of vertical fluxes with depth separately for PAR and NIR. It is clearly seen that the role of NIR increases consistently with the depth of stand. While in the upper layers the total NIR flux Q is 1.4 times larger than the mean flux of PAR, at the depth $L=5$ the ratio is five to one. The flux Q^T of PAR decreases monotonically with depth; the flux Q^T of NIR first increases reaching its maximum at about $L\approx 2$ and then decreases with the increasing L. Fluxes Q^T and Q° differ considerably. At the depth $L=5$ the fluxes in gaps (Q°) are 50 times as large as those in shaded areas (Q^T) in the case of PAR, and only 5 times in the case of NIR.

It is interesting to note that the total relative downward flux Q^0/Q_0 in sunny gaps in upper layers of the stand exceeds unity in the NIR region and is less than unity in the PAR region. This difference, which is due to complementary scatter-

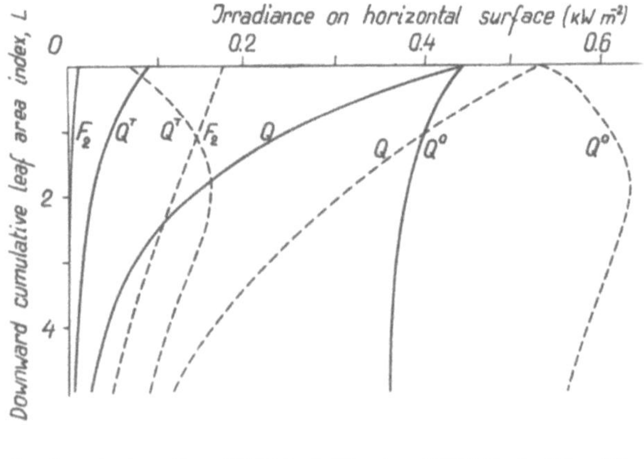

Fig. 101. Calculated vertical profiles of PAR (solid lines) and NIR (dashed lines) fluxes within a stand with uniformly oriented. $S_0=805$, $D_0=147$, $h_\odot=60°$, $C_S=0.43$, $C_D=0.60$, $\omega_L=0.15$ for PAR and $\omega_L=0.85$ for NIR.

Table II.7.1 *Depth L reached by 10% of incident PAR flux*

Orientation	$h_\odot$ 15	30	45	60
Horizontal	2.2	2.2	2.2	2.2
Vertical	1.5	2.5	3.9	5.2
Uniform	1.6	2.5	3.3	4.1
Maize	1.7	2.5	3.2	3.5
Sunflower	1.8	2.6	2.9	3.1

Table II.7.2 *Relative fluxes of direct radiation a_S, diffuse sky radiation a_D and complementary field from direct radiation downward a_{S1}^H and upward a_{S2}^H for specified values of ω_L=0.10, 0.25 and 0.85 and different plant stand structures at $h_\odot$ = 15° and 60°*

L	a_S	a_D	ω_L=0.10		ω_L=0.25		ω_L=0.85	
			a_{S1}^H	a_{S2}^H	a_{S1}^H	a_{S2}^H	a_{S1}^H	a_{S2}^H
Horizontal orientation								
0	1.000	1.000	0.000	0.026	0.000	0.072	0.000	0.442
0.4	0.670	0.670	14	18	37	51	186	378
1.0	368	368	19	10	53	30	311	300
1.5	223	223	18	6	50	20	336	247
2.0	135	135	15	4	42	13	326	204
3.0	50	50	8	2	24	5	263	138
5.0	7	7	2	0	6	1	138	64
Vertical orientation								
$h_\odot$ = 15°								
0	1.000	1.000	0.000	0.046	0.000	0.098	0.000	0.528
0.4	0.387	0.617	32	19	66	43	292	333
1.0	92	384	31	5	67	14	364	210
1.5	28	282	23	2	50	6	330	161
2.0	9	216	15	1	34	3	281	129
3.0	1	119	6	0	15	1	194	86
5.0	0	73	1	0	3	0	90	40
$h_\odot$ = 60°								
0	1.000	1.000	0.000	0.014	0.000	0.040	0.000	0.296
0.4	0.863	0.718	6	12	16	36	92	298
1.0	692	502	10	10	28	30	185	287
1.5	576	393	11	8	31	25	230	272
2.0	479	316	11	7	31	21	253	254
3.0	332	218	9	5	26	15	262	215
5.0	159	123	5	2	15	7	205	137

Table II.7.2 (*Continued*)

L	a_S	a_D	$\omega_L=0.10$		$\omega_L=0.25$		$\omega_L=0.85$	
			a_{S1}^H	a_{S2}^H	a_{S1}^H	a^{HS2}	a_{S1}^H	a_{S2}^H
Uniform orientation								
$h_\odot=15°$								
0	1.000	1.000	0.000	0.034	0.000	0.092	0.000	0.511
0.4	0.462	0.623	22	16	58	47	266	353
1.0	145	360	24	6	65	18	359	233
1.5	55	240	19	2	52	9	340	178
2.0	21	164	13	1	37	5	296	142
3.0	3	80	5	0	17	2	209	94
5.0	0	22	1	0	3	0	97	43
$h_\odot=60°$								
0	1.000	1.000	0.000	0.019	0.000	0.054	0.000	0.366
0.4	0.794	0.695	9	16	24	44	140	353
1.0	566	438	14	11	38	33	236	311
1.5	419	307	14	8	40	25	297	285
2.0	314	218	13	6	38	20	315	254
3.0	177	113	10	4	28	12	287	192
5.0	56	33	4	1	12	4	189	104
Sunflower								
$h_\odot=15°$								
0	1.000	1.000	0.000	0.032	0.000	0.085	0.000	0.488
0.4	0.545	0.632	19	18	50	25	235	370
1.0	219	345	23	8	62	23	347	260
1.5	103	220	19	4	52	12	345	202
2.0	48	140	14	2	40	7	312	161
3.0	11	59	6	1	20	2	228	106
5.0	0	11	1	0	4	0	108	48
$h_\odot=60°$								
0	1.000	1.000	0.000	0.023	0.000	0.063	0.000	0.407
0.4	0.736	0.680	11	17	30	48	159	370
1.0	464	399	17	11	46	32	282	314
1.5	316	260	16	8	46	23	320	270
2.0	215	170	14	5	41	16	323	230
3.0	100	75	9	3	27	8	280	164
5.0	21	15	3	1	9	2	161	80

ing inside the stand, has also been revealed in the work of Varlet-Grancher and Bonhomme (1971), which presents spectral curves of the relative fluxes $Q_\lambda^0/Q_{0\lambda}$ at various heights in a maize crop in France.

For a better estimation of the effect of geometrical structure on a_Q, Table II.7.1 gives the values of depth L reached by 10% of the incident flux of PAR. It is seen from this table that in the range of solar elevations 30–40° the penetration of total downward PAR flux is practically independent of the orientation of leaves. It is largest at low solar elevations in the case of horizontal leaves, and at great elevations in the case of vertical leaves.

The presence of snow below the stand (a coniferous forest in winter) increases the total downward flux $a_D + a_{D1}^H$ by 10–20%.

Table II.7.2 presents the relative fluxes a_S, a_D, a_{S1}^H and a_{S2}^H. When the conditions of incident radiation are known, the fluxes of PAR and NIR as also of the shortwave radiation at different depths in the stand can be calculated from this Table.

II.7.4 ANGULAR DISTRIBUTION OF TOTAL RADIATION FLUX

As regards the radiation regime in a foliage canopy the field of irradiances determined from formulae (II.1.78) and (II.1.79) is a suitable characteristic of the total radiation field. In comparison with the field of intensities the field of irradiances $I(L, r_L)$ is smoothed and has neither an abrupt jump in the direction of the Sun, nor a discontinuity plane at $L=0$ and $\theta=\pi/2$.

In the general case the dependence of I_1 and I_2 on the zenith angle is complicated in character. But assuming that the complementary fields i_S and i_D and also the diffuse sky radiation are isotropic, then in accordance with (II.1.4d), (II.1.78) and (II.1.79) we obtain the following expressions for the irradiances of the total field

$$I_1(L, r_L) = S(L, r_S)\,|\cos \widehat{r_L r_S}| + D(L)\frac{1+\cos \vartheta_L}{2} + \\ + \frac{H_1(L)+H_2(L)}{2} + \frac{H_1(L)-H_2(L)}{2}\cos \vartheta_L \tag{II.7.9}$$

$$I_2(L, r_L) = D(L)\frac{1-\cos \vartheta_L}{2} + \frac{H_1(L)+H_2(L)}{2} - \\ - \frac{H_1(L)-H_2(L)}{2}\cos \vartheta_L. \tag{II.7.10}$$

In these formulae H_1 and H_2 denote the sums $H_{S1}+H_{D1}$ and $H_{S2}+H_{D2}$. Formulae (II.7.9) and (II.7.10) are exact for a stand with horizontal leaves and give the Schwarzschild approximation for a stand of any geometrical structure.

On a sunny day the role of H_{D1} and H_{D2} is slight in comparison with the other components and can be neglected. On a cloudy day

$$I_1(L, r_L) = \frac{D(L) + H_{D1}(L)}{2}(1 + \cos\vartheta_L) + $$

$$+ H_{D2}(L)\frac{1 - \cos\vartheta_L}{2}, \qquad \text{(II.7.11)}$$

$$I_2(L, r_L) = \frac{D(L) + H_{D1}(L)}{2}(1 - \cos\vartheta_L) + $$

$$+ H_{D2}(L)\frac{1 + \cos\vartheta_L}{2}. \qquad \text{(II.7.12)}$$

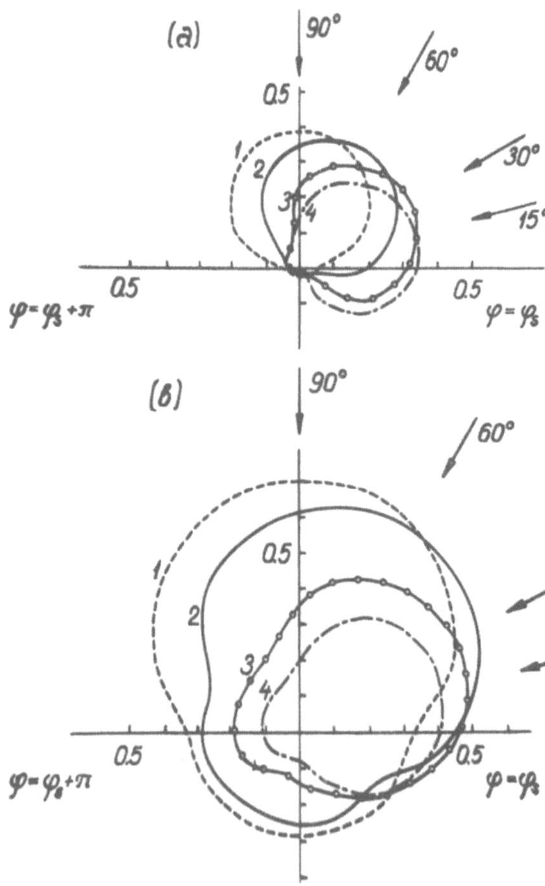

Fig. 102. Calculated angular distribution of irradiance of total PAR (a) and NIR (b) in solar vertical plane for different solar elevations at depth $L=1$. Uniformly oriented leaves, $Q_0=1$ and $L_0 \to \infty$. $T_L=0.06$ and $R_L=0.09$ for PAR; $T_L=0.50$ and $R_L=0.35$ for NIR;
$1 - h_\odot=90°$, $2 - h_\odot=60°$, $3 - h_\odot=30°$, $4 - h_\odot=15°$.

If on a sunny day the radiation field in a plant stand is divided into gaps and shaded areas, it is advisable to use formulae (II.7.9) and (II.7.10) substituting $S(L, r_S)$ by $S_0(r_S)$ in the case of gaps, and formulae (II.7.11) and (II.7.12) substituting H_{D1} and H_{D2} by H_{S1} and H_{S2}, respectively, in the case of shaded areas.

The above formulae indicate that the Schwarzschild approximation for the radiation field allows considerable simplifications for the calculation of irradiances.

The angular dependence of the irradiance $I(L, \vartheta_L)$ of PAR and NIR as calculated for horizontal leaves from formulae (II.7.9) and (II.7.10) is presented in our work (Ross & Nilson 1967). Formulae (II.6.15) and (II.6.16) have been used for $H_1 = \pi i_1$ and $H_2 = \pi i_2$.

Figure 102 shows the angular dependence of the irradiance of PAR (a) and NIR (b) in the solar vertical plane at the depth $L = 1$ for different solar elevations. It will be seen that the direction of the Sun r_S is the determining factor for the angular dependence of irradiances. Also, the angular dependence of irradiances changes but little in its general character when passing from the uppermost level of the stand to the depth $L = 1$. Figure 103 shows experimental data (Laisk 1965b) of the angular dependence of the irradiances of PAR on a cloudy (a) and a sunny day (b) for a maize crop at the tassel flowering stage obtained with the Laisk spectrophotometer (Laisk 1964).

It will be seen from Figure 102 that in the presence of sunlight the hodograph of irradiances of PAR resembles a spherical body of rotation whose axis is nearly coincident with the direction of sunrays. The sphericity is due to the predominance of the first term $S(L, r_S)\,|\cos r_L r_S|$ in expression (II.7.9). In the case of NIR

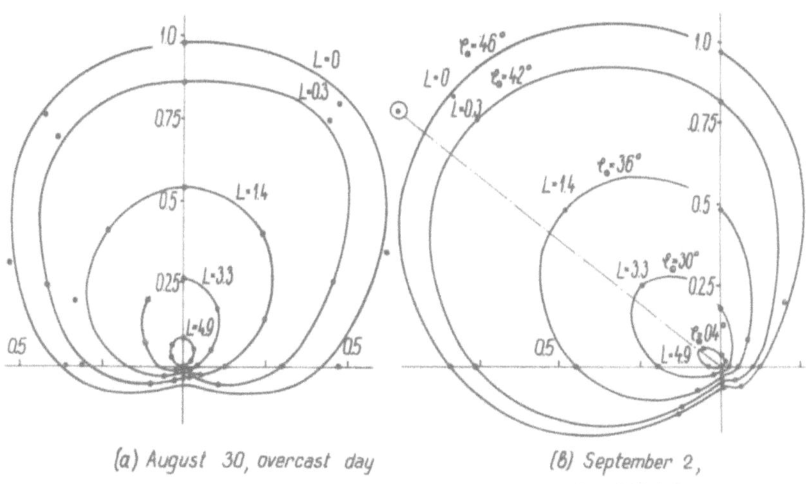

Fig. 103. Angular distribution of irradiance of total PAR in a plane perpendicular to rows at different depths L in a maize crop in units of incident flux ($Q_0 = 1$). Experimental data of Laisk (1965b), experiment No. 6, Jõgeva, 1963.

$\varphi_\odot$ – Sun's azimuth relative to measurement plane.

(Figure 102B) the hodograph is more complicated in shape because of the increasing role of the complementary field i_S.

The effect of foliage area orientation on the hodograph of irradiances is small in upper layers, but it increases with depth. As depth increases the symmetry axis of the hodograph is increasingly deflected from the direction r_S towards the vertical in connection with the increasing role of near-zenith intensities. On a cloudy day and in shaded areas inside the stand on a clear day the symmetry axis of the hodograph coincides with the vertical direction.

The great eccentricity of the hodograph of irradiances relative to the zero point is explained by the fact that the downward fluxes exceed the upward ones by 0.5–1.5 orders of magnitude.

It can be stated in conclusion that the angular dependence of irradiances inside a stand is determined mainly by the spatial structure of incident radiation and is

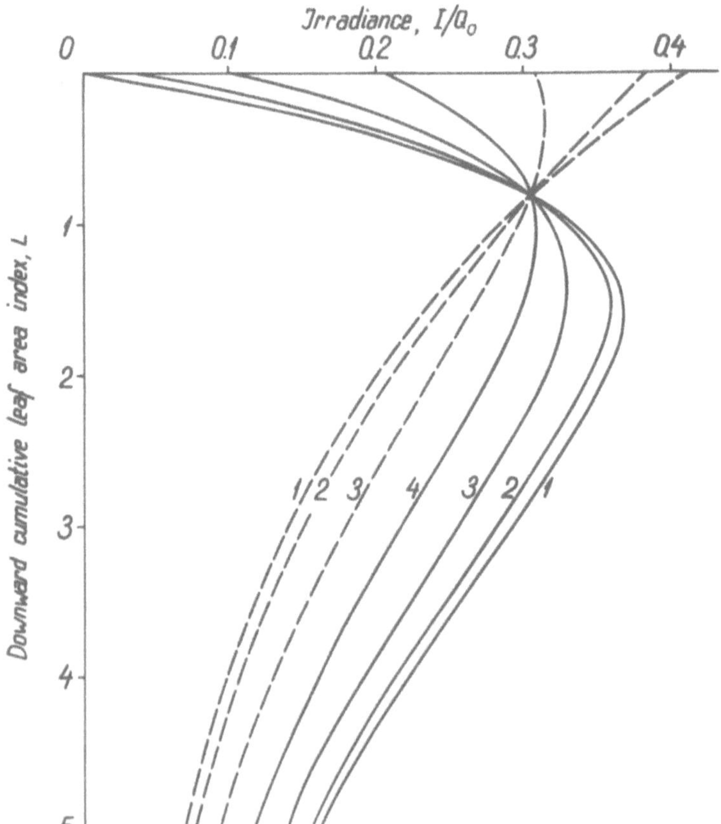

Fig. 104. Calculated vertical profiles of irriadiances $I_1(L_1\vartheta)$ (solid lines) and $I_2(L,\vartheta)$ (dashed lines) of complementary PAR at different angles ϑ in a stand with horizontal leaves. $Q_0=1$, $T_L=0.06$, $R_L=0.09$. $L_0\rightarrow\infty$.

$1-\vartheta=0°$, $2-\vartheta=30°$, $3-\vartheta=60°$, $4-\vartheta=90°$.

less dependent on the geometrical structure of the stand and the optical properties of leaves.

The vertical profiles of the irradiances $I_1(L)$ and $I_2(L)$ of the complementary field at different angles ϑ are shown in Figure 104 for the PAR region. The calculation was performed from formulae (II.7.8) and (II.7.9). It is seen from Figure 104 that $I_2(L)$ decreases monotonically with depth, while $I_1(L)$ reaches its maximum at $L \approx 1$. As the inclination angle increases the maximum shifts to higher layers. There is a peculiar isotropy zone at a depth of $L \approx 0.8$ where the irradiances $I_1 = I_2$ and do not depend on the inclination angle ϑ.

On the whole, an analogous situation exists in the NIR region, where the absolute values of irradiance are 10–15 times in excess of those in the PAR region. The maximum of downward irradiances $I_1(L)$ is shifted to deeper layers and is at $L \approx 1.5$.

II.7.5 LEAF ABSORPTION IN TOTAL RADIATION FIELD

In the general case the leaf absorption function in the total radiation field $\Pi_L(L, r_L)$ is defined by formulae (II.1.88)–(II.1.93).

For approximate calculations, when the radiation field is specified by the vertical fluxes and the diffuse sky radiation is isotropic, the total of irradiances $I_1(L, r_L) + I_2(L, r_L)$, with (II.7.9), (II.7.10) and (II.1.84) taken into account, is equal to

$$I_1(L, r_L) + I_2(L, r_L) = S(L, r_L)\,|\cos \widehat{r_L r_S}| + D(L) + H_{S1}(L) + H_{S2}(L) + H_{D1}(L) + H_{D2}(L) = S(L, r_L)\,|\cos \widehat{r_L r_S}| + Q^T(L) + F_2(L). \quad \text{(II.7.13)}$$

and the absorption function, with (II.1.88) taken into account, is equal to

$$\Pi_L(L, r_L) = \kappa(S\,|\cos \widehat{r_L r_S}| + Q^T + F_2) = \kappa(Q + F_2). \quad \text{(II.7.14)}$$

On a clear day the leaf absorption function of total solar radiation in gaps is

$$\Pi_L^0(L, r_L) = \kappa[S_0\,|\cos \widehat{r_L r_S}| + Q^T(L) + F_2(L)], \quad \text{(II.7.15)}$$

and the leaf absorption function of total solar radiation in shade is

$$\Pi_L^T(L) = \kappa[Q^T(L) + F_2(L)]. \quad \text{(II.7.16)}$$

The leaf absorption function of total solar radiation on a cloudy day is

$$\Pi_L^C(L) = \kappa[D(L) + H_{D1}(L) + H_{D2}(L)]. \quad \text{(II.7.17)}$$

It is clear from the above expressions that leaf orientation must be taken into account in approximate calculations of the absorption of radiation by a leaf in the presence of direct solar radiation, i.e. in gaps. The leaf absorption function Π_L

is isotropic in shade and on a cloudy day. This is due to the total of irradiances I_1+I_2 being independent of the leaf orientation r_L, though the irradiances I_1 and I_2 are angle dependent as defined by formulae (II.7.8) and (II.7.9).

II.7.6 RADIATION IN A PLANT STAND WITH HORIZONTAL LEAVES

Among all special cases of leaf orientation a stand with horizontal leaves is most interesting from the standpoint of theory, since in this case the transfer equation has an exact solution which coincides with the Schwarzschild approximation. This special case allows a comparatively simple, analytical estimation of the approximation methods. Though the estimates thus obtained are numerically true for horizontal leaves alone, they can be applied for a qualitative analysis of other plant stand structures.

As was shown in II.6.3, the complementary radiation field $i(L, r_L)$ is isotropic in the case of horizontal mat leaves. Consequently, the angular dependence of the total radiation field is determined by the spatial structure of the total radiation incident on the stand. If diffuse sky radiation is assumed to be isotropic, the intensities of the total field inside the stand are the same in all directions, except that of the Sun. Regularities of the attenuation of direct and diffuse sky radiation $S(L)$ and $d(L)$ are also similar.

The expressions for vertical fluxes are given by formulae (II.7.6) and (II.7.7). Let us analyze some special cases.

1. *A dense and high stand, $L_0 \to \infty$.* From (II.7.5)–(II.7.7) we obtain

$$\left.\begin{aligned} Q(L) &= Q_0 \exp(-aL), \\ F_2(L) &= Q_0 A' \exp(-aL), \end{aligned}\right\} \qquad \text{(II.7.18)}$$

and

$$A_Q = A' = \frac{R_L}{1-T_L+a} = 1 - \frac{1-\omega_L}{R_L}\left[\left(1+\frac{2R_L}{1-\omega_L}\right)^{\frac{1}{2}} - 1\right]. \qquad \text{(II.7.19)}$$

It is seen from these formulae that in a dense and high stand with horizontal leaves both the downward (Q) and upward (F_2) fluxes, as well as direct (S) and diffuse (D) radiation fluxes decrease with depth exponentially. The multiplier $kG_S/\sin h_\odot$ in the exponent, which in the case of direct solar radiation characterizes the geometrical structure of the stand and the path length of the beam, is in this case equal to unity, but the multiplier in the exponent of total flux is equal to $a=[(1-T_L)^2-R_L^2]^{\frac{1}{2}}$. In this case a does not characterize the geometrical structure, but depends on the optical parameters of the leaf. In the case of pure absorption $R_L=T_L=0$ and $a=1$, while in the case of pure scattering $R_L+T_L=1$ and $a=0$. Consequently, the coefficient a characterizes the degree of absorption of radiation by the leaf and can be called *the generalized absorption coefficient in a layer of horizontal leaves of unit depth.* The absorption coefficient is generally understood to mean $\kappa=1-T_L-R_L$; κ and a are related as $a=[\kappa(\kappa+2R_L)]^{\frac{1}{2}}$. Consequently, a and κ coincide in the absence of reflection when $R_L=0$.

The physical sense of the parameter

$$A' = \frac{1-T_L-a}{R_L} = \frac{1-T_L}{R_L}\left\{1-\left[1-\left(\frac{R_L}{1-T_L}\right)^2\right]^{\frac{1}{2}}\right\}$$

is obvious from the expression for the albedo. It is the albedo of a dense and high stand. As the regularities of the variation of radiation with depth are the same for direct and diffuse radiation, $A_Q = A_S = A_D = A'$ and is independent of depth L in this case. Since the upward intensities do not depend on the direction, the stand reflects as a mat surface. Consequently, in this case the albedo of the stand does not depend on the conditions of plant stand illumination, corresponds to a mat surface and does not change with depth. It can be shown that the case of $L_0 \to \infty$ is equivalent to the case of $A' - A_{B'}$.

It should be noted that with a constant scattering coefficient $\omega_L = R_L + T_L$ the coefficient a can vary depending on various combinations of R_L and T_L. If $R_L = 0$, then $A' = 0$ and $a = 1 - \omega_L$ at any T_L. If $R_L = T_L = \omega_L/2$, then

$$A' = \frac{1-\frac{\omega_L}{2}-(1-\omega_L)^{\frac{1}{2}}}{\omega_L/2}$$

and $a = (1-\omega_L)^{\frac{1}{2}}$. If $R_L = \omega_L$ and $T_L = 0$, then $A' = [1-(1-\omega_L^2)^{\frac{1}{2}}]/\omega_L$ and $a = (1-\omega_L^2)^{\frac{1}{2}}$. It follows from these expressions that when $\omega_L = \text{const}$, the albedo increases with the leaf reflection coefficient R_L.

2. *A comparatively dense stand with*

$$A' \frac{A' - A_B}{1 - A_B A'} \exp(-2aL_0) \ll 1.$$

Then from (II.7.6) and (II.7.7) we obtain

$$\left.\begin{aligned} Q(L) &= Q'(L) + \Delta Q(L), \\ F_2(L) &= F_2'(L) + \Delta F_2(L), \end{aligned}\right\} \qquad \text{(II.7.20)}$$

when

$$\left.\begin{aligned} \Delta Q(L) &= Q_0 A' \frac{A_B - A'}{1 - A' A_B} e^{aL} - e^{-aL}) e^{-2aL_0}, \\ \Delta F_2(L) &= Q_0 \frac{A_B - A'}{1 - A' A_B} (e^{aL} - A'^2 e^{-aL}) e^{-2aL_0}. \end{aligned}\right\} \qquad \text{(II.7.21)}$$

Here Q' and F_2' are the fluxes for $L_0 \to \infty$, while ΔQ and ΔF_2 are corrections to the solution for finite L_0. The signs of corrections are determined by that of the difference $A_B = A'$, i.e. the difference between albedo of the ground surface and of the dense stand. Since A' ranges from 0 to 0.4, and the ground surface albedo A_B from 0.1 to 0.25, the corrections can have both positive and negative values. In the PAR the corrections are positive, with $\Delta Q \approx 0$ and $\Delta F_2 \approx A_B Q_0$.

3. *Strongly absorbing leaves*: T_L and R_L values are small. In this case single scattering may be assumed in the transfer equation, and the terms containing

integrals discarded from equation (II.6.11). Then instead of the set (II.6.11) we obtain two usual differential equations, whose solution, with the boundary conditions (II.6.12) taken into account, gives

$$\left.\begin{aligned} H_1(L)&=Q_0T_Le^{-L},\\ Q(L)&=Q_0(1+T_LL)e^{-L},\\ F_2(L)&=Q_0\left[\frac{R_L}{2}(e^{-L}-e^{-2L_0+L_0})+A_B(1+T_LL)e^{-2L_0+L}\right] \end{aligned}\right\} \tag{II.7.22}$$

It is clear from the expression for $H_1(L)$ that in the case of single scattering the downward intensities are only due to transmission of radiation by leaves and are proportional both to the transmission coefficient T_L and to the downward cumulative leaf area index L. The term e^{-L} in the expression for $H_1(L)$ accounts for the attenuation of once penetrated radiation due to shading by the overlying leaves.

4. *Pure scattering:* $\omega_L=R_L+T_L=1$. Substituting $1-T_L=R_L$ in the set of equations (II.6.11a) and taking into account (II.6.12) we obtain the following solutions:

$$\left.\begin{aligned} H_1(L)&=Q_0\left[\frac{1+R_L(1-A_B)(L_0-L)}{1+R_L(1-A_B)L_0}-e^{-L}\right],\\ Q(L)&=Q_0\left[1-\frac{R_L(1-A_B)L}{1+R_L(1-A_B)L_0}\right],\\ F_2(L)&=Q_0\frac{A_B+R_L(1-A_B)(L_0-L)}{1+R_L(1-A_B)L_0}. \end{aligned}\right\} \tag{II.7.23}$$

In this case the total vertical fluxes are linear functions of L_0-L, i.e. the depth of vegetable layer between the soil and the given level L. In the given case the variation of fluxes is only due to the absorption of radiation by the soil. When $L_0\to\infty$, $Q(L)=F_1(L)=F_2(L)=Q_0$.

II.7.7 ERRORS OF THE APPROXIMATE METHODS OF CALCULATION

A proper calculation of the complementary integral radiation field in the regions of PAR, NIR and short-wave radiation necessitates, first, the determination of monochromatic intensities $i_\lambda(L,r)$ by solving the transfer equation (II.1.53) and, second, the integration of intensities over wavelengths in the corresponding spectral range. The dependence of intensities i_λ on λ is expressed in the formulae by the spectral scattering coefficient of the leaf $\omega_\lambda=T_\lambda+R_\lambda$ and by the spectral composition of the radiative fluxes $S_{0\lambda}$ and $D_{0\lambda}$ incident on the stand.

Since calculation using the proper method is very laborious work, it is usually replaced by an approximate method which contains three simplifications.

1. Instead of integrating over wavelengths a certain effective scattering coefficient $\omega_E = T_E + R_E$ is used, which is determined from the condition

$$i(\omega_E) = \int_{\lambda_1}^{\lambda_2} i_\lambda(\omega_\lambda)\mathrm{d}\lambda. \qquad \text{(II.7.24)}$$

Since, however, the monochromatic intensities depend not only on λ but also on the depth L and direction r_L, the effective coefficient ω_E is, strictly speaking, also dependent on them. The larger the variation of ω_λ in the interval $\lambda_2 - \lambda_1$, the stronger the dependence of ω_E on the depth L and direction r_L.

2. Multiple scattering is not taken into account. The lower the scattering coefficient ω_E, the more valid the simplification.

3. The finite depth of the stand is not taken into account ($L_0 \to \infty$). This assumption simplifies the calculation formulae for intensities and fluxes considerably and is valid in the case of a dense stand.

In atmospheric optics, the dependence of integral optical properties of the atmosphere on its optical depth at an invariable physical state of the atmosphere is known as the Forbes effect: as the optical depth increases the transmission coefficient also increases and the radiation spectrum grows 'redder'. The Forbes effect in a plant stand is expressed by the dependence of the effective scattering coefficient ω_E on the depth L of stand and reveals itself in variation of the spectral quality of radiation, with an increase in the green and particularly near infrared.

An attempt to determine the variation of ω_E with the depth L, ω_E corresponding to the whole short-wave spectral region, has been made by the author jointly with Tooming (Ross & Tooming 1964). Expression (II.7.5) for $a_Q(L)$, where a_S, a_{S1}^H, a_D and a_{D1}^H are defined respectively by formulae (II.4.4), (II.6.39), (II.5.12a) and (II.6.40), were treated as an equation for the determination of $\omega_E(L)$. Generalized experimental data of the penetration of total short-wave radiation in maize stands collected during expeditions in Estonia (1962–1963) (experiments Nos 3–6) were used for $a_Q(L)$. In these experiments the orientation of leaves is approximated well by means of a uniform distribution. This calculation proceeds on the assumption that all discrepancies between experimental and calculated data of $a_Q(L)$ are due to the dependence $\omega_E(L)$, and therefore the values of ω_E are not expected to be highly accurate. The results of calculation (Figure 105) indicate that the Forbes effect does exist in the stand and ω_E depends strongly on the depth L, with ω_E increasing approximately twofold as the depth increases from 0.5 to 6.0. The value $\omega_L = 0.65$ which was assumed as a basis for the calculation of radiation in the short-wave spectral region gives correct results only at $L = 1$–2. In view of theoretical considerations it can be supposed that actually the dependence $\omega_E(L)$ in Figure 105 is less pronounced. At small L values, when the accuracy of determining ω_E is not high, single scattering is predominant and therefore ω_E must correspond to the integral scattering coefficient of a maize leaf, equal to 0.35–0.40. It is clear from Figure 105 that this is true. If it is assumed that at large L integral radiation consists only of NIR, for which $\omega_\lambda \leqslant 0.85$, then $\max \omega_E < 0.85$. But it is seen from Figure 105 that at large L values the value of ω_E

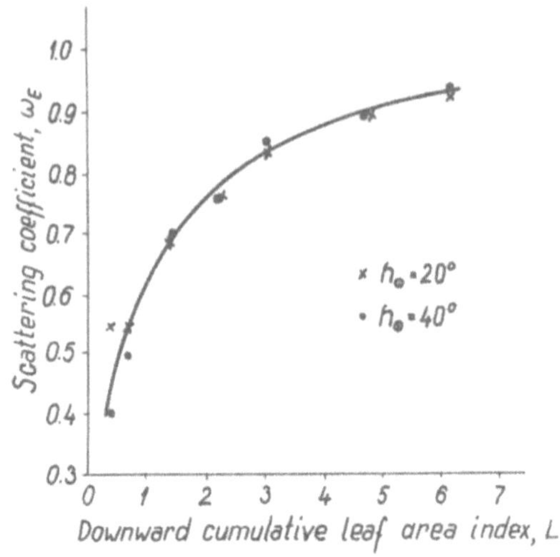

Fig. 105. Leaf's effective scattering coefficient ω_E as a function of depth L.

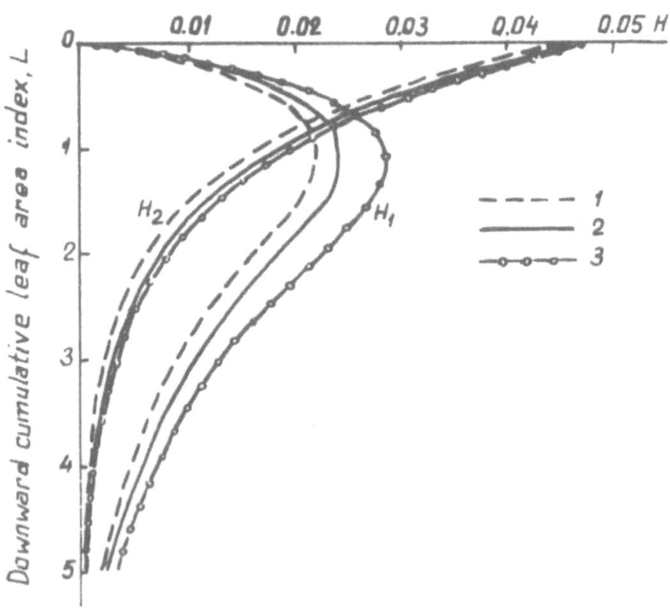

Fig. 106. Vertical profiles of complementary PAR fluxes $H_1(L)$ and $H_2(L)$ in a stand with horizontal leaves. Calculated by three methods at $Q_0 = 1$, $L_0 \rightarrow \infty$. See text.

exceeds 0.9, which is obviously not true. Still, it can be concluded from the curve of $\omega_E(L)$ that when a single value of ω_E for all depths L is used for the calculation of the integral short-wave radiation, the values of a_{S1}^H and a_{D1}^H are overestimated by 20–50% at small L values and underestimated by the same amount at large L values.

In the PAR and NIR the influence of the Forbes effect must be less strong, since ω_λ varies in a narrower range.

Figure 106 shows the vertical profiles of complementary PAR fluxes $H_{S1}(L)$ and $H_{S2}(L)$ in a stand with horizontal leaves calculated by three methods. Curves 1 correspond to the single scattering approximation at $R_E=0.09$ and $T_E=0.06$; curves 2, to an exact solution with multiple scattering taken into account, at $R_E=0.09$ and $T_E=0.06$; curves 3 correspond to an exact solution in the case of monochromatic radiation and to subsequent integration over the spectrum in the PAR region, i.e.

$$H_{S1}(L)=\int_{380}^{710} H_{S1}(L,\lambda)\mathrm{d}\lambda \quad \text{and} \quad H_{S2}(L)=\int_{380}^{710} H_{S2}(L,\lambda)\mathrm{d}\lambda.$$

The difference between curves 2 and 3 is characteristic of the influence of the Forbes effect. It is seen that using the effective values ω_E, R_E and T_E results in underestimated values of fluxes. In the case of the downward flux, the Forbes effect is greatest at the depth $L\approx 1.5$ and its influence does not exceed 30%.

Since in the case of PAR the influence of H_{S1} and H_{D1} in the total flux $Q(L)$ is slight, the influence of the Forbes effect must be small.

Calculations show that the influence of the Forbes effect on the downward PAR flux $Q(L)$ does not exceed 0.5%, and the use of effective coefficients ω_E, T_E and R_E is fully justified. The absolute error of the upward fluxes $F_2(L)$ are of the same order, but the relative error is an order higher and amounts to 3–9%. It is also confirmed by experimental data (Gulyavev et al. 1967) that the Forbes effect in the PAR region can be neglected.

Let us proceed to an estimation of the influence of multiple scattering on the radiant fluxes in stands. Since as ω increases the influence of multiple scattering also increases, this influence is least in the case of PAR and greatest in the case of NIR.

Calculations show that failing to take account of multiple scattering in the PAR leads to an error amounting to 5–7% for the fluxes of the complementary radiation field i. It will be seen from Figure 106 that for downward fluxes $H_{S1}(L)$ the error due to not taking multiple scattering into account is only half the error due to the Forbes effect and does not exceed 15%. A reverse situation exists for upward fluxes $H_{S2}(L)$: the error due to not taking multiple scattering into account is twice as large as that due to not taking into account the Forbes effect and amounts to 10–15%.

A different situation exists in the NIR region. In the downward flux the influence of multiple scattering is not considerable in the topmost layer $L=0$–5, but starting from the depth $L\approx 2$ it becomes the determining factor. Multiple scattering predominates in the upward flux in all cases.

Table II.7.3 *Downward near infrared radiation fluxes* ($\lambda = 0.760\ \mu m$) *scattered by vegetation n times and their percentage in total flux of multiple scattering as a function of leaf area index* ($L_0 \to \infty$, *horizontal leaves*)

	n=0		n=1		n=2		n=3		n=4			
L	H_1^0	H_1^0/Q, %	H'_1	H'_1/Q, %	H_1^2	H_1^2/Q, %	H_1^3	H_1^3/Q, %	H_1^4	H_1^4/Q, %	β	$Q, n \to \infty$
0	1.000	100	0	0	0	0	0	0	0	0	0	1.0000
0.4	0.6703	77.2	0.1342	15.5	0.0298	3.4	0.0124	1.4	0.0067	0.8	1.7	0.8670
1.0	0.3679	52.5	0.1840	26.3	0.0685	9.8	0.0302	4.3	0.0164	2.3	4.8	0.6997
2.0	0.1353	27.6	0.1353	27.6	0.0841	17.2	0.0474	9.7	0.0278	5.7	12.2	0.4896
3.0	0.0498	14.5	0.0748	21.8	0.0652	19.0	0.0464	13.5	0.0308	9.0	22.2	0.3426
4.0	0.0183	7.6	0.0366	15.2	0.0411	17.1	0.0356	15.3	0.0274	11.4	33.4	0.2398
5.0	0.0067	4.0	0.0162	9.6	0.0222	13.2	0.0229	13.7	0.0201	12.0	47.5	0.1676

Table II.7.3 characterizes the influence of the order of scattering on the total downward flux of PAR at different depths L in the case of horizontal leaves. It follows from this Table that in the upper layer the NIR flux $Q(L)$ is mainly due to the penetrated radiation ($n=0$). At a depth $L=2 \div 3$ the orders of scattering 0, 1 and 2 are of predominant importance, but at deeper depths the fraction of penetrated radiation, for example, amounts only to 4%, and the maximum contribution (14%) is due to the third order scattering. The part contributed by first orders (up to $n=4$) at $L=5$ to the total flux is 52%. Consequently, the contribution of higher orders ($n>4$) is equal to 48%. The contribution of single scattering is greatest at the depth $L=2$; that of double scattering, at the depth $L=3$; that of triple scattering, at $L=4$, etc. The quantity $\beta = 100(Q - \sum_{n=1}^{4} H_1^n)/Q$ defines the contribution of higher order scattering.

As seen from the comparison of formulae (II.6.39) and (II.6.42) or formulae (II.6.56), (II.6.61) and (II.6.62), neglecting the reflection from the ground surface below the stand simplifies the formulae noticeably. It is interesting to estimate the influence of the albedo A_B on the vertical fluxes. Of course, this influence must be greater in the case of upward fluxes in a sparse stand. As seen from formula (II.7.2), it also increases with the increasing difference $A_B - A'$. In the case of a sparse stand ($L_0 < 3$) an increase of A_B from 0.15 to 0.30 increases the upward PAR flux 1.5–2 times, but in the case of a dense stand ($L_0 > 4$) this influence is felt only in the lower layers and is practically inconsiderable. Figure 107 illustrates the influence of the albedo A_B on the vertical profiles of fluxes $H_1(L)$ and $H_2(L)$ in the NIR region. Because of the high value of multiple scattering the influence of A_B in this case is larger than for PAR. It is seen that the influence of A_B on the downward flux H_1 increases with depth, but on the whole is considerably less

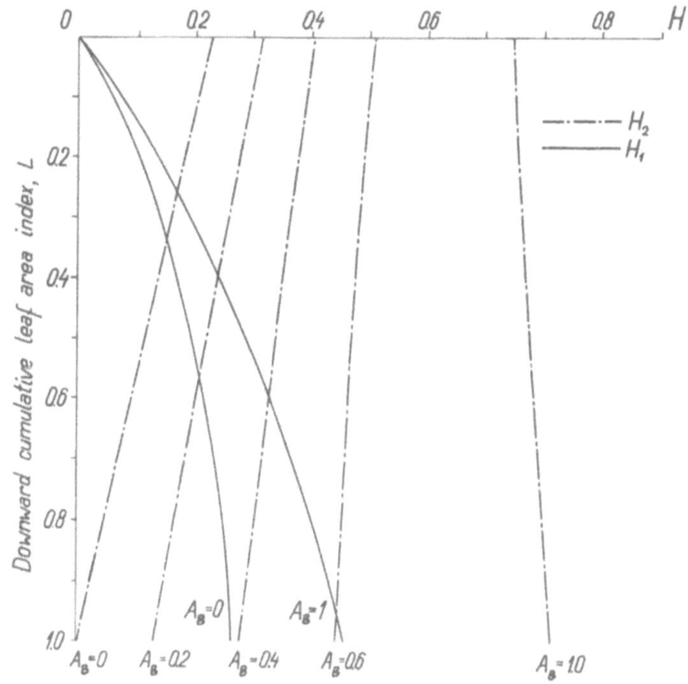

Fig. 107. Influence of reflection coefficient A_B of ground surface on the vertical profiles $H_1(L)$ and $H_2(L)$ of the complementary NIR fluxes. A sparse stand at $Q_0 = 1$, $L_0 = 1$, $T_L = 0.50$, $R_L = 0.30$.

than on H_2. In the NIR region below the stand $A_B \approx 0.2$, while the albedo of the stand at $L_1 \to \infty$ is equal to 0.41.

In the case of a sparse stand a snow cover below the stand increases the complementary short-wave radiation flux by approximately 30%. The following can be stated to summarize the influence of the albedo A_B of the ground surface below the stand on the vertical fluxes.

1. In the PAR region the influence of A_B on the downward fluxes is immaterial and can be neglected.

2. In the case of upward fluxes the influence of A_B should be taken into account in a sparse stand but can be neglected in a dense stand.

3. In the NIR region the influence of A_B is more substantial due to multiple scattering. But taking into account that A_B of many types of ground surface below the stand is equal to 0.20–0.30, its influence on the downward fluxes can also be neglected.

4. The influence of A_B on upward NIR fluxes is substantial and should be taken into account if $L_0 < 4$.

The results in this chapter and their analysis show that the total radiation field averaged horizontally has a complicated structure and depends on a plurality of factors. But this problem becomes still more complicated in converting from averaged values to statistical characteristics of the total field. Some interesting results of experimental and theoretical studies of the statistical character of the attenuation of radiation in plant stands have been obtained by Laisk (1964, 1965a. b, c, 1968) at our laboratory.

An important problem arises in connection with the total radiation field, viz. that of the optimum radiation regime in a plant stand. This problem is intimately linked with the photosynthesis and yield in stands, and is therefore outside the scope of this book. We will only mention that this is a complicated problem wanting an ultimate solution. Some of its aspects and special solutions are dealt with in the works of Verhagen et al. (1963), Tooming and Nilson (1967), Kuroiwa (1968a, b, 1970), Nilson (1968b), Ross and Bichele (1968, 1969), Tooming (1968a, b) and Ross (1970).

II.7.8 NEW THEORIES

A number of new, up-to-date mathematical models of the radiation regime in plant stands, distinguished by an extensive use of the computer art, have been developed during and after the publication of this book. A brief review of the most instructive models, which reflect the recent progress in theoretical phytoactinometry, is given below. An excellent critical review of radiation models for estimating the short-wave regime of plant stands until 1974 is given by Lemeur and Blad (1974).

Idso and de Wit's model (de Wit 1965, Idso and de Wit 1970). This is the first computer model in which the irradiances of penetrated direct $S'(L)$ and diffuse $D(L)$ solar radiation are treated separately, with due regard for leaf orientation. The penetration is calculated by a binomial formula similar to (II.4.13), the fluxes $H_1(L)$ and $H_2(L)$ are calculated approximately using iteration by order of scattering, assuming that one half of the radiation scattered by a horizontal layer goes up and the other half down. Instead of the G-function Idso and de Wit take account of leaf orientation by means of a cumulative light distribution function $S(I)$ defined as the probability that a light ray is intercepted by leaves with a cosine of its normals angle to the light equal to or smaller than $\cos \widehat{r_L r}$. This model has been used for calculating the vertical profiles of radiation, albedo and the photosynthesis of crops.

Goudriaan's theory. Idso and de Wit's model was further developed and generalized by Goudriaan in his monograph (Goudriaan 1977). The numerical experiments performed by Goudriaan yielded results that essentially coincide with our data. Much attention is paid in Goudriaan's theory to developing approximate formulae obtained by approximation of computer calculations by simple expressions and convenient for practical application. Thus, for example, vertical profiles of radiation may often be described by exponental formulae similar to $\exp(-K_Q L)$, where K_Q has the form $K_Q = 0.0353 + 0.9462(1-\omega_L)^{0.5} K$, in which $K = G/\sin h_\odot$.

Goudriaan has also shown that for a wide class of leaf orientations the $G(\vartheta)$ function may be given by an empirical formula using the leaf inclination index χ_L

$$G(\vartheta)=G_O(\chi_L+0.877(1-2G_O(\chi_L)]\cos\vartheta,$$

where

$$G_O(\chi_L)=0.5-0.633\chi_L-0.33\chi_L^2, \quad -0.4<\chi_L<0.6.$$

Lemeur's theory. During the period from 1971 to 1974 Lemeur (1971, 1973a, b), Lemeur and Blad (1974) derived a theory for radiative transfer in plant stands, which is quite similar to that of ours. Lemeur also used the concept of G-function, individually discussed the transmission and absorption of direct and diffuse solar radiation and also the complementary radiation field due to scattering by phytoelements. Vertical profiles of all radiation field components were calculated for different orientations of leaves at different solar elevations. Lemeur's results essentially coincide with the results presented in this book.

The Suits theory. This theory (Suits 1972a, b) is a further development of the model proposed by Allen et al. (1970), achieved by proceeding from a single-layer to a multilayer stand, including the effect of solar elevation and describing the optical characteristics of the layer by means of its architecture and the coefficients of reflection R and transmission T of phytoelements. A simplifying method is used for taking account of the orientation of phytoelements: each phytoelement within the canopy is replaced by a horizontal and vertical panel of the same area as the projected area of the leaf on a horizontal plane and two orthogonal vertical planes. By shifting projected leaf areas from horizontal to vertical, one can simulate a change in leaf slope. Three types of radiant flux are identified in the plant canopy: the downward complementary H_1, the upward complementary H_2 and the downward specular flux density S' directed along r_S. Lambertian phytoelements and their random orientation in azimuth are accepted. Under these assumptions the complementary radiation field is azimuthally isotropic and no contribution to azimuthal variation in infinitesimal layer radiance can be expected from the complementary radiation field. The azimuthal variation in radiance is due entirely to the contribution of the specular flux S^1.

The considering of the 'hot spot' effect is an important feature of the Suits theory. In all previous theories formulae for reflected radiation included a term $a(L, r_S)a(L, r)=\exp[-(K_S+K)L]$, where K_S and K are the extinction coefficients of incident and reflected beams, respectively. The function $a(L, r_s)$ can be treated as the probability of achieving the line of sight r_S at least at the level L for specular flux, and $a(L, r)$ as the probability of achieving the line of sight r from the level L to the region outside the stand. The product of these probabilities means that both of them are independent. But it is obvious that, when the directions r_S and r differ a little, these probabilities are not independent. When r_S and r coincide, then if the specular flux can penetrate to the level L along the direction r_S, it is certain that the line of sight is achieved outward from the level L in the same direction and the joint probability is not equal to $\exp[-(K_S+K)L]$ but to $\exp(-K_S L)$. The important assumption by Suits is that the joint probability may be expressed by $\exp(-K_S-fK)L$, where $f=1-\exp[-\beta(\theta_S-\vartheta)^2-\gamma(\varphi_S-\varphi)^2]$ and β and γ are constants characteristic of the coarseness or fineness of the stand structure. This expression has the property of reducing the joint probability to $\exp(-KL)$ when $r_S=r$ and smoothly changes to $\exp[-(K_S+K)L]$ for large values of $(\theta_S-\vartheta)^2$ or $(\varphi_S-\varphi)^2$ when independent parts of the stand are involved for specular flux and viewing direction r. Originally the Suits theory was derived for the calculation of the spectral directional relfectance of a plant stand, but later was used by Chance and LeMaster (1978) for the calculation of the crop absorption also. An experimental test of the theory in different spectral intervals of short-wave radiation for wheat and cotton stands performed by Chance and LeMaster (1977) has shown that (i) model predictions for percentage light absorption as a function of L_0 are in agreement with experimental data. (ii) The Suits theory exhibits qualitative agreement with experimental data for spectral reflectance, but overestimates reflectance at all wavelengths. (iii) The theory shows poor agreement with experimental results for wheat reflectance as a function of solar elevation.

The Norman-Jarvis theory. This comprehensive and powerful theory (Norman & Jarvis 1975) is elaborated for coniferous forests for both a random and nonrandom plant stands. The canopy is assumed to be divided into horizontal layers, and the clumping of foliage is considered at two levels of grouping: first, randomly located horizontal whorls composed of branches and shoots and, second, within the confines of a whorl, randomly located shoots comprising one year's growth of needles. All shoots inside the layer are modelled as cylinders and assumed to lie in a horizontal plane being uniformly distributed about the azimuth. In a given layer the structure is approximated by the actual number of whorls, but all the whorls are assumed to be of the same dimension representative for that layer. Similarly, all the shoots which compose the whorls in a layer are assumed to have the same dimensions for that layer. The radiation field inside the stand is determined by the following terms: the downward and upward diffuse fluxes above a layer, the solar beam flux (S') penetrating unintercepted through the stand to a given level and the total downward and upward diffuse source fluxes arising from reflection and transmission of intercepted beam radiation in a given layer. The main radiation characteristics of the layer are the layer reflection and transmission coefficients. These layer properties are derived from leaf spectral characteristics, the angular distribution and the layer architecture. First in phytoactinometry a comprehensive treatment of the penumbra effect and sunflecks size distribution is included in the theory using the works of Miller and Norman (1971). Results of the theory are compared to measurements in PAR and NIR. The best agreement between theory and measurements was obtained with the grouping theory which includes stems and branches. The calculated PAR transmission distribution based on the grouping theory agreed well with theory. It is probable that the most important penumbral contribution for photosynthesis arises from gaps between needles and shoots.

Ondok's model. This semiempirical simulation model (Ondok 1975) is derived for the system which consists of a *Phragmites communis Trin.* stand above the water surface. Two submodels (one for direct and the other for diffuse sky radiation) are introduced, each containing three components: PAR intercepted by foliage, PAR intercepted by water below the stand and PAR reflected to the atmosphere by both the foliage and water. The model inputs are the hourly sums of PAR, and the outputs – the daily course of radiant energy content in different parts of the system.

In the above-described models the plant stand was treated as a horizontally uniform layer with a random or nonrandom distribution of phytoelements. But in recent years models for the calculation of the radiation regime in a single plant and a row crop have also been developed. We will discuss some of them.

Charles-Edwards and Thornley (1973) proposed a *light interception model for an isolated tree.* The crown of a tree is presented as an ellipsoid which is cut below at some height by a horizontal plane. Varying the cutting height various shapes of crowns can be obtained. If the exponential light attenuation inside the crown is assumed, the main objective is to calculate the beam path length from the ellipsoid surface to a specified point inside the crown in a given direction. Integration over all directions allows transition from the specular beam to the diffuse radiation.

Allen's hedge-row model (1974). A canopy planted in rows is treated as a series of infinitely long horizontal quadrangles placed in parallel, with a given height, width and distance from one another. Inside the hedge-row the parabolic leaf area distribution in the vertical direction and an exponentail light attenuation are assumed. The main objective is to calculate the beam path length as a function of the beam inclination angle and azimuth. The model predicts the average direct-beam penetration to any depth as a function of distance from the row and its orientation.

The Fukai-Loomis model (1976). This model is a further development of Allen's model, and it takes account of the nonuniform distribution of leaf area, leaf inclination and azimuth angle. The spatial nonrandomness is approached by dividing the plant stand horizontally and vertically into a number of small rectangular tubes parallel to the row. The leaf distribution functions may differ for different tubes of a particular row but the same functions are assumed for the corresponding tubes at identical positions in adjacent rows. The model calculates light interception within a particular tube from the same position and the distribution for leaf area and inclination and azimuth among tubes. Leaf orientation is taken into account by methods discussed in this book. The penetration of radiation is described by the binomial relation (II.4.13). The direct and diffuse radiation fluxes are calculated separately. In calculating the complementary radiation field it is assumed that all sunlit leaves reflect light upwards and transmit light downwards: thus no scattered light is considered to move horizontally from tube to tube. The model was tested with experimental data of sunlit-area profiles, light-flux profiles and frequency distribution of light flux density of the leaf surface for cotton and soybean crops.

A number of new up-to-date mathematical models of the radiation regime in plant stands distinguished by an extensive use of the computer art have been developed during and after the publication of this book. This new field was pioneered by Japanese workers. In the first work of Tanaka (1969) the sunlit area in a tobacco plant community was calculated by the Monte Carlo method. The plant model, leaf arrangement and orientation simulated a real plant. Each leaf was divided into four to eight triangles, and each point of a triangle was shown in cartesian coordinates. 1000 random number were produced to hit each triangle, and the sunlit area was estimated by the random number that reached a triangle without being disturbed by upper triangles. Calculations were performed for three values of solar elevation: 30°, 60° and 77°.

More time-consuming calculations were performed by Oikawa and Saeki (1972, 1977) and Oikawa (1977a, b). Their first model of canopy included 500 randomly distributed black leaves of the same size, their individual areas being just 1/100 of the unit land area, i.e. the leaf area index L_0 was equal to 5. The model of leaves had an elliptical shape with a major axis to minor axis ratio 2:1. Four leaf inclination angles were considered: 0°, 30°, 45° and 60°. The leaves were arranged at equal vertical intervals from top to bottom and the leaf area density was $u_L(z)=5$ throughout the population, the height of stand being equal to 100 cm. The beam emission experiments were conducted for two different stand types. One type termed the 'infinite' plant stand, was extended in the horizontal directions by an infinite repetition of unit plant stands. The other type, termed the 'patchlike' plant stand included unit plant stands placed apart from one another without interacting with neighbouring populations, which could be treated as a solitary plant stand. The beam emission corresponded to the uniform overcast sky and each beam radiated into the plant stand was followed until it was absorbed by any of the leaves or by the soil surface.

Concurrently with the Monte Carlo experiment, the penetration function $a_S(L, h_\odot)$ was calculated from the exponential formula of Monsi and Saeki. The results obtained by the two methods agree well for all leaf inclination angles. Additional calculations were performed to evaluate the influence of the boundary effect in the case of the patchlike stand. A calculation of the variance of irradiance of individual leaves located at the same height showed that the variance decreased with the depth L.

Oikawa's second plant stand model (1977a, b) has the same height, 100 cm, leaf area index $L_0=5$, and the same shape of leaves as the first model, but the leaves are arranged about the plant vertical axis, the plants being square-planted with different planting densities. Each constituent plant with senile leaves has a spiral phyllotaxis, its divergence being $\frac{1}{2}$, $\frac{1}{3}$ and $\frac{1}{5}$ and a constant leaf inclination angle having values of 0°, 30°, 45° and 60° in different numerical experiments.

An increase in planting density N_p from 4 to 36 plants per m^2 in such a model stand means a decrease in the distance between neighbouring plants from 50 cm down to 16.7 cm and, at a constant $L_0 = 5$, a decrease in the foliage area of a single plant $\hat{S}_L$ from 125 to 13.9 dm^2. Each plant volume is represented by a vertical cylinder, 100 cm high, with an average diameter varying with the size and inclination angle of leaves as well as with the divergence of the spiral phytotaxis. The leaf area density inside the plant volume $\hat{S}_L = \text{const}$, but at a constant L_0 it increases with the increasing planting density and the leaf inclination angle ϑ_L and with the decreasing single leaf area S_L. If N_p is small, a packing effect is present in the stand, i.e. a large number of leaves are located around the plant axis and at a considerable distance from one another. The relative variance of distance between neighbouring leaves averaged over all directions $V_a' \gg 1$, and leaves in such a stand have a clumped distribution. If N_p and the single leaf area S_L increase and ϑ_L decreases, the packing effect will decrease, $V_a' \to 1$ and the stand will become random. *If* N_p is large, then S_L is small, and overlapping of individual plants begins in the presence of large horizontal leaves. In such stands the leaf distribution is semiregular and $V_a' < 1$.

Varying different parameters of the plant stand architecture Oikawa performed numerous numerical experiments to determine the coefficient of penetration $a_D(L)$ for the uniform overcast sky.

The results of calculation may be summarized as follows.

If the planting density N_P decreases, the number of leaves per plant increases. This causes an increasing packing effect within the plant volume, an increase in the variance $V' < 1$ and an increase in the penetration a_D in comparison with a random distribution of leaves.

An increase in the inclination ϑ_L of leaves around the stem from 0° to 60° decreases the plant volume, increases the relative variance $V' > 1$, increases the packing effect and, consequently, increases the penetration a_D in comparison with a random leaf distribution.

An increase in leaf divergence from $\frac{1}{5}$ to $\frac{1}{2}$ causes a decrease in the plant volume filled with leaves, increases the packing effect and the penetration a_D in comparison with a random leaf distribution. However, this effect is not great.

The penetration $a_D(L)$ is greater in a forb-type canopy, in which $u_L(z)$ increases with height, in comparison with a herb-type canopy, in which $u_L(z)$ decreases with height.

At a constant single plant leaf area $\hat{S}_{LP}$ an increase in the leaf area with a corresponding decrease in the number of leaves causes an increase in the plant volume. That means that the packing effect decreases and penetration increases. At $\hat{S}_{LP} = 500$ cm and $S_L < 10$ cm^2 $a_D(L)$ is greater than a_D for a random distribution of leaves, but at $S_L > 20$ cm^2 the situation is reversed.

Oikawa also studied the influence of the leaf area density $u_L(z) = u_L$ on the penetration function $a_D(L)$ at a constant L_0. An increase in u_L was achieved by moving leaves from bottom layers of the stand to upper layers. This procedure shortens the cylinder length at the bottom, and the leaf area density $\hat{S}_L$ within the cylinder increases accordingly. When the depth of stand is contracted to the limit, the stand becomes a monolayer of square-planted spherical formations, within which assumes extremely high values. The penetration a_D in this case increases tending to max $a_D \neq 0$.

It should be noted that a principally different situation, not discussed by Oikawa, will exist where the distribution of plants deviates from a random to a regular one. A decrease in plant stand depth at a constant L_0 also causes an increase in u_L, but a_D will decrease. In the extreme case we have a monolayer, the penetration of which is extremely small. Thus an increase in the leaf area density u_L at a constant L_0 causes a decrease in a_D in the case of a semiregular ($V' < 1$) and an increase of a_D in the case of a clumped ($V' > 1$) distribution of leaves in the stand.

The pattern in which the leaves of a square-planted stand are arranged was examined by the point quadrat method. The relative variance V' calculated only in a vertical direction was always greater than unity, even when $a_D(L)$ was smaller than in the case of a random leaf distribution. This discrepancy is due to the fact that in determining a_D the penetration is summed over all directions from the upper hemisphere, and V' found for the vertical direction only is not equal to V'averaged over all directions.

The model of Szwarcbaum and Shaviv (1976). A spherical plant of a diameter R on which vertical beams fall at randomly chosen points of the sphere is discussed. The distance a between neighbouring leaves is given by a distribution function $P(a)$. An isotropic and uniform distribution of leaves inside the tree is assumed, i.e. $P(a)$ does not depend either on the direction or on the initial point a. The optical properties of leaves are specifified by the transmission T_L and reflection R_L coefficients. All leaves have the same inclination angle ϑ_L and are uniformly distributed in azimuth. Specular reflection alone is considered. The vertical profiles of absorption and albedo at varying inclination angles and $T_L = 0.1$ and $R_L = 0$ and 0.1 are calculated during numerical experiments. The path of beam following multiple reflection and transmission is traced until a specified intensity is achieved. Calculations show that at $R_L = 0$ the vertical profile of absorption is described well by an exponential relation. The dependence of the albedo on ϑ_L is strange because of taking into consideration only the specular reflection, a limitation which never occurs in reality.

The model of Smith and Oliver (1974) assumes that the horizontal plant stand is composed of nonhomogeneous layers of Lambertian surfaces of known optical properties, statistical composition and geometric arrangement. Total radiation which reaches the top of stand is composed of direct solar and diffuse sky radiation, and the interaction with the canopy of each of these initial radiation sources is treated independently. The direct solar radiation component is treated as a point source, whereas the diffuse fraction is divided into source sectors by partitioning the hemisphere into 10° inclination bands and further subdividing these bands to form 20° azimuthal sectors. The interaction with the canopy of each of these initial radiation sources is treated independently. The orientation of the leaf in each layer is determined by sampling from the inclination distribution and a uniform azimuthal distribution. The interaction probabilities within each layer for both incoming and outgoing fluxes at each specified source direction are given by formula (II.4.13). The leaf's reflectance and transmittance are treated as a Lambertian surface. The flux which passes through a gap or is reflected downward from an upper layer of the canopy interacts with the next lower layer. The flux which reaches the soil surface is reflected into each of the upward directed source bands. The upward directed flux from the top layer escapes from the canopy into each of the bands and is separately accumulated. The iteration procedure continues until the level of flux in any source direction within any layer is below a threshold value.

The model was used in calculating the spectral brightness for a *Bouteloua gracilis* canopy with a leaf are index of 6.5. The angular dependence of the brightness coefficient was calculated for different wavelengths in the PAR region.

In summary it should be stressed that the Monte Carlo method is a promising new method for modelling the radiation regime of plant stands, and that its advantages are best realized for the following problems:

(i) Investigation of the influence of architecture parameters in the case of a nonrandom distribution of leaves. First results in this field have been obtained in excellent works of Oikawa (1977a, b).

(ii) Investigation of the statistical structure of the radiation field inside a stand, where besides average values of intensities higher statistical moments and distribution functions of intensities are required. Serious results in this field have been scarce so far.

II.8 Semiempirical formulae for total radiation fluxes

II.8.1 GENERAL

In order to evolve a quantitative theory of radiative transfer in plant stands attempts have been made to approximate experimental data by means of empirical or semi-empirical formulae on the basis of a simplified treatment of the general regularities of the attenuation of radiation in stands. The development of a quantitative theory of radiative transfer has permitted deriving theoretical formulae for the calculation of radiation fluxes, which offer a more correct treatment of the physical essence of the problem. But as we have seen in Chapters II.6 and II.7 these formulae are very cumbersome and not suitable for practical purposes. Also, because of a number of simplifications in the derivation their application to different types of vegetation is limited.

Some semiempirical formulae for the calculation of total short-wave radiation, PAR and NIR in plant stands have been developed at our laboratory. Some of the initial theoretical formulae have been simplified, and some of the functions they contain have been replaced by empirical constants (Tooming 1969; Tooming & Ross 1969). This has enabled some comparatively simple formulae to be derived, which on the whole reflect the principal regularities of the attenuation of radiation in stands and, owing to the empirical constants they contain, are in better agreement with experimental data than theoretical or empirical formulae.

The first part of this chapter gives a review and an estimation of the different empirical and semiempirical formulae which are used in the literature. The second part of the chapter contains new semiempirical formulae and estimates their accuracy.

Bouguer's law. The law that defines the attenuation of radiation in a homogeneous absorbing and scattering medium is expressed in physics as Bouguer's law $I = I_0 \exp(-kx)$, where x is the path length traversed by the beam in the medium, and k is the attenuation coefficient.

Bouguer's formula was first used to describe the attenuation of total radiation in stands in the work of Monsi and Saeki (1953), where the beam path length in the stand is expressed by means of the foliage area index L. According to Monsi and Saeki the total downward flux of PAR in the stand is described by formula

$$Q(L) = Q_0 \exp(-k_Q L), \tag{II.8.1}$$

where k_Q is determined experimentally. These authors measured radiation with a photocell, and the foliage area by the stratifying clip method. Profiles of radiation were studied in many types of stands and, according to the authors, agreed satisfactority with the exponential law of attenuation. The attenuation coefficient varied in a wide range from 0.3 to 1,5, its low values being characteristic of stands with nearly vertical leaves (grasses), and high values characteristic of stands with large, mainly horizontal leaves. Monsi and Saeki also proposed the binomial formula

$$Q(L) = Q_0 \left(1 - \frac{k_Q L}{N}\right)^N, \tag{II.8.2}$$

where N is the number of independent horizontal layers in the stand.

In contrast to the one-parameter exponential formula the binomial formula is a two-parameter expression, in which both k_Q and N are determined experimentally.

In a later Japanese work (Kasanaga & Monsi 1954) the binomial formula was completed by taking into account the leaf transmission coefficient T_L so that

$$Q(L) = Q_0 \left[1 - k_Q(1 - T_L)\frac{L}{N}\right]^N. \tag{II.8.3}$$

If $N \to \infty$, formula (II.8.3) changes into an exponential formula

$$Q(L) = Q_0 \exp[-k_Q(1 - T_L)L]. \tag{II.8.4}$$

Formulae (II.8.1) and (II.8.2) can be treated theoretically assuming that total radiation consists of vertical beams. Then they are derived analogously to the formulae for direct solar radiation in II.4.3 at $h_\odot = \pi/2$, and the attenuation coefficient k_Q assumes the meaning of $G(\pi/2) = G_2$ i.e. it is equal to the cosine of the effective angle of leaf inclination.

Formulae (II.8.3) and (II.8.4) are derived analogously assuming that the leaves transmit vertically a part T_L of the incident radiation.

To calculate the phytosynthetic activity of nonhorizontal leaves we require to know the irradiance on the leaf surface rather than the downward flux. For this reason Saeki (1960a) modified formula (II.8.1) and proposed

$$Q^*(L)=Q_0\frac{k_Q}{1-T_L}\exp(-k_Q L) \tag{II.8.5}$$

in view of the following considerations: as is known, $-dQ/dt$ represents the radiation Q_{at} attenuated in a unit layer of the stand for any leaf orientation. In this case the attenuation includes reflection and absorption of radiation by leaves, since the transmission is taken into account in the flux Q. Consequently, $Q^*=Q_{at}+Q_T$, where $Q_T=T_L Q^*$ is the radiation penetrating through the unit layer and

$$Q^*=-\frac{1}{1-T_L}\frac{dQ}{dL}. \tag{II.8.6}$$

This formula, however, is not substantiated theoretically. If we proceed from the above mentioned theoretical treatment of the formulae, Q is the mean irradiance on a horizontal surface at the depth L, k is the cosine of the effective leaf inclination angle. Consequently, the irradiance on the leaf surface must be equal to

$$Q^*=kQ. \tag{II.8.6a}$$

Saeki's inferences are logically incorrect, since while he takes into account the transmission of leaves in formula (II.8.6), he uses formula (II.8.1) for Q, which does not take into account the transmission of leaves, to determine Q^*. If formula (II.8.4) for Q is used, then (II.8.6) gives (II.8.6a).

The fact that the above formulae do not take into account the influence of solar elevation on the penetration of total radiation is one of their drawbacks. That is why the empirical constant k_Q depends considerably on solar elevation $h_\odot$. In order to reduce this dependence Tooming and Ross (1964) proposed the following modified exponential formula:

$$Q(L)=Q_0\exp\left(-\frac{k_Q^* L}{\sin h_\odot}\right), \tag{II.8.7}$$

where the empirical constant k_Q^* is less variable in comparison with k_Q in formula (II.8.1). Formula (II.8.7) implies that the attenuation of total short-wave radiation follows the laws that govern the penetration of direct solar radiation. It can therefore be expected that at high solar elevations, when the penetration of S exceeds that of D formula (II.8.7) will overestimate values and underestimate them at low solar elevations. A test of formula (II.8.7) in a maize stand (experiments Nos. 3, 4, 5) has shown a comparatively good agreement in the range of solar elevations from 20 to 50° at $k_Q=0.5$. That means that total short-wave radiation penetrates in a maize stand as direct solar radiation in a model stand with uniformly oriented leaves.

The applicability of formula (II.8.7) to total PAR was verified by Tooming (1967) in a maize stand and, in greater detail, by Gulyayev et al. (1967, 1969) in stands of maize, sugar beet and winter wheat. In the case of winter wheat the

Table II.8.1 *Coefficient* k_Q^* *for different crops* (*Gulyayev* et al. 1969)

Crop	Solar elevation		
	50–60°	35–45°	15–25°
Maize			
well fertilized	0.37	0.40	0.21
poorly fertilized	0.60	0.60	0.30
Sugar beet	0.70	–	–
Winter wheat			
middle of growing season	0.41	0.42	0.32
end of growing season	0.50	–	–

downward cumulative leaf area index L was replaced by the total effective area of phytoelements $L^* = L_0 + (L_{GE} + L_{RE})\cos h_{\odot}$, where L_{CE} and L_{RE} are the relative vertical section areas of stems and spikelets, respectively. The authors have concluded that formula (II.8.7) provides a satisfactory description of the penetration of total PAR in these crops at $h_{\odot} > 20°$. The coefficients whose values are given in Table II.8.1 are still dependent on solar elevation, which indicates that the laws of penetration for direct and total radiation are different on the whole.

Referring to table II.8.1 note the large difference of penetration into a maize stand between well fertilized and poorly fertilized areas in the case of $L_0 \leqslant 3$. It can be supposed on the basis of Figure 71A that the high k_Q^* values characterize the regularity of distribution of leaves in a sparse crop. The near values of k_Q^* for maize and wheat indicate that the penetration of total PAR is similar notwithstanding the different structure of these crops. The high values of k_Q^* for sugar beet are probably due to the large proportion of horizontal leaves in this crop.

Budagovsky has proposed a simple empirical formula for the penetration of total short-wave radiation

$$Q(L) = \frac{Q_0}{1 + k_Q L}, \tag{II.8.8}$$

a theoretical analysis of which has been performed by the author jointly with Tooming (Tooming & Ross 1965).

Budagovsky's formula for short-wave radiation was verified for validity by Tooming and the author (Tooming & Ross 1965) and satisfactory results were obtained in a maize crop (experiments Nos. 3–5, 10 and 11). It was found that the coefficient k_Q is somewhat dependent on solar elevation. At $h_{\odot} = 20$, 40 and 60° it is equal to 0.50, 0.69 and 0.63, respectively. For approximate calculation, k_Q can be assumed as constant and equal to 0.65. As Gulyayev et. al. (1967) points out, Budagovsky's formula does not sufficiently approximate experimental data for total PAR.

A modified binomial formula was proposed by Monteith (1965a) proceeding from the following assumptions:

(a) a plant stand, L_0 in depth, is divided into L_0 'elemental' layers of $\Delta L = 1$ thickness;

(b) no mutual shading of leaves exists within the 'elemental' layer;

(c) the relative area of gaps s in an 'elemental' layer is equal for all layers and constant during the day;

(d) the distribution of leaves in a single 'elemental' layer does not depend on the distribution of leaves in other layers;

(e) reflection from leaves is not taken into account, while transmission is taken into account by means of effective transmission coefficients $T = \int Q_\lambda T_\lambda \mathrm{d}\lambda / \int Q_\lambda \mathrm{d}\lambda$ (effective transmission coefficient of single leaf) and $T^L = \int Q_\lambda T_\lambda^L \mathrm{d}\lambda / \int Q_\lambda \mathrm{d}_\lambda$ (effective transmission coefficient of L leaves). In these assumptions s corresponds to the penetration of radiation through an 'elemental' layer without interacting with leaves, and $(1-s)T$ to the penetration through a single layer of leaves. The total penetration through an elemental layer is $s+(1-s)T$, and after penetrating L layers

$$Q(L) = Q_0[s+(1-s)T]^L, \qquad \text{(II.8.9)}$$

when $0 \leqslant L \leqslant L_0$.

The main advantage of the Monteith formula is that it allows approximate division of fluxes of different orders of scattering in accordance with the terms in the binomial formula. Thus, the radiation which has penetrated the stand without interception is equal to s; and the radiation transmitted by leaves once, $Ls^{L-1}(1-s)T$; etc. The total area of unshaded leaves is

$$A_0 = 1 + s + s^2 + \ldots + s^{L_0 - 1} = \frac{1 - s^{L_0}}{1 - s}.$$

Since T^L values can be determined independently, from the optical properties of leaves, formula (II.8.9) is essentially a one-parameter expression. The values of s determined by actinometric measurement are equal to 0.6–0.8 for barley and 0.4–0.6 for cabbage (Monteith 1965a). The Monteith formula, however, has some drawbacks which result from simplified initial assumptions. Assumptions b, c and e are obviously incorrect. Actually the parameter $s \approx 1 - G(h_\odot)/\sinh_\odot$ depends on solar elevation. Also, the formula does not distinguish between direct and diffuse radiation. The drawbacks of the Monteith formula are analyzed in greater detail in the work of Warren Wilson (1967a).

II.8.3 NEW SEMIEMPIRICAL FORMULAE

Failing to take proper account of the different attenuation of penetrated and scattered radiation is the major drawback common to all above formulae. Not only the intensity of total radiation but also its spatial structure and spectral quality change with solar elevation. This determines the variability of the empirical parameters contained in the formulae.

On the other hand, the theoretical formulae are complicated and also approximate because of simplifications made either when evolving the mathematical model or when solving the transfer equation. For that reason the author jointly with Tooming (Ross & Tooming 1968; Tooming 1967; Tooming & Ross 1965) have derived semiempirical formulae to describe the short-wave radiation regime in plant stands. We proceeded from the theoretical formulae set forth in the preceding chapters, simplifying some of their terms or replacing theoretical functions with empirical constants.

Expression (II.7.5) for the penetration function of total radiation was taken as a basis. Instead of the theoretical formula (II.4.4) for the penetration function of direct solar radiation a_s we used a semiempirical formula (II.4.60) in which the empirical constant k_S characterizes the influence of the geometrical structure of the stand on penetration. Formula (II.5.2) in which $G(h_\odot)$ was replaced k_S, i.e.

$$a_D=\int_0^{2\pi} d\varphi' \int_0^{\pi/2} d_0(r') \exp\left(-\frac{k_s L}{\cos\theta'}\right)\cos\theta' \sin\theta' d\theta' \approx 2\bar{d}_0 E_3(k_s L)$$

was used as initial expression for the penetration function of diffuse sky radiation a_D. Values of $\bar{d}_0$ were determined from Põldmaa's (1963) experimental data on the zonal distribution of diffuse sky radiation. Calculations show that $\bar{d}_0$ ranges from 0.7–1.0 depending on solar elevation and the depth L. Approximate calculations may use $\bar{d}_0=0.9$.

As seen from formula (II.6.41), the downward complementary radiation flux H_{S1} consists of the difference of two exponents. Following the general form of this formula we proposed the following expression for

$$a_{S1}^H(L)=C_2\left[\exp\left(-C_3\frac{k_S L}{\sinh_\odot}\right)-\exp\left(-\frac{k_S L}{\sinh_\odot}\right)\right]. \qquad \text{(II.8.10)}$$

In this formula the empirical constant C_2 defines the part of scattered radiation in the total flux, and C_3 characterizes the intensity of scattering on phytoelements and depends on their optical properties. As follows from (II.8.10), at $L=0$ $a_{S1}^H(L)=0$ and reaches its maximum value

$$C_2\frac{1-C_3}{3}\exp\left(-\frac{k_S L}{\sin h_\odot}\right) \quad \text{at the depth} \quad L^*=\frac{\ln C_3}{C_3-1}\frac{\sin h_\odot}{k_S}.$$

Under a clear sky the role of a_{D1}^H in expression (II.7.5) is small, and we can assume $a_{D1}^H \approx a_{S1}^H$. For the penetration function of total short-wave radiation below a clear sky we obtain the following formula:

$$a_Q=\frac{\frac{S_0'}{D_0}e^{-k_S L/\sin h_\odot}+2d_0 E_3(k_S L)}{1+S_0'/D_0}+C_2\left(e^{-C_3 k_S L/\sin h_\odot}-e^{-k_S L/\sin h_\odot}\right). \qquad \text{(II.8.11)}$$

On a cloudy day it is suitable to take

$$a_Q^C = 2E_3(k_S L) = C_2\left[\exp\left(-\frac{C_3 k_S L}{\sin 40^\circ}\right) - \exp\left(-\frac{k_s L}{\sin 40^\circ}\right)\right]. \tag{II.8.12}$$

In the solar elevation range 20° to 50° $a_S \approx a_D$, $a_{S1}^H \approx a_{D1}^H$ and from (II.8.11) we obtain

$$a_Q = (1 - C_2)\exp\left(-\frac{k_S L}{\sinh_\odot}\right) + C_2 \exp\left(-\frac{C_3 k_S L}{\sin h_\odot}\right). \tag{II.8.13}$$

The last formula clearly explains the physical essence of the semiempirical formula. Total radiation inside a stand can be treated as consisting of two parts different in the character of attenuation, the part $1 - C_2$ of the total radiation being attenuated in accordance with one exponential law, and C_2 in accordance with another. The first term characterizes the part of total radiation penetrating to the depth L without interception; the other term denotes the radiation generated in the overlying layer in the course of scattering of total radiation on phytoelements.

The formulae derived were tested in stands of maize (experiments Nos. 3–5, 11 and 14) and sorghum (experiment No. 15). The coefficients $k_S = 0.5$, $C_2 = 0.3$ and $C_3 = 0.15$ provide good approximation for short-wave radiation (Figure 108) and are practically the same in a row and a square-sown crop if $L_0 \geqslant 3$. Within the accuracy of experimental data they are also independent of geographical lo-

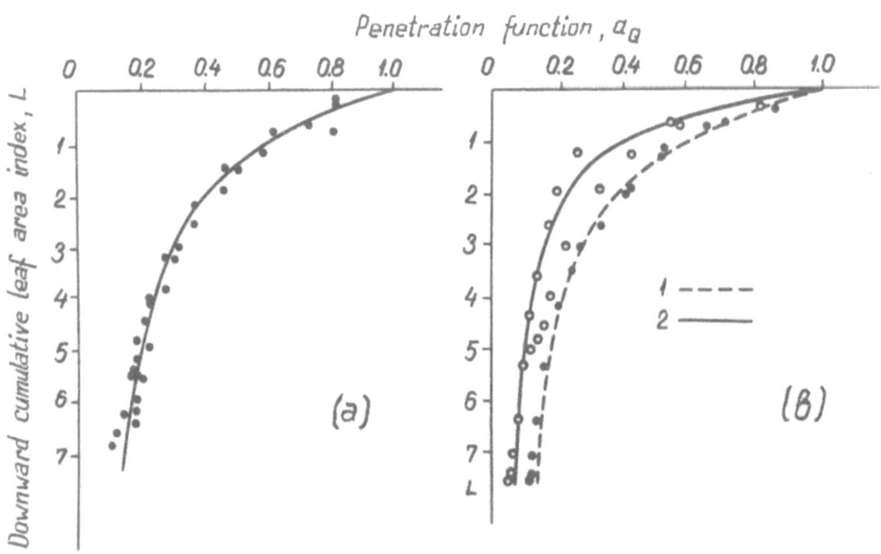

Fig. 108. Penetration function of total short-wave radiation a_Q as a function of depth L in maize stands.
(a) – cloudy sky, stand depth 100–160 cm (curve calculated from equation (II.8.12)); (b) – cloudless sky, stand depth 140–160 cm: 1 – $h_\odot = 40^\circ$, 2 – $h_\odot = 20^\circ$ (curves calculated from equation (II.8.13)).

cation. As it follows from the work of Abashina (1968), formula (II.8.11) is valid within a 10–20% error for the calculation of short-wave radiation fluxes in barley.

Measurements of the penetration of total short-wave radiation in a stand of cotton (Tadjikistan, experiment No. 18) failed to reveal any dependence of a_Q on the solar elevation $h_\odot$ ranging from 25 to 65°.

The experimental data for cotton are approximated by means of this formula

$$a_Q(L) = (1 - C_2)\exp(-k_S L) + C_2 \exp(-C_3 k_s L) \tag{II.8.14}$$

with coefficients $k_S = 0.65$, $C_2 = 0.3$ and $C_3 = 0.15$.

The proposed semiempirical formulae allow determining the mean total short-wave radiation flux in stands of maize, sugar beet, sorghum and cotton within a 15–20% error when the incident flux of total short-wave radiation Q_0, solar elevation and the vertical distribution of foliage area are specified.

Formulae (II.8.11)–(II.8.14) are three-parameter expressions, though the parameter k_S can be determined independently of others from data on the penetration of direct solar radiation. The drawback of the formula is the absence of a simple method to determine the parameters C_2 and C_3.

Validity of formula (II.8.11) for a barley crop has been varified by Abashina and Sirotenko (1973). The meansquare error of the calculation of a_Q^C is reduced from 0.0130 to 0.0105 if $k_S = 0.25 + 0.6z \times 10^{-3}$. The first term in the expression for k_S takes account of the orientation of leaves; and the second, the effect of the nonrandomness of leaf distribution in space.

Validity of formulae (II.8.11) and (II.8.14) in the PAR and NIR regions has been discussed by Tooming (1967) and Laisk. Values proposed by Tooming for PAR and NIR are given in Table II.8.2. The coefficients for NIR need further experimental verification.

For attenuated total PAR the role of the terms a_{S1}^H and a_{D1}^H in expression (II.7.5) is less important than the terms a_S and a_D. Therefore, approximately

$$a_{QP} = \frac{a_D + a_S \dfrac{S'_{0P}}{D_{0P}}}{1 + \dfrac{S'_{0P}}{D_{0P}}}. \tag{II.8.15}$$

At large values of L, however, the role of a_{S1}^H and a_{D1}^H increases noticeably and formula (II.8.15) is certain to give underestimated values.

Table II.8.2 *Values of empirical coefficients in different spectral regions*

Type of radiation	k_S	C_2	C_3
Short-wave radiation	0.5	0.30	0.15
PAR	0.5	0.04	0.15
NIR	0.5	0.60	0.15

Laisk has derived the following semiempirical formulae for a_{S1}^H and a_{D1}^H which take account of scattering in the stand:

$$a_{S1}^H = \frac{C_S}{C_S + C_D}\frac{\omega_L}{2}[\exp(-C_3 L) - \exp(C_D L)]$$

and

$$a_{D1}^H = C_D \frac{\omega_L}{2} L \exp(-C_D L),$$

where

$$c_S = \frac{1}{2\sin h_\odot} \qquad C_D = \frac{1}{2\sin 40^\circ},$$

a_{QP} being determined from formula (II.7.5).

Exponential formula (II.8.7) is invalid at $h_\odot < 35^\circ$ when the penetration of diffuse sky radiation exceeds that of direct solar radiation. As one should expect, formula (II.8.15) underestimates values in lower layers of the stand where the role of a_{S1}^H and a_{D1}^H increases. A formula of the type (II.7.5) provides the best agreement with experimental data. But at $L > 3$ all calculated curves underestimate values in comparison with experiment. This, apparently, is due to the theoretical value $(1/2)\sin h_\odot$ adopted for k_S in Laisk's semiempirical formulae. This value corresponds to a uniform orientation and does not take into account the clumping of leaves which occurred in the maize stands.

Measurements of total PAR and NIR fluxes in stands need special instruments and devices not always available. A procedure has been developed by Tooming (1967) to convert from the penetration function of short-wave radiation in a stand to penetration functions of the total PAR and total NIR. This procedure is as follows: dividing the flux of total short-wave radiation into two parts, PAR and NIR, we obtain the relation

$$a_Q(L) = C_Q a_{QP}(L) + (1 + C_Q) a_{QN}(L), \qquad \text{(II.8.16)}$$

where a_{QP} and a_{QN} are the penetration functions of PAR and NIR, respectively, and C_Q is the PAR conversion factor for total short-wave radiation given by formula (II.2.12).

Functions a_{QP} and a_{QN} must depend in some way on the effective scattering coefficients of leaves ω_P and ω_N, respectively, for PAR and NIR. These latter can be determined independently. As shown in Chapter II.3, for living leaves, they have certain, sufficiently stable values. Also, a_{QP} and a_{QN} must decrease with the depth L. For a_{QP} and a_{QN}, Tooming adopted empirical formulae

$$\left.\begin{array}{l} a_{QP} = \omega_P^{n(L)}, \\ a_{QN} = \omega_N^{n(L)}, \end{array}\right\} \qquad \text{(II.8.17)}$$

where $n(L)$ is a certain function characterizing the mean order of scattering for a layer, L in depth.

Taking into account (II.8.16) and (II.8.17) we obtain

$$a_Q(L) = C_Q \omega_P^{n(L)} + (1 - C_Q)\omega_N^{n(L)} \tag{II.8.18}$$

Using experimental data for a_Q expression (II.8.18) may be treated a transcendental equation to determine the unknown function $n(L)$.

A more practicable way, however, is to prepare a table in which every value of a_Q has corresponding values of a_{QP} and a_{QN} (Table II.8.3).

Laisk compared experimental data obtained by him in a maize stand with calculated results obtained by using Tooming's procedure (Figure 109). The points plotted on the graph correspond to Laisk's experimental data obtained in a maize stand in Moldavia, the curve of short-wave radiation corresponds to the semiempirical formula (II.8.11), and Tooming's curve is obtained from the shortwave radiation curve and Table II.8.3. Niilisk's curve was obtained from monochromatic radiation measurements (1964) with a spectral photometer (Niilisk 1962, 1965) in maize stands in Estonia (experiments Nos. 4 and 5). Laisk's

Table II.8.3 *Penetration functions for total short-wave radiation and corresponding penetration functions for total PAR and NIR*

a_Q	0.90	0.80	0.70	0.60	0.50	0.40	0.30	0.20	0.15	0.10	0.05
a_{QP}	0.86	0.73	0.60	0.48	0.36	0.24	0.15	0.06	0.03	0.01	0.002
a_{QN}	0.94	0.89	0.82	0.75	0.67	0.58	0.47	0.34	0.28	0.19	0.10

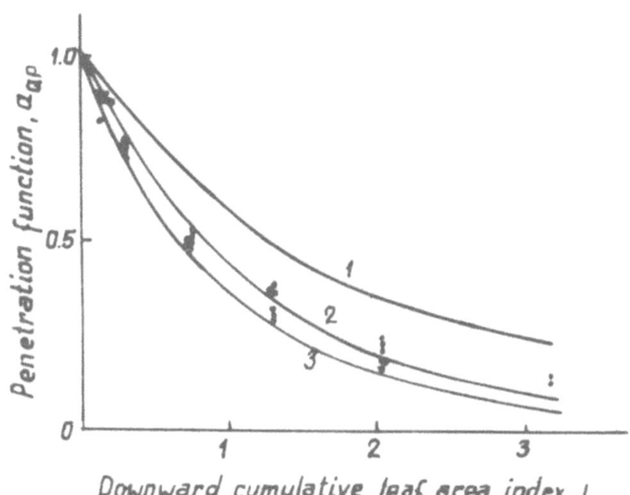

Fig. 109. Penetration function a_{QP} of total PAR as a function of depth L in a maize stand (data of Laisk).

1 – short-wave radiation, 2 – PAR according to Tooming; 3 – PAR according to Niilisk, PAR points according to Laisk. $h_\odot = 40°$.

experimental data show a better coincidence with Tooming's curve than with Niilisk's curve. No suitable experimental data to verify the Tooming procedure for conversion from short-wave radiation to NIR are available at this date.

Thus, where no high accuracy is required, fluxes of PAR inside a stand can be determined from fluxes of shortwave radiation by means of the conversion Table II.8.3.

The comparatively simple new semiempirical formulae derived in this Chapter from complicated theoretical ones allow determining total fluxes of short-wave radiation, PAR and NIR in a stand within a 10–20% accuracy if the solar elevation, total incident radiation flux Q_0 and the vertical distribution of leaf area are specified.

Anderson in her work (1969) compares Saeki's formula (II.8.5) with our simplified formulae on the basis of experimental data obtained in a sunflower stand in England and in a wheat stand in Australia. She came to the conclusion that our formulae are in better agreement with experimental data than the Japanese formulae. It should be remarked that Anderson did not use our simplified formulae quite accurately. A still better agreement with Anderson's experimental data is obtained by using our formulae of the type $a_A \approx a_S + a_{S1}^H$ where a_S and a_{S1}^H are given by formulae (II.4.4) and (II.6.55), respectively.

II.9 Albedo of plant stand

II.9.1 GENERAL

Among the numerous optical characteristics of the plant stand albedo is one of greatest importance. It is widely used in geophysics, geography, aerial photography, etc. In most cases the reflection coefficient (albedo) of a stand is smaller than that of the leaves composing it. The difference is caused by the mutual shading of leaves and by multiple scattering within the stand, which leads to the additional absorption of radiation. This 'black body' effect increases when the absorption coefficient of the leaves and the degree of multiple scattering increase. Since the mean degree of scattering of a beam is determined by its incident angle, geometrical structure of the stand, and also depends on the scattering coefficient, the albedo of a plant stand also depends on these factors. Since the spectral coefficients of reflection and transmission of living leaves show typical variations, the spectral reflection coefficient (albedo) follows the same type of variation (see, for example, the work of Alkseyev (1969)).

In this chapter we derive theoretical formulae for the albedo of the stand in PAR, NIR and short-wave radiation on the basis of general formulae for intensities and fluxes. Also, the dependence of albedo on the main factors of its variability is discussed.

II.9.2 FORMULAE FOR THE ALBEDO AND BRIGHTNESS COEFFICIENT

General formulae. Formula (II.1.20) defined the reflection function for the ground surface below the stand. The reflection function for the system 'plant stand – soil' $\gamma(r', r)$ can be defined analogously. The reflection function $\gamma_S(r_S, r)$ of plant cover for direct solar radiation (bidirectional reflectance) is equal to the upward relative intensity at the upper level of the stand $L=0$ and coincides essentially with the brightness coefficient for direct solar radiation $b_S(r_S, r)$, i.e.

$$\gamma_S(r_S, r) = b_S(r_S, r) = \frac{i_{S2}(O, r)}{S'_0}. \tag{II.9.1}$$

The brightness coefficient of plant cover for diffuse sky radiation, taking into account (II.1.21) and (II.2.5), is equal to

$$b_D(r) = \frac{i_{D2}(O, r)}{D_0} = \int_{\Omega_1} b_S(r', r) d'_0(r') \cos \vartheta' d\Omega'. \tag{II.9.2}$$

The brightness coefficient of plant cover for total radiation is

$$b_Q(r) = \frac{i_{S2}(O, r) + i_{D2}(O, r)}{Q_0} = \frac{b_D(r) + b_S(r)\dfrac{S'_0}{D_0}}{1 + \dfrac{S_0}{D_0}}. \tag{II.9.3}$$

Taking into account (II.1.21), (II.1.7), (II.1.72) and (II.1.73) we can convert the brightness coefficient of plant cover to the albedo.

The reflection coefficient (albedo) for direct solar radiation is

$$A_S(r_S) = \frac{H_{S2}(\mathrm{O}, r_S)}{S_0} = a^H_{S2}(\mathrm{O}, r_s) = \int_{\Omega_1} b_S(r_S, r) \cos \vartheta \mathrm{d}\Omega. \tag{II.9.4}$$

and for diffuse sky radiation, taking into account (II.2.7) also,

$$A_D = \frac{H_{D2}(\mathrm{O})}{D_0} = a^H_{D2}(\mathrm{O}) = \int_{\Omega_1} b_D(r) \cos \vartheta \mathrm{d}\Omega = \\ = \int_0^{\pi/2} A_S(\vartheta') D'_{0z}(\vartheta') \mathrm{d}\vartheta'. \tag{II.9.5}$$

In the case of isotropic sky brightness,

$$A_D = 2\int_0^{\pi/2} A_S(\vartheta') \cos \vartheta' \sin \vartheta' \mathrm{d}\vartheta'. \tag{II.9.5a}$$

For total radiation,

$$A_Q = \frac{F_2(\mathrm{O})}{Q_0} = \frac{A_D + A_S\dfrac{S'_0}{D_0}}{1 + \dfrac{S'_0}{D_0}}. \tag{II.9.6}$$

Calculation of the reflection coefficient from the formulae derived requires the upward intensities and fluxes at the uppermost level of the stand. The corresponding general formulae which serve as a basis for calculation were derived in Chapter II.6.

Albedo of a stand with horizontal leaves. From (II.6.15) taking into account (II.9.1) we derive

$$A_S = A_D = A_Q = \pi b_S = \pi b_D = \pi b_Q = \qquad \text{(II.9.7)}$$
$$= \frac{A'(1 - A'A_B) - (A' - A_B)\exp(-2aL_0)}{1 - A'A_B - A'(A' - A_B)\exp(-2aL_0)},$$

where according to (II.7.19) $A' = R_L/(1 - T_L + a)$ is the albedo of the stand at $L_0 \to \infty$.

If $R_L = T_L = \frac{\omega}{2}$, then

$$A' = \frac{1 - \sqrt{1 - \omega_L}}{1 + \sqrt{1 - \omega_L}}.$$

In the case of single scattering $A' = R_L/2$. Consequently, the albedo of a stand consisting of randomly disposed horizontal leaves is half as large as that of a horizontal monolayer. This is caused by the mutual shading of leaves, as a result of which the radiation reflected from the lower layers is absorbed by the overlying leaves and does not reach the uppermost level of the stand (the 'black body' effect). From the condition $A' = R_L$ we obtain a combination of R_L and T_L, which provides the same amount of reflection from a dense and high stand ($L_0 \to \infty$) as from a monolayer. This condition gives $R_L = (1 - 2T_L)^{\frac{1}{2}}$. At $R_L = 0.09$ the compensating transmission should be equal to $T_L = 0.496$. At $R_L = 0.25$, $T_L = 0.469$. Real leaves have smaller transmission, we can, therefore, state that for real stands $R_L/2 < A' < R_L$, i.e. the albedo of the stand for PAR is somewhat lower than that of a monolayer (reflection coefficient of leaves). This conclusion may be supposed to be valid not only for a stand with horizontal leaves but for any deep stand. It is not valid, however, at large values of ω_L, for example, in the case of NIR, when at low solar elevations the albedo of a stand can exceed that of a monolayer.

From formulae (II.7.20) and (II.7.21) we obtain the following expression for the albedo of a comparatively dense stand

$$A = A' + \triangle A' \qquad \text{(II.9.8)}$$

where

$$\triangle A = \frac{(A_B - A')(1 - A'^2)}{1 - A'A_B}\exp(-2aL_0). \qquad \text{(II.9.9)}$$

The sign of the correction $\triangle A$ is determined by the difference $A_B - A'$ and can either be positive or negative.

Formulae (II.7.22) allow determining the albedo of a stand for strongly absorbing leaves (e.g. in PAR) with single scattering taken into account

$$A = \frac{R_L}{2}[1 - \exp(-2L_0)] + A_B\exp(-2L_0). \qquad \text{(II.9.10)}$$

The first term in expression (II.9.10) takes account of reflection from the leaves; the second, the radiation which reaches the ground surface, is reflected from it and escapes upwards without interacting with leaves on the whole path length.

Table II.9.1 *Albedo of a deep stand ($L_0 \to \infty$) with horizontal leaves at different values of reflection R_L and transmission T_L coefficients in the case of single (A'_0) and multiple (A')scattering*

T_L	0.00	0.02	0.06		0.12		0.50		0.35
R_L	0.05	0.13	0.09		0.13		0.35		0.50
ω_L	0.05	0.15	0.15	0.15	0.25	0.65	0.85	0.85	0.85
A'_0	0.025	0.065	0.045	0.0375	0.065	0.162	0.175	0.212	0.250
A'	0.0255	0.067	0.048	0.041	0.074	0.256	0.410	0.442	0.470
$\frac{A'}{A'_0}$	1.015	1.03	1.07	1.09	1.14	1.57	2.35	2.06	1.88
A_D				0.037	0.065	0.236		0.417	

In the case of pure scattering we obtain the following from (II.7.23)

$$A = \frac{A_B + R_L(1-A_B)L_0}{1+R_L(1-A_B)L_0}. \tag{II.9.11}$$

Since $A_B < 1$, the albedo of the stand in this case increases with the increasing depth of stand L_0, and at $L_0 \to \infty$, $A = 1$. The albedo also increases with increasing reflection ($R_L = 1 - T_L$).

It is seen from Table II.9.1 that the albedo of a stand with horizontal leaves ranges from 0.04 to 0.07 in the PAR, from 0.40 to 0.50 in the NIR, and from 0.25 to 0.30 in the short-wave radiation region.

Albedo of a stand for direct solar radiation. Taking into account (II.6.46b) we obtain an expression for the brightness of a stand b_S in the case of direct solar radiation from (II.9.1)

$$b_S(r_S, r) = \frac{1}{\pi}\Big\{ b'_S(r_S, r) - [b'_S(r_S, r) - A_B] \times$$

$$\exp\left[-L_0\left(\frac{G_S}{\mu_0} + \frac{G}{\mu}\right)\right]\Big\}, \tag{II.9.12}$$

where according to formula (II.6.48)

$$\frac{1}{\pi} b_S(r_S, r) = \frac{1}{\pi} \frac{\Gamma(r_S, r)}{G_S\mu + G\mu_S}$$

is the brightness coefficient at $L_0 \to \infty$.

The brightness coefficient can be converted to albedo in accordance with formula (II.9.4), but since $b(r_S, r)$ is not integrable in elementary functions over r, we cannot obtain an analytical expression for A_S. Instead we shall use the expression for the flux H_{S2} (0) in the Schwarzschild approximation. When $L_0 \to \infty$, taking into account (II.6.41) we obtain the following expression from (II.9.4)

$$A'_S = \frac{\omega_L}{1+\alpha} \frac{1}{1+\alpha \frac{\sin h_\odot}{G(r_S)}} \tag{II.9.13}$$

in the single scattering approximation $\alpha = 1$ and

$$A'_{S0} = \frac{\frac{\omega_L}{2}}{1+\alpha \frac{\sin h_\odot}{G(r_S)}}. \tag{II.9.14}$$

For uniformly oriented leaves

$$A'_S = \frac{1-\alpha}{1+2\alpha \sin h_\odot} \tag{II.9.15}$$

and for vertical leaves

$$A'_S = \frac{1-\alpha}{1+\frac{\pi}{2}\alpha \tan h_\odot}. \tag{II.9.16}$$

In I.6.3 for the G-function was proposed an empirical formula (II.6.2) $G(h_\odot) = G_1 + (G_2 - G_1)\sin h_\odot$, where G_1 and G_2 are functions of the inclination index of foliage area χ_L. Including (II.6.2) in (II.9.13) we obtain a semiempirical formula for albedo

$$A'_S = (1-\alpha) \frac{1 + \frac{G_2 - G_1}{G_1} \sin h_\odot}{1 + \frac{\alpha + G_2 - G_1}{G_1} \sin h_\odot} \tag{II.9.17}$$

which defines the dependence of albedo on three factors: solar elevation $h_\odot$, the leaf scattering coefficient $\omega_L = 1 - \alpha^2$ and the orientation of leaves.

If other plant organs must be taken into account besides leaves, the G-functions in the formulae for albedo should be replaced with the $\bar{G}$-function in accordance with formula (II.1.61a).

In grasses, for example, we must take into account spikelets and stalks, for which $\bar{G}$ is given by formula (II.1.6a), with $G_C = G_R = \cos h_\odot / \pi$.

According to Table I.6.2 the orientation of leaves in numerous cereal crops is rather close to a uniform orientation. Taking into account (II.1.62b) we obtain the following for the albedo of grasses (cereals) from (II.6.13)

$$A'_S(h_\odot) = \frac{1-\alpha}{1+\frac{2\alpha \sin h_\odot}{\beta_L + \frac{2}{\pi}(\beta_C + \beta_R)\cos h_\odot}}. \tag{II.9.18}$$

When the stand has a finite depth, general expressions for albedo will be very complicated, e.g., with uniformly oriented leaves the albedo of a stand in the Schwarzschild approximation will be (Ross 1962)

$$A_S=\frac{A'_S(1-C_1A_B)-C_3(C_1-A_B)e^{-2\alpha L_0}+\mu_8(1-C_1A_B)e^{-L_0[\alpha+(1/\mu_S)]}}{1+C_1A_B-(C_1-A_B)e^{-2\alpha L_0}}, \tag{II.9.19}$$

were

$$C_1=\frac{1-\alpha}{1+\alpha},\quad C_3=\frac{1-\alpha}{1-2\alpha\mu_S}$$

$$\mu_8=\frac{4\alpha}{(1+\alpha)^2}\left[A_B\left(1-\frac{\omega_L}{2}\frac{1+2\mu_S}{1-4\alpha^2\mu_S^2}\right)-\frac{\omega_L}{2}\frac{1-2\mu_S}{1-4\alpha^2\mu_S^2}\right].$$

Albedo for diffuse sky radiation. In the general case the calculation of albedo from formulae (II.9.5) and (II.9.2) is complicated and laborious. Only in the special case of isotropic sky brightness and uniform leaf orientation, taking into account (II.9.15) we obtain a simple expression for A_D from (II.9.5)

$$A_D=\frac{1-\alpha}{\alpha^2}\left[\alpha-\frac{1}{2}\ln(1+2\alpha)\right]. \tag{II.9.20}$$

If the stand depth is finite, using the same assumptions and the Schwarzschild approximation we can obtain the following expression (Ross 1962)

$$A_D=\frac{(A'_D-C_4)(1-C_1A_B)-C_5(C_1-A_B)e^{-2\alpha L_0}+\mu_g e^{-\alpha L_0}}{1-C_1A_B-(C_1-A_B)e^{-2\alpha L_0}}, \tag{II.9.21}$$

where

$$\left.\begin{aligned}
C_4&=\frac{1-\alpha}{\alpha^2}E_1\left[(1+2\alpha)\frac{L_0}{2}\right],\\
C_5&=F_1\frac{L_0}{2}-\frac{1-\alpha}{\alpha^2}\left(1+\frac{1}{2}\ln|1-2\alpha|\right),\\
\mu_g&=\frac{C_1}{\alpha}\left[(1+A_BE_1\left(\frac{L_0}{2}\right)-2(1-A_B)E_2\left(\frac{L_0}{2}\right)+\right.\\
&\left.+\frac{8\alpha^2A_B}{\omega_L}E_3\left(\frac{L_0}{2}\right)\right],
\end{aligned}\right\} \tag{II.9.22}$$

and C_1, $E_n(x)$ and $F_1(x)$ are given by the formulae at (II.6.40).

When calculating A_D from A_S according to formula (II.9.5) individual values of $A_S(\vartheta')$ are taken into account as the weight of $D_{oz}(\vartheta')$. As seen from Figure 82 the largest weight is contributed at zenith angles from 40 to 50°. For approximate calculations we can assume $A_D=A_S(h_E)$ where h_E is a certain effective solar elevation at which albedo for direct solar radiation and that for diffuse sky radiation are equal. Table II.9.2 gives values of A'_D for uniformly oriented leaves, as calculated from formula (II.9.21), and effective solar elevations determined from formula (II.9.15).

Table II.9.2 *Albedo for diffuse sky radiation, and effective solar elevations h_E at which $A'_D = A'_S(h_E)$*

ω	0.15	0.25	0.65	0.85
A'_D	0.037	0.065	0.236	0.417
h_E	38	38	38.5	37.5

It will be observed from table II.9.2 that in the case of a deep stand with uniformly oriented leaves the effective solar elevation is equal to 38° and is practically the same in all regions of the spectrum. A similar situation is expected to exist for other leaf orientations.

Albedo of a stand in the case of single scattering. Single scattering may be assumed for PAR. Taking into account (II.6.56) we derive an expression for albedo for direct solar radiation from (II.9.4)

$$A_{SO} = A'_{SO} + (A_B - A'_{SO}) \exp\left[-L_0 \left(1 + \frac{G_S}{\mu_S}\right)\right], \qquad \text{(II.9.23)}$$

where A'_{SO} is given by formula (II.9.14).

A somewhat more exact formula is obtained by replacing A'_{SO} in (II.9.23) with A'_S. Then

$$A_S = A'_S + (A_B - A'_S) \exp\left[-L_0 \left(1 + \frac{G_S}{\mu_S}\right)\right]. \qquad \text{(II.9.24)}$$

This formula is valid for PAR and short-wave radiation, but not sufficiently exact for NIR.

In a sparse stand,

$$A_S = A_B + (A_B - A'_S)\left(1 + \frac{G(r_S)}{\sin h_\odot}\right) L_0. \qquad \text{(II.9.25)}$$

The above derived formulae define the dependence of the albedo of the stand on the solar elevation $h_\odot$, the depth L, the leaf orientation and scattering coefficient, and the albedo of the ground surface A_B. Let us discuss the dependence of albedo on the above factors on the basis of these formulae.

II.9.3 ALBEDO AND ITS DEPENDENCE ON VARIOUS FACTORS

Dependence of albedo on the leaf scattering coefficient. Consider Figure 89 which presents this dependence for diffuse sky radiation. It is seen that albedo increases with the increasing leaf scattering coefficient ω_L and that this increase is more rapid at large values of ω_L (NIR region). Consequently, slight changes in the optical properties of phytoelements in the NIR region can cause noticeable changes in the albedo of the stand in this spectral region. This dependence is more pronounced in a deep stand, where the role of multiple scattering is greater. The albedo A_D of a dense stand is not large in PAR, ranging from 0.03 to 0.07, as compared with A_D=0.4–0.5 in the NIR region and A_D=0.23–0.27 in the short-wave radiation region. Analogous relations exist for albedo in the case of direct solar radiation.

Dependence of albedo of a system 'plant stand – soil' on the albedo of the soil and the depth of stand. If the stand is not dense, the albedo of this system is determined, in accordance with formula (II.9.23), by reflection from leaves and reflection from the ground surface below the stand. The influence of this latter increases with decreasing depth of the foliage layer above the ground and with increasing albedo of the soil. Figure 110 presents the albedo of the system for diffuse short-wave radiation of the sky as a function of L_0. The different curves correspond to different values of the soil reflection coefficient (albedo) A_B. Largest variations of the albedo of the system occur as the depth L_0 varies from 0 to 2, i.e.

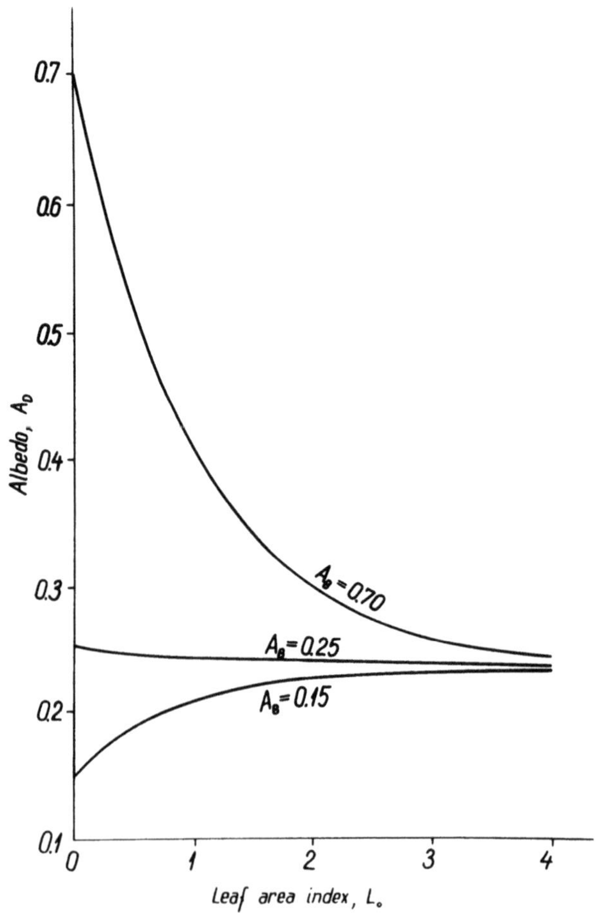

Fig. 110. Reflection coefficient (albedo) of plant cover for diffuse short-wave radiation as a function of stand depth L_0 for different values of the reflection coefficient of the ground surface A_B. Calculated from equation (II.9.21), ω_L=0.65. A_B=0.15, corresponds to bare ground, A_B=0.25, to grassy undergrowth or bright litter, A_B=0.70, to snow cover.

exactly in a sparse stand. The direction of this variation, according to formula (II.9.23), depends on the term $A_B - A'_D$ characteristic of the contrast between leaves and soil. If the soil is darker than leaves, as when it is wet, with $A_B = 0.15$, the albedo A_D increases with the increasing depth L_0. If $A_B = A'_D$, albedo does not depend on L_0. Under natural conditions the highest contrast exists with a snow cover ($A_B = 0.70$) below the green plants. In this case albedo decreases twofold if L increases from 0 to 1.5. The contrast with snow is still higher in the PAR region. In the case of bare soil, dead undergrowth or grass below the stand the contrasts in the PAR region between the ground surface and the stand are small, and formulae (II.9.13) and (II.9.18) can be used for A'_S and A'_D, respectively. A different situation exists in the NIR region. If the soil below the stand is not covered with green plants, a noticeable contrast exists between plants and soil. This case, discussed by Nilson (1968a), is shown in Figure 111, which presents the dependence of A_S on L_0 at three specified solar elevations for different leaf orientations. According to this graph, the albedo of maize at high solar elevations in the NIR region is lower than that of sunflower. An inverse situation occurs at low solar elevations. It should be noted that this difference is only due to the different leaf orientation, with the same leaf scattering coefficient $\omega_L = 0.85$.

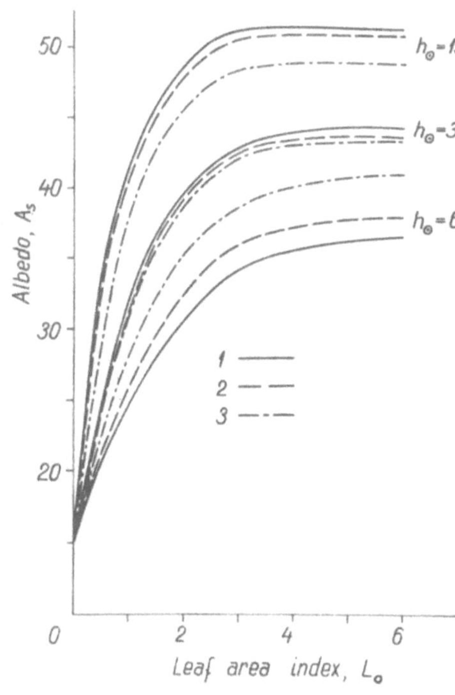

Fig. 111. Reflection coefficient (albedo) for direct NIR as a function of stand depth L_0 at different solar elevations $h_\odot$ for various foliage orientations. Calculated using Nilson's procedure (1968a), $A_B = 0.15$, $\omega_L = 0.85$.
1 – uniformly oriented leaves, 2 – maize, 3 – sunflower.

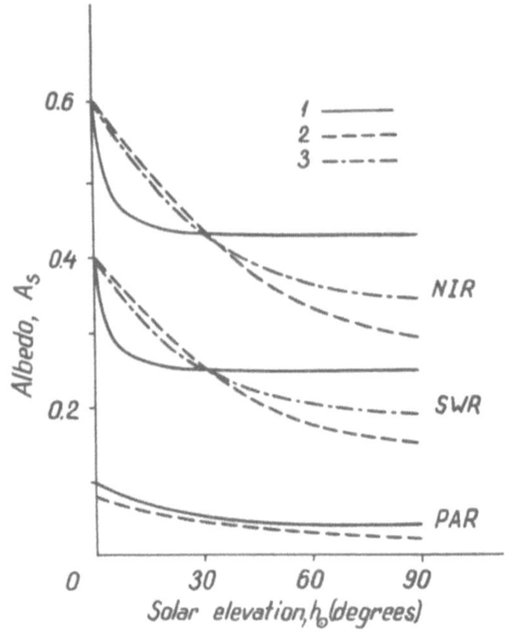

Fig. 112. Reflection coefficient (albedo) of a dense stand for direct PAR ($\omega_L = 0.15$), NIR ($\omega_L = 0.85$), and short-wave radiation ($\omega_L = 0.65$) as a function of solar elevation. Calculated from semiempirical equation (II.9.17) $L_0 \rightarrow \infty$.
1 – most horizontal leaves $\chi_L - 0.65$; 2 – most vertical leaves $\chi_L \approx 0.6$, 3 – *uniformly oriented leaves.*

Dependence of albedo on solar elevation and leaf orientation. Some general conclusions as to the dependence of albedo on solar elevation and leaf orientation can be drawn from Figure 112 which shows the dependence of the albedo A'_S of a dense stand on solar elevation as calculated from semiempirical formula (II.9.17).

1. The dependence of albedo on leaf orientation in PAR is insignificant and can be neglected. The dependence of albedo on solar elevation is weak and can also be neglected for approximate calculations.
2. In the NIR and short-wave radiation regions the effect of leaf orientation is substantial at high and low solar elevations. At high solar elevations the albedo of a dense stand A'_S increases as leaves become more horizontal and decreases as they become more vertical. The situation is reversed at low solar elevations.
3. In the solar elevation range $h_\odot = 20$–$40°$ albedo is practically independent of leaf orientation.
4. The dependence of albedo A'_S on solar elevation becomes less pronounced as leaves become more horizontal.
5. The region between curves 1 and 2 characterizes maximum variations of the albedo A'_S due to variations of the geometrical structure of real stand types. These variations are largest in the NIR and smallest in the PAR region.

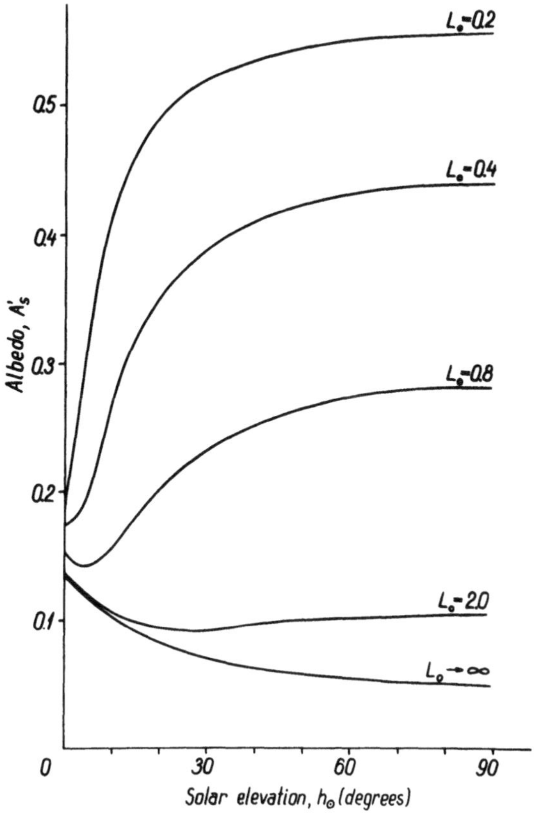

Fig. 113. Reflection coefficient (albedo) of system 'green stand above snow cover' for direct PAR, plotted against solar elevation for different stand depths L_0. Calculated from equation (II.9.20), $\omega_L=0.25$, $A_B=0.70$.

The examples discussed show that albedo decreases with increasing solar elevation in summer time when $A_B - A'_S < 0$. But if the soil below the stand is covered with snow (a coniferous forest in winter), this dependence is more complicated as shown in Figure 113. In a sparse stand albedo increases with increasing solar elevation, but it decreases in a dense stand. In the interval of depths $L_0 = 1$–3 the dependence of albedo on solar elevation is weak.

The albedo of a stand for total radiation is defined by formula (II.9.6) and also depends on solar elevation. Figure 114 shows examples of this dependence. Shown in the upper part of the graph is the dependence of the albedo of a sparse stand above a cover of snow for direct and total short-wave radiation, and the lower part shows corresponding dependences for a dense stand. According to this graph, the albedo for total radiation at high solar elevations differs but slightly from the albedo for direct solar radiation. A difference appears only at $h_\odot < 20°$. The albedo of a dense stand for total radiation has a maximum at approximately $h_\odot = 10°$. The presence of a maximum is also confirmed by the experimental data of Tooming (1960). In the case of the system 'sparse stand above snow' the albedo for total radiation has a minimum at approximately $h_\odot = 5°$.

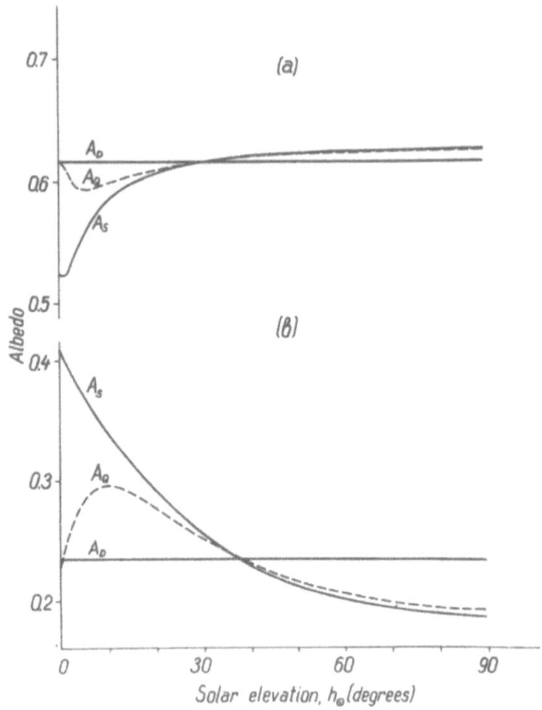

Fig. 114. Reflection coefficient (albedo) of plant stand as a function of solar elevation, for total (A_Q), direct (A_S) and diffuse (A_D) short-wave radiation. Calculated from equations (II.9.6), (II.9.20) and (II.9.21); $\omega_L = 0.65$, uniformly oriented leaves.

(a) – sparse green vegetation above snow cover, $L_0 = 0.2$, $A_B = 0.70$; (b) – dense stand $L_0 \to \infty$.

II.9.4 COMPARISON OF CALCULATED AND EXPERIMENTAL DATA

Most of the experimental data on the albedo of plant stands relate either to total short-wave radiation or to diffuse short-wave radiation of a cloudy sky. Data on the albedo for direct solar radiation are comparatively scarce, since direct measurement is impossible. The albedo A_S is determined indirectly from formula (II.9.6) by consccutively measuring A_Q, with the solar disk not covered by clouds, and A_D, with the solar disk covered by clouds, at the same solar elevation.

Plotted as points in Figure 115 are experimental data of the albedo A_S of a rye field for direct solar short-wave radiation (Ross 1957). The curve in this figure was calculated from formula (II.9.18) at $\omega_L = 0.65$. It is seen that within the spread of experimental data the theoretical curve describes the variation of albedo very well.

Figure 115B presents experimental data on the albedo of a maize stand (Tooming & Ross 1964) at $L_0 = 3$–7. The curve was calculated from formula (II.9.6) at $\omega_L = 0.65$ and $L_0 \to \infty$; A_D was calculated from formula (II.9.20), and $A'_S(h_\odot)$ from formula (II.9.15). It is seen that the agreement is good. It can be

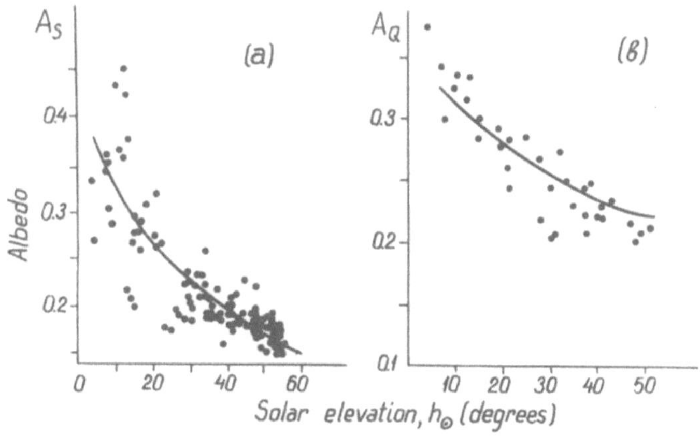

Fig. 115. Reflection coefficient (albedo) of crops as a function of solar elevation. Plotted points correspond to experimental data.
(a) – for direct solar radiation, in a rye field at heading and flowering stage; (b) – for total solar radiation in a maize stand at tassel flowering stage, $L_0 = 3$–7.

improved if in addition to leaves the stalks and tassels of maize are taken into account in accordance with formula (II.9.18).

It can be stated in summary that the proposed formulae for the albedo of a plant stand, derived from a mathematical model of the radiation regime in stands, coincide very well with experimental data, thus proving the general theory developed in the present work.

A number of experimental data on the albedo A_Q of grass and a maize crop has been obtained by Tooming (1960) at our laboratory. According to his data the albedo of a green stand is asymmetrical about midday; after midday the albedo is 5–20% higher than at the corresponding hours before midday. This asymmetry can be attributed to a change in optical properties of leaves, particularly, in the NIR region. Tooming has also found that as S_0'/D_0 increases, the albedo of the stand A_Q also increases at low solar elevations and somewhat decreases at high elevations.

II.10 Calculation of long-wave radiation in a stand

Calculation of long-wave radiation fluxes in stands is sometimes simpler and sometimes more complicated than that of short-wave radiation. A simplifying circumstance is the fact that the scattering coefficient of phytoelements in the long-wave region $\omega = 0.05$, and therefore the integral scattering term in transfer equation (II.1.53) may be neglected. Then the integral-differential equation (II.1.53) reduces to a usual differential equation in the form

$$\frac{\mathrm{d}i(L,r)}{\mathrm{d}L}\cos\vartheta = -G(L,r)i(L,r) + \eta(L,r) \qquad \text{(II.10.1)}$$

or, taking into account (II.1.2) and (II.1.3),

$$\left.\begin{aligned} \frac{\mathrm{d}i_1(L,r)}{\mathrm{d}L}\cos\theta &= -G(L,r)i_1(L,r) + \eta(L,r) \\ \frac{\mathrm{d}i_2(L,r)}{\mathrm{d}L}\cos\vartheta &= G(L,r)i_2(L,r) - \eta(L,r) \end{aligned}\right\} \qquad \text{(II.10.2)}$$

with the boundary conditions

$$\left.\begin{aligned} &i, (0,r) = \varepsilon_A(r), \\ &i_2(L_0, r) = \varepsilon_B + \frac{A_{BJ}}{\pi}\int_{\Omega_1} i_1(L_0, r')\cos\theta' \mathrm{d}\Omega', \end{aligned}\right\} \qquad \text{(II.10.3)}$$

where ε_A is the intensity of long-wave radiation of the atmosphere, ε_B that emitted from the ground surface below the stand, $\eta(L, r)$ is the emission coefficient of the medium, defined by formula (II.1.50), and $A_{BJ} = 1 - \delta_B$ is the albedo of the ground surface for long-wave radiation.

The long-wave radiation of the ground surface is usually assumed to be isotropic. Then analogously to (II.1.32),

$$\varepsilon_B(r) = \frac{1}{\pi}E_B = \frac{\delta_B}{\pi}\sigma(273 + t_B)^4, \qquad \text{(II.10.4)}$$

where t_B, E_B and δ_B are the temperature, irradiance of long-wave radiation emitted by the ground surface and the emissivity of the ground surface, respectively.

Assuming $G(L, r) = G(r)$ and $\eta(L, r)$ to be known we obtain the solution of (II.10.2) in the form

$$i_1(L, r) = \varepsilon_A(r) \exp\left(-\frac{LG}{\cos\theta}\right) + \\ + \int_0^L \frac{\eta(L', r)}{\cos\theta} \exp\left[-\frac{(L-L')G}{\cos\theta}\right] \mathrm{d}L', \qquad (11.10.5)$$

$$i_2(L, r) = \frac{E_B}{\pi} \exp\left[-(L_0 - L)\frac{G}{\cos\vartheta}\right] + \\ + \int_L^{L_0} \frac{\eta(L', r)}{\cos\vartheta} \exp\left[-(L'-L)\frac{G}{\cos\vartheta}\right] \mathrm{d}L' + \\ + A_{BJ} \exp\left[-(L_0 - L)\frac{G}{\cos\vartheta}\right] \int_{\Omega_1} i_1(L_0, r') \cos\theta' \mathrm{d}\Omega'. \qquad (II.10.6)$$

Formulae (II.10.5) and (II.10.6) are initial expressions for the calculation of long-wave radiation in stands. The first term in expression (II.10.5) for i_1 denotes the long-wave radiation of the atmosphere which has penetrated the stand to the depth L without interception. The second term corresponds to the long-wave radiation from plant parts generated in the layer $(0, L)$.

In expression (II.10.6) for i_2 the first term denotes the long-wave radiation from the ground surface which has penetrated the layer (L, L_0) without interacting with plant parts. The second term corresponds to the long-wave radiation of plant parts generated in the layer (L, L_0), the third term describes the part of the downward flux reflected from the ground surface, which has reached the level L without attenuation.

The angular dependence of long-wave radiation intensities is determined by the angular dependence of the long-wave radiation of the atmosphere and the orientation of phytoelements in the expression for $\eta(L_1 r)$.

The greatest difficulties of calculating long-wave radiation from formulae (II.10.5) and (II.10.6) are associated with determining the temperature of phytoelements, which governs their long-wave radiation and is included in the expression for $\eta(L, r)$ in accordance with formula (II.1.50). The temperature of plant organs is determined by the total effect of their radiative and water-and-heat exchange between them (leaves) and the environment and can theoretically be determined by solving the corresponding set of equations. Under steady-state conditions the leaf temperature can be determined from its heat balance equation.

Considerable simplifications, however, are permissible in a number of cases. If the water supply is not a limiting factor and air humidity is moderate, then temperature differences between separate leaves are small and leaf temperature may be assumed to be equal to air temperature. In the shortage of water leaf temperature increases in excess of air temperature and correlates with the amount of short-wave radiation absorbed by the leaf. In this case the temperature of leaves in the shade can also be considered as equal to that of the air, but the temperature of the sunny leaves is higher and depends on the angle between the leaf normal and the direction of sunrays.

Consider in greater detail a special case when the leaf temperature is the same in a horizontal layer and only depends on the depth L. Then according to (II.1.31) and (II.1.32) the leaf emission coefficient is

$$\eta_L(L) = \frac{\delta_L}{\pi}\,\sigma[273 + t_L(L)]^4 = \frac{E_L(L)}{\pi}. \tag{II.10.7}$$

According to (II.1.50) the emission coefficient of the medium is

$$\eta_L(L) = \eta_L(L) G(r) = \frac{E_L(L)}{\pi} G(n) \tag{II.10.8}$$

and assuming $\varepsilon_A = \varepsilon_A(\theta)$ we obtain the following expressions for the irradiances $E_1(L)$ and $E_2(L)$ from (II.10.5) and (II.10.6):

$$\begin{aligned} E_1(L) &= \int_{\Omega_1} i_1(L_1 r) \cos\theta \,\mathrm{d}\Omega = \\ &= 2\pi \int_0^{\pi/2} \varepsilon_A(\theta) \exp\left[-\frac{LG(\theta)}{\cos\theta}\right] \cos\theta \sin\theta \,\mathrm{d}\theta + \\ &+ \int_0^L E_L(L')\{2\int_0^{\pi/2} G(\theta) \exp\left[-(L-L')\frac{G(\theta)}{\cos\theta}\right] \sin\theta \,\mathrm{d}\theta\}\mathrm{d}L', \end{aligned} \tag{II.10.9}$$

$$\begin{aligned} E_2(L) &= \int_{\Omega_2} i_2(L, r) \cos\vartheta \,\mathrm{d}\Omega = \\ &= 2E_B \int_0^{\pi/2} \exp\left[-(L_0 - L)\frac{G(\vartheta)}{\cos\vartheta}\right] \cos\vartheta \sin\vartheta \,\mathrm{d}\vartheta + \\ &+ \int_L^{L_0} E_L(L')\left\{2\int_0^{\pi/2} G(\vartheta) \exp\left[-(L'-L)\frac{G(\vartheta)}{\cos\vartheta}\right] \sin\vartheta \,\mathrm{d}\vartheta\right\}\mathrm{d}L' + \\ &+ 2\pi(1-\delta_B) \int_0^{\pi/2} \exp\left[-(L_0 - L)\frac{G(\vartheta)}{\cos\vartheta}\right] \cos\vartheta \sin\vartheta \,\mathrm{d}\vartheta \times \end{aligned} \tag{II.10.10}$$

$$\begin{aligned} &\times 2\pi \int_0^{\pi/2} \varepsilon_A(\theta) \exp\left[-L_0 \frac{G(\theta)}{\cos\theta}\right] \cos\theta \sin\theta \,\mathrm{d}\theta + \\ &+ 2\pi(1-\delta_B) \int_0^{\pi/2} \exp\left[-(L_0 - L)\frac{G(\vartheta)}{\cos\vartheta}\right] \cos\vartheta \sin\vartheta \,\mathrm{d}\vartheta \times \\ &\times \int_0^{L_0} E_L(L)\left\{2\int_0^{\pi/2} G(\theta) \exp\left[-(L_0 - L')\frac{G(\theta)}{\cos\theta}\right] \sin\theta \,\mathrm{d}\theta\right\} dL'. \end{aligned}$$

In expression (II.10.9) for the downward flux of long-wave radiation the first term defines that part of long-wave radiation of the atmosphere which penetrates the layer $(0, L)$ of the stand and reaches the level L, the second term defines the downward flux of the long-wave radiation of leaves in the layer $(0, L)$, which reaches the level L. In expression (II.10.10) for the upward flux the first term defines the part of long-wave radiation of the soil which penetrates the layer (L, L_0) and reaches the level L, the second term characterizes the upward flux of long-wave radiation of leaves in the layer (L, L_0) which reaches the level L, and the third term characterizes that part of long-wave radiation of the atmosphere which penetrates the whole layer of the stand, is reflected from the soil and, penetrating through the layer (L, L_0), reaches the level L. The last term characterizes that part of the downward flux of radiation of the leaves of the whole stand which reaches the soil, is reflected from it, penetrates the layer (L, L_0) and reaches the level L. Formulae (II.10.9) and (II.10.10) for irradiances at $\delta_B=1$ were first derived in our work (Budagovsky et al. 1968).

Since $1-\delta_B \approx 0.05$, the last two terms in expression (II.10.10) E_2 can be neglected as a first approximation.

Above the stand $(L=0)$

$$E_1(0)=E_A=2\pi\int_0^{\pi/2} \varepsilon_A(\theta)\cos\theta\sin\theta d\theta \tag{II.10.11}$$

and

$$E_2(0)=E_{20}=E_B a_D(L_0)+\int_0^{L_0} E_L(L')a_G(L')dL', \tag{II.10.12}$$

where a_D is given by formula (II.5.7a), and

$$a_G(x)=2\int_0^{\pi/2} G(\vartheta)\exp\left[-\frac{xG(\vartheta)}{\cos\vartheta}\right]\sin\vartheta d\vartheta. \tag{II.10.13}$$

The first term in expression (II.10.12) characterizes the part of the soil, and the second defines the part of the stand in the total upward long-wave radiation flux of the system 'soil-plant stand'.

On the soil surface $(L=L_0)$

$$E_1(L_0)=E_A a_D(L_0)+\int_0^{L_0} E_L(L')a_G(L_0-L')dL' \tag{II.10.14}$$

and

$$E_2(L_0)=E_B.$$

The difference of fluxes $E_1(L)$ and $E_2(L)$ is termed the net long-wave radiation B_1

$$B_I(L)=E_1(L)-E_2(L). \tag{II.10.15}$$

Above the stand

$$B_I(0)=B_{IO}=E_A-E_{20}, \tag{II.10.16}$$

below the stand

$$B_I(L_0)=E_1(L_0)-E_B. \tag{II.10.17}$$

In the case of isotropic long wave-radiation of the atmosphere $\varepsilon_A=E_A/\pi$ and instead of (II.10.9), (II.10.10), and (II.10.15) we obtain

$$E_1(L)=E_A a_D(L)+\int_0^L E_L(L')a_G(L-L')\mathrm{d}L', \tag{II.10.9a}$$

$$E_2(L)=E_B a_D(L_0-L)+\int_L^{L_0} E_L(L')a_G(L'-L)\mathrm{d}L' \tag{II.10.10a}$$

$$B_I(L)=E_A a_D(L)-E_B a_D(L_0-L)+$$

$$+\int_0^L E_L(L')a_G(L-L')\mathrm{d}L'-\int_L^{L_0} E_L(L')a_G(L'-L)\mathrm{d}L'. \tag{II.10.15a}$$

In a special case when the temperature of phytoelements is the same for the whole stand, $E_L(L)=E_L$ and instead of (II.10.9a), (II.10.10a) and (II.10.15a) we obtain, taking into account

$$\int_L^{L_0} a_G(L-L')\mathrm{d}L'=1-a_D(L)$$

and

$$\int_L^{L_0} a_G(L'-L)\mathrm{d}L'=1-a_D(L_0-L)$$

$$E_1(L)=E_L-(E_L-E_A)a_D(L), \tag{II.10.9b}$$

$$E_2(L)=E_L-(E_L-E_B)a_D(L_0-L), \tag{II.10.10b}$$

$$B_I(L)=-(E_L-E_A)a_D(L)-(E_B-E_L)a_D(L_0-L). \tag{II.10.15b}$$

From the last expression it is seen that if $E_B>E_L$, then always $B_I<0$. $B_I=0$ if $(E_L-E_A)/(E_B-E_L)=a_D(L_0-L)/a_D(L)$. At the uppermost level of the stand

$$B_{I0}=-(E_L-E_A)-(E_B-E_L)a_D(L_0). \tag{II.10.16a}$$

For a dense stand $E_B\approx E_L$, $E_2=E_L$ and

$$B_I(L)=-(E_L-E_A)a_D(L). \tag{II.10.18}$$

As a rule, $E_L>E_A$. Consequently in a dense stand net long-wave radiation is negative and proportional to the difference between the long-wave radiation of the leaves and that of the atmosphere, the coefficient of proportionality $a_D(L)$ depending only on the stand architecture. The absorption of long-wave radiation by the whole stand

$$\Pi_{LI}(L_0)=B_I(0)-B_I(L_0)=(E_A+E_B-2E_L)[1-a_D(L_0)]. \tag{II.10.19}$$

It is seen from expression (II.10.19) that the long-wave radiation absorbed by the stand is proportional to the difference $(E_A+E_B)-2E_L$. If $E_A+E_B<2E_L$, the stand will emit long-wave radiation rather than absorb it. As the depth of stand increases the coefficient of proportionality $1-a_D(L)$ also increases.

It is important to note that in the above formulae denotes the long-wave radiation coming only from one side of the leaf. The emission of the whole leaf is $2E_L$. Let us discuss two special cases of leaf orientation.

1. *Uniform orientation*, $G=\frac{1}{2}$. In accordance with (II.5.12a)

$$a_D(L)=2E_3\left(\frac{L}{2}\right)$$

and

$$E_1(L)=E_L-2E_3\left(\frac{L}{2}\right)(E_L-E_A),$$

$$E_2(L)=E_L-2E_3\left(\frac{L_0-L}{2}\right)(E_L-E_B),$$

$$B_I(L)=-2E_3\left(\frac{L}{2}\right)(E_L-E_A)-2E_3\left(\frac{L_0-L}{2}\right)(E_I-E_L),$$

$$\Pi_L(L_0)=\left[1-E_3\left(\frac{L_0}{2}\right)\right](E_A+E_B-2E_L).$$

2. *Horizontal orientation*, $G=\cos\theta$,

$$a_D(L)=\mathrm{e}^{-L}$$

$$E_1(L)=E_A\,\mathrm{e}^{-L}+E_L(1-\mathrm{e}^{-L}),$$

$$E_2(L)=E_B\,\mathrm{e}^{-L}+E_L(1-\mathrm{e}^{-(L_0-L)}),$$

$$B_I(L)=-(E_L-E_A)\,\mathrm{e}^{-L}+(E_L-E_B)\,\mathrm{e}^{-(L_0-L)},$$

$$\Pi_L(L_0)=(E_A+E_B-2E_L)(1-\mathrm{e}^{-L_0}).$$

Determining from the condition $\mathrm{d}B_I/\mathrm{d}L=0$ the value of L^* at which B_I reaches its maximum we obtain the following at $E_B>E_A$

$$L^*=\frac{L_0}{2}+\frac{1}{2}\ln\frac{E_L-E_A}{E_B-E_A}$$

and

$$B_I(L^*)=-2[(E_B-E_L)(E_L-E_A)]^{\frac{1}{2}}\mathrm{e}^{-L_0/2}.$$

If $E_B<E_L$, B_I increases monotonically with the depth L. As a rule, $B_I<0$. Only at $E_B<E_L$ the net radiation B_I passes the zero position at the point

$$L'=\frac{L_0}{2}\ln\frac{E_L-E_A}{E_L-E_B}$$

An approximate theory of long-wave radiation transfer in plant stands, in which the penetration function $a_D(L)$ is given as an empirical function, was first expounded in Saito's work (1964). In his work Groen (1973) derived formulae (II.10.9a) and (II.10.10a) for a uniform orientation of leaves. Groen also discussed the case in which the isotropic scattering coefficient of leaves is equal to $1-\delta_L$. It appears that in this case the stand depths L_0 and $\delta_L L$ should be replaced with the effective quantities $\delta_L L_0$ and $\delta_L L$ and formulae (II.10.9a), (II.10.10a) and (II.10.15a) will remain valid. Groen also proposed a formula for the long-wave radiative heating of plant stands

$$\left(\frac{\partial T_L}{\partial t}\right)_{JR} = \frac{1}{C_L^* \sigma_L} \frac{\mathrm{d}B_I}{\mathrm{d}L},$$

where C_L^* is the specific heat of plant material.

Rauner (1972) proposed an exponential formula for the net long-wave radiation B_I in the form

$$B_I\left(\frac{z}{z_0}\right) = B_I(0) \exp\left(-a_R \frac{z}{z_0}\right),$$

where z_0 is the height of stand. For a birch-tree canopy $a_R = 0.25$, for a cotton crop or a steppe-and-meadow community $a_R = 0.20$.

By way of example, Figure 116 shows the vertical profiles of E_1, E_2 and B_I at two values of L_0 in a stand with horizontal leaves. The calculation used $E_A = 3.83$,

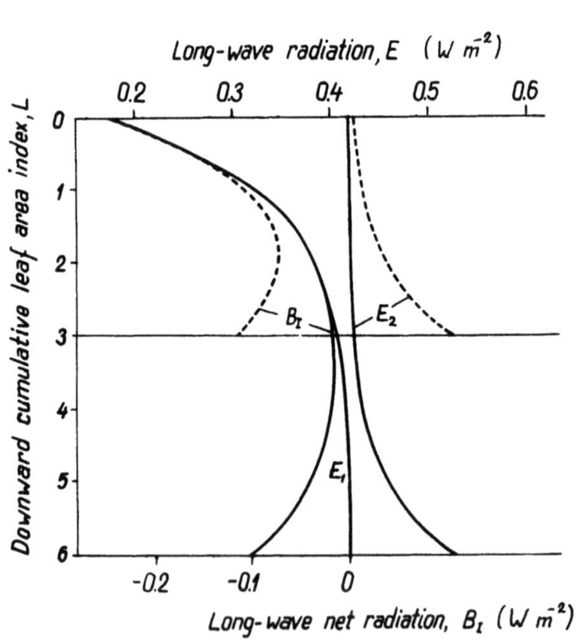

Fig. 116. Vertical profiles of downward (E_1) and upward (E_2) long-wave fluxes and net long-wave radiation B_I in model stand with horizontal leaves for $L_0 = 3$ (dashed lines) and $L_0 = 6$ (solid lines).

$E_L=4.18$ and $E_B=4.53\,\mathrm{W\,m^{-2}}$ characteristic of a cloudless summer day in Tadjikistan. It is seen from Figure 116 that as the depth L increases the downward flux E_1 increases rapidly in the upper layers from E_A to E_L and then remains unchanged. An inverse situation is observed for the upward flux E_2: it decreases in the lower layers of the stand from E_B to E_L. The net longwave radiation $B_I<0$ at the stand boundaries is close to zero in the middle of the stand.

In the case of a dense stand with an adequate water supply the vertical variations of long-wave radiation fluxes are not large and their importance for the radiation regime is not great either. The importance of long-wave radiation fluxes for the process of photosynthesis increases in a sparse stand when the soil below the stand is hot and the role of E_B increases, or in the shortage of moisture when the leaf temperature t_L increases noticeably due to the reduction or even cessation of transpiration. The influence of long-wave radiation on plant life, however, is expressed by means of the thermal regime and is not to be compared to the vast importance of solar radiation for plant life.

Analysis of experimental data (e.g. Monteith & Szeicz 1961) shows that the diurnal variation of B_I is not caused by the variation of E_A but by that of E_B, i.e. by the course of the ground surface temperature, which in turn depends on the diurnal course of absorbed solar radiation $(1-A_Q)Q_0$. Barashkova et al. (1961) have found that in the case of a grass stand there is a linear empirical relation

$$-B_{I0}=a_1(1-A_Q)Q_0+B \qquad \text{(II.10.20)}$$

where $a_1=0.42$.

Häckel (1975) calculated the net long-wave radiation of a long and thin cylinder whose axis makes an angle ψ with a horizontal plane. Depending on ψ the parts of E_A and E_B incident on the cylinder vary. The effect of horizon closure on the decrease in the incoming E_A was evaluated, and it was found that E_A decreases linearly until the closure angle reaches 40°.

II.11 Net radiation in plant stands

Net radiation within a plant stand $B(L)$ is the difference of all downward and upward fluxes of solar and long-wave radiation, i.e.

$$B(L) = Q(L) - F_2(L) + B_I(L) \tag{II.11.1}$$

or taking into account (II.1.80) and (II.1.81)

$$\begin{aligned} B(L) &= S'(L) + D(L) + H_{S1}(L) + H_{D1}(L) - H_{S2}(L) - H_{D2}(L) + E_1(L) \\ &- E_2(L) = \\ &= [a_S(L) + a^H_{S1}(L) - a^H_{S2}(L)]S'_0 + \\ &+ [a_D(L) + a^H_{D1}(L) - a^H_{D2}(L)]D_0 + \\ &+ E_1(L) - E_2(L). \end{aligned} \tag{II.11.2}$$

Net radiation is an important energy characteristic of the stand, and the quantity dB/dL defines the amount of radiant energy which is converted into other forms of energy within a unit layer of the stand. Net radiation as a function of the depth L determines the degree of radiant exchange within the stand in its different layers and is associated with the plant stand microclimate. Net radiation above the plant stand, i.e. the quantity $B(0) = B_0$ is an important term in climatological and geophysical studies.

Taking into account (II.11.1), (II.11.2), (II.9.6) and (II.10.16) we derive the formula for net radiation above the stand

$$\begin{aligned} B_0 &= Q_0 - F_2(0) + B_{I0} = (1 - A_0)Q_0 + B_{I0} = \\ &= (1 - A_S)S'_0 + (1 - A_D)D_0 + E_A - E_{20}. \end{aligned} \tag{II.11.3}$$

Under the stand, and also taking into account (II.10.17)

$$\begin{aligned} B_0(L_0) &= Q(L_0) - F_2(L_0) + B_I(L_0) = \\ &= (1 - A_B)Q(L_0) + E_1(L_0) - E_B = \end{aligned}$$

$$=(1-A_B)[a_S(L_0)+a_{S1}^H(L_0)]S_0'+$$

$$(1-A_B)[a_D(L_0)+a_{D1}^H(L_0)]D_0+E_1(L_0)-E_2. \qquad \text{(II.11.4)}$$

It follows from an analysis of formula (II.11.2) that net radiation in plant stands is determined by the total effect of three groups of factors:

(a) Weather conditions above the stand. These primarily include the fluxes of direct solar radiation (S_0') incident on the stand and of diffuse radiation (D_0), and also the vertical profiles of temperature and humidity at the interface between atmosphere and stand, which determine the long-wave radiation of the atmosphere E_A. The presence of clouds is an important factor also.

(b) Plant stand architecture and optical properties of the stand and the soil. Their influence on net radiation is expressed by the penetration functions a_S, a_{S1}^H, a_{S2}^H, a_D a_{D1}^H, a_{D2}^H and the albedo A_B of the soil.

(c) The temperature of the stand and soil in the expression for B is expressed in terms of the long-wave radiation of the system soil-stand E_2 and partially in terms of E_1. The temperature of both the stand and the soil is determined by the combined effect of all components of the heat balance of the system, and thus the net radiation of the stand is associated, through the long-wave fluxes E_1 and E_2, with the energy balance of the ground surface below the stand. Besides radiation, the temperature of the stand depends considerably on its water supply. When the water supply is sufficiently plentiful, the stomata of leaves are open, a considerable amount of absorbed solar radiation is spent for transpiration, and leaf temperature differs only a little from air temperature. In the shortage of water in the soil the stomata are half closed and a reduced transpiration leads to an overheating of leaves to an amount of 10 centigrade degrees. Leaf temperature is also influenced to a certain extent by wind velocity, particularly at low values of it.

Thus, net radiation in plant stands depends on a great variety of factors.

The theoretical formula for the net radiant flux in plant stands (II.11.2) is complicated and unwieldy for practical purposes. Attempts have, therefore, been undertaken to develop empirical and semiempirical formulae. It follows from experimental data and from an analysis of formula (II.11.1) that the total radiation Q is the most substantial component of net radiation, which determines its temporal and spatial variability.

Taking into account (II.11.3) and empirical formula (II.10.15) we obtain a semiempirical formula for net radiation above the stand

$$B_0=a(1-A_Q)Q_0-b, \qquad \text{(II.11.5)}$$

where $a=1-a_1$.

Formula (II.11.4) may be presented in a different form.

Denoting

$$a'=a(1-A_Q) \qquad \text{(II.11.6)}$$

we obtain from (II.11.4)

$$B_0 = a'Q_0 - b. \tag{II.11.5a}$$

Formula (II.11.5a) was first proposed by Fleischer (1953). The empirical constant $a' = \mathrm{d}b_0/\mathrm{d}Q_0$ characterizes the rate at which net radiation varies with the varying total radiation. When $Q_0 = 0$, $B_0 = -b$, whence it follows that the empirical constant b is equal to the effective radiation at sunrise or sunset.

Monteith and Szeicz (1961) modified formula (II.11.4).

Eliminating the term $(1-A_Q)Q_0$ from (II.11.3) and (II.11.4) gives $-B_{I0} = b/a + B_0(1-a)/a$.

Denoting

$$\beta = \frac{1-a}{a} = \frac{\mathrm{d}B_I}{\mathrm{d}B_0} \tag{II.11.7}$$

we obtain instead of (II.11.4)

$$B_0 = \frac{1-A_Q}{1+\beta} Q_0 - b. \tag{II.11.8}$$

In the daytime $B > 0$ and $B_1 < 0$, therefore $\beta = \mathrm{d}B_I/\mathrm{d}B > 0$ and it characterizes the rate of heating of the ground surface. That is why Monteith and Szeics called β the heating coefficient.

Numerous studies have been performed in recent years in order to ascertain whether formulae (II.11.4), (II.11.5a) and (II.11.8) are valid for different types of ground surface below the stand.

Table II.11.1 gives a summary of coefficients a', β, A_Q and b for a cloudless day, which were obtained by different authors.

Table II.11.1 *Values of a, a', β, A_Q and b in empirical equations $B_0 = a(1-A_Q)Q_0 - b$ and $B_0 = a'Q_0 - b$ for various types of ground surface on a cloudless day*

1	2	3	4	5	6	7	8
Type of surface	a	a'	β	A_Q	$-b$(W/m^2)	Remarks	References
Fish pond	1.015	0.90	−0.015	0.11	84	Israel	Stanhill et al. (1966)
Lava	0.97	0.92	0.03	0.05	174	Havaii. Mauna Loa Slope observ.	Ekern (1965)
	0.93	0.88	0.07	0.05	174	3400 m elev.	
Bare soil	0.74		0.35	0.17	53	Frequently cultivated powdery	
	0.72		0.39	„	61	mulch with negligible evaporation	Monteith & Szeicz (1961)
	0.71		0.41	„	80	England	
	0.73		0.37	„	69		Ekern (1965)
	0.97	0.60	0.03	0.08	105	Hawaii	
	0.93		0.07	„	105		
	0.825	0.59	0.21	0.28	84	USA. Minnesota	Idso et al. (1969)
		0.58			39		Kalma (1972)

Table II.11.1 (*Continued*)

1	2	3	4	5	6	7	8
Type of surface	a	a'	β	A_Q	$-b$(W/m^2)	Remarks	References
Desert	1.18	0.64	−0.15	0.26	126	Soviet Central Asia. Data of Aizenschtat	Gay (1971)
Desert wadi vegetation	1.04		−0.035	0.37	105	Israel. 0.2–4 m tall	Stanhill et al. (1966)
Stressed awn-grass		0.65		0.35	63	South Africa	Glover (1972)
Short grass	0.82		0.20		63	England. 40% timothy + 60% meadow fescue	Monteith & Szeicz (1961)
	0.80		0.25	0.25	53	USA. Prairie grass	
	0.99		0.01	„	43		Berger-Landefeld (1964)
	0.915		0.09	0.26	66	Sod	Idso et al. (1969)
Pasture	0.735	0.57	0.36	(0.18)	84	Israel. Nonirrigated. Grass and herbaceous cover	Stanhill et al. (1973)
Semi-steppe hillside	1.04		−0.04	0.20	105	Batha communities. Israel. 10–60 cm tall	Stanhill et al. (1966)
Corynephoretum cover	0.82		0.22		63		Berger-Landefeld (1964)
Americo-festucetum cover	0.82		0.22		18		Berger-Landefeld (1964)
Caricetum-gracilis cover	0.87		0.16		30		Berger-Landefeld (1964)
Annual grasses *Brachiaria ramosa* and *Digitaria ascendens*		0.615		6			Kalma (1972)
Townsville stylo		0.645			45		Kalma (1972)
Alfalfa	0.93	0.75	0.075	0.24		Irrigated. USA Arizona	Idso et al. (1969)
Sugar beet	0.83		0.21		68	July–Sept. England Dense stand	Monteith & Szeicz (1961)
Beans	0.925		0.09		82		Monteith & Szeicz (1962)
	0.82	0.69	0.22		49		
Soybean	„		„		21		Impens & Lemeur (1969a)
	0.865				5.72 MJ/m^2 day	L_0 3 USA. July	Kanemasu et al. (1976)
	0.725				2.11 MJ/m^2 day	L_0 0	

Table II.11.1 (*Continued*)

1	2	3	4	5	6	7	8
Type of surface	a	a'	β	A_Q	$-b$(W/m²)	Remarks	References
Wheat	0.87		0.15		92	Spinny wheat. May–July. England	Monteith & Szeicz (1961)
	0.925		0.09		82		Monteith & Szeicz (1962)
		0.70			81	Australia. Deniliquin	Denmead (1976)
		0.80			124		
		0.68			89		Paltridge et al. (1972) (ref. after Denmead 1976)
	0.96	0.84	0.044	0.21			Idso et al. (1969)
Barley	0.925		0.09		82		
	0.82	0.655	0.25	0.23	84		Monteith & Szeicz (1962)
Rice	1.10	0.86	−0.09	0.22	56	Fully grown crop	Fritschen (1967)
Cotton		0.80			35		Uchijima (1976a) Uchijima (1976b)
Sunflower	0.93	0.74	0.075	0.22			Idso et al. (1969)
	1.06	0.87	0.06	0.18	91	Irrigated. Soviet Central Azia. Data of Aizenschtat	Gay (1971)
	0.79		0.26		54		Impens & Lemeur
		0.65			7		(1969a)
	0.785		0.27		21		Impens & Lemeur
		0.66			42		(1969a)
Maize	0.78–		0.15–	0.15–	35–	Irrigated. Israel.	Stanhill et al.
	0.87		0.28	0.21	126	$L_0 = 1.3$–7.3	(1968)
	1.01	0.82	0.01	0.19	90	Fully grown crop	Uchijima (1976a)
Sorghum	0.93	0.75	0.075	0.21		L_0 3 USA July	Idso et al. (1969)
Sorghum	0.837				5.50 MJ/ m² day		Kanemasu et al. (1976)
	0.725				2.11	L_0 3	
Sugar cane	0.97		0.03	0.15	105	Hawaii	Ekern (1965)
Various crops		0.775		0.225	70	Irrigated. South Africa	Glover (1972)
Various crops	0.725		0.38		28	Canada, June, August	Davies & Buttimor (1969)
Various crops	0.875		0.14		21	Cloudy	Impens & Lemeur (1969a)
		0.69			35		
Dwarf-desert shrubs	0.98		0.015	0.305	105	Israel. 11–15 cm tall	Stanhill et al. (1966)
Evergreen magnis shrubs	1.00		0.00	0.16	84	Israel. 2–4 m tall	Stanhill et al. (1966)
		0.63			97	Israel. 3–4 m tall	Stanhill et al.
		0.0006			3.9	Overcast. Israel. 3–4 m tall	(1973)
Apple tree	0.86		0.165	0.17	42	Canada. June–Oct. 10-year-old	Proctor et al.
		0.80			98	McIntoch	(1972)

Table II.11.1 (*Continued*)

1	2	3	4	5	6	7	8
Type of surface	a	a'	β	A_Q	$-b$(W/m^2)	Remarks	References
Pineapple plantation	0.97		0.03	0.12	105	Hawaii 4.3 plants per m^2	Ekern (1965)
Orange orchard	1.07		0.07	0.17	98	Irrigated. Israel. 4 m tall	Stanhill et al. (1966)
	0.95		0.05	0.16	63	Irrigated. Israel 4.5 m tall. Annual mean	Kalma & Stanhill (1969)
Oak forest. *Quercus*	0.96		0.04	0.175	70	Israel. 5m tall	Stanhill et al. (1966)
ithaburensis *Styrax*	0.85		0.18		19	Dense canopy 16–21 m tall	Galoux & Grulois,
officinalis association	0.78		0.28		60	Belgium	(1968)
Loblolly pine plantation	0.97	0.87	−0.03	0.13	112	USA North Carolina 24 m tall	Gay (1971)
Pine forest	0.99		0.015	0.125	84	Evergreen. Israel. 6.5 m tall	Stanhill et al. (1966)
Coniferous forests		0.71–0.94		0.05 0.20	-6–110		
Bare soil	0.83				35	For climatological calculations	Barashkova et al. (1961)
Grass:							
South regions of USSR		0.70			56	"	Dyatchenko (1958)
Middle region of USSR		0.65			42	"	"
Humid regions	0.9				48 MJ/(m^2 month)	Monthly totals for climatological calculations	
Arid regions	0.7				84 MJ/(m^2 month)		Berlyland & Muchenberg (1963)
Other regions	0.8				63 MJ/(m^2 month)		

Analysis of this table shows that the comparatively great variations of empirical coefficients a, a', and b are ambiguously related to different types of surface or albedo values. From general considerations it may be concluded that the quantity b, i.e. the net long-wave radiation of the ground surface below the stand at $Q_0=0$ must be systematically higher at sunset than at sunrise (assuming constant water supply conditions and a cloudless sky), since the ground surface temperature is higher in the afternoon than in the morning. In the presence of clouds b must decrease. In the shortage of water in the soil b must increase, but it must decrease with the increasing albedo A_Q. Therefore, the coefficients a, a', β and b in empirical formulae (II.11.4), (II.11.6) and (II.11.8) describe the type of ground surface ambiguously, since they depend on the energy parameters of the system 'soil–vegetation–atmosphere'. Stanhill et al. (1966, 1973), Davies and Buttimor (1969) and Idso et al. (1969) have observed that diurnal and seasonal variations of a, a', β and b are not less than their differences in the case of different surfaces, and conclude, with good reason, that formulae (II.11.4) and (II.11.8) are not theoretically more advantageous than the simplest formula (II.11.6). Besides, Idso (1968) has shown that in a crop soon after watering $\beta<0$ and is devoid of its initial physical meaning. Discrepancy between data contributed by different authors may be partially attributed to differences in instrumentation and methods of measurement. It is well known that the accuracy of measurement and calibration of net radiometers is noticeably inferior to the accuracy of short-wave radiation meters. An appropriate scale of space average is also an important methodical problem when measurements within a plant stand are concerned. These measurements will systematically overestimate data of penetration because of the finite dimensions of instruments.

Formulae (II.11.4) and (II.11.6) have been widely used by Soviet radiation climatologists for calculations of the geographical distribution and annual variation of net radiation as a component of the thermal balance (Barashkova et al. 1961; Berlyand & Mukhenberg 1963; Gagua 1968). The work of Gagua (1968) gives monthly mean values of the coefficients a and b for different regions of the Earth. The values of a vary in the range from 0.78 to 0.93, those of b from $+35$ in polar regions to -70 W/m^2 in arid southern regions.

Barashkova (1961b) proposed the formula $B_0=98\,(h_0-10)$ in the case of bare soil, Barashkova et al. (1961) proposed the formula for the daily totals of net radiation $\sum_d B_0=c(h_\odot^*-10)^2-d$, where $h_\odot^*$ is solar elevation at midday. On the basis of an analysis of a large number of data on various surfaces (snow cover, water, meadown, tundra, forest, bare soil) Pavlov (1962) found that the coefficient a is a universal function of albedo, and that $a=0.81-1.39\;A_Q+0.45A_Q{}^2+0.34A_Q{}^3$, but later investigations by Idso et al. (1969) have not confirmed this universality. For an oak stand below an overcast sky Galoux and Grulois (1968) have recommended the formula $B_0=aQ_0^n+C$.

Bohren and Thorud (1973) have proposed an approximate formula for net radiation under the stand

$$B(L_0)=\frac{(1-A_B)(1-a_D)}{1-a_D A_B A_Q}Q_0+(E_A-E_L)+(E_L-E_B)a_D(L_0) \qquad \text{(II.11.9)}$$

treating the stand and the ground surface under it as two horizontal surfaces engaged in radiative exchange. The first term in expression (II.11.9) takes account of the net short-wave radiation of the ground surface below the stand. From an analysis of formula (II.11.9) Bohren and Thorud have concluded that $B(L_0)$ is a monotonically increasing or decreasing function of the penetration function A_D. These authors have used their model for calculating the process of snow thawing below a forest canopy.

The vertical profile of net radiation $B(L)$ on a cloudless day is primarily determined by the vertical profile of direct solar radiation $S(L)$. Since the latter quantity varies according to an exponential law, it might be expected that as a first approximation $B(L)$ also follows an empirical relation. That was first established by Uchijima (1961) for a rice stand.

$$B(L)=B_0\exp(-C_B L), \qquad \text{(II.11.10)}$$

where C_B is an empirical constant.

A series of investigations have proved a satisfactory agreement between formula (II.11.10) and experimental data, the empirical constant C_B depending on crop species, solar elevation and other factors.

Table II.11.2 gives a summary of coefficients C_B. It is seen that $B(L)$ is attenuated in a forest more rapidly than in a crop. According to data of Kalma and Stanhill (1969) in an orange plantation the coefficient C_B depends considerably on solar elevation. Formula (II.11.10) is valid in the interval $26° < h_\odot < 38°$, but about midday ($h_\odot = 80°$) C_B is nearly equal to zero. The deviation of

Table II.11.2 *Values of coefficient C_B in formula $B(L)=B(0)\exp(-C_B L)$*

Canopy	C_B	Remarks	Author
Maize	0.66		Uchijima (1961)
”	0.53	Clear sky	Denmead et al. (1962)
	0.40	Overcast sky	”
	0.53		Brown & Covey (1966)
	0.5–0.6	$h_\odot < 25$	Budagovsky et al. (1968)
Rice	0.45–0.65		Uchijima (1976a)
Sunflower	0.72	$L_0=2$	Saugier (1976)
Oak forest	0.35–0.55	C_B decreases with increasing $h_\odot$	Tajchman (1972)
Sitka spruce	0.98	$L_0=0$–2, $h_\odot=42°$	
	1.2	$L_0=2$–4.2, $h_\odot=42°$	
	0.60	$L_0=0$–2, cloudy	Jarvis et al. (1976)
	0.69	$L_0=2$–4.2, cloudy	
	1.1	$L_0=4.2$–7.0, cloudy	
	4.3	$L_0=7.0$–8.3, cloudy	

the attenuation of $B(L)$ from the exponential law is due to the orientation of leaves. In upper layers of an orange tree the orientation of leaves is nearly vertical, and the attenuation of $B(L)$ is insignificant. Samarina (1965) proposed an empirical relation between C_B and solar elevation: $C_B = 0.51 - 0.0046\, h_{\odot}$, which was obtained for maize and wheat crops. Budagovski (1964) tabulated the dependence $C_B(h_{\odot})$ for various crops as follows

$h_{\odot}$	10°	20°	30°	40°	50°	60°	70°
$C_B(h_{\odot})$	0.97	0.54	0.40	0.34	0.30	0.29	0.29

Budagovski et al. (1968) have proposed the following formula for a maize stand at $h_0 > 25°$:

$$B_L = B_0 \exp\left(-\frac{0.24L}{\sin h_{\odot}}\right). \tag{II.11.11}$$

According to Tajchman (1972) in a pine forest better results in comparison with (II.11.10) are achieved with the formula $B(L) = B_0 \exp(-C_B L^2)$. In a row crop of maize the attenuation of net radiation, the azimuth effect and the direction of rows can be taken account of (McChaughey & Davis 1974) by means of formula

$$B(L) = B_0 \exp\left[-\frac{0.436L}{(\sin h_{\odot})^{z/z_0}} \cos\left(\psi - \frac{\pi}{2}\right)\right], \tag{II.11.13}$$

where ψ is the difference of azimuth angles of the solar beam and the row direction; and z_0 is the height of crop. Rauner (1966), Rauner and Ananyeva (1967) and Rudnev (1977) have proposed formulae like

$$B(\tilde{z}) = B_0 \exp\left[-\beta(h_{\odot}) \int_0^{\tilde{z}} \omega(\tilde{z}) d\tilde{z}\right].$$

or $B(\tilde{\omega}) = B_0 \exp(-\beta\tilde{\omega}^n)$, where $\tilde{z} = 1 - z/z_0$ and $\tilde{\omega} = \omega/\omega_0$, ω – upward cumulative leaf area index, i.e. $\omega = L_0 - L$ and $\omega_0 = L_0$.

Introduction of a relative height in this case, however, is not justified and is only confusing. Besides, in using relative phytometric characteristics in these works the authors fail to observe the principle of like dimensions. In the latest work of Rauner (1976) the author proposed a new formula

$$B(L) = B_0 \exp(-C_B L)^m \tag{II.11.14}$$

where the parameter C_B depends noticeably on $h_{\odot}$ and has a maximum value at $h_{\odot} = 15$ to 20°.

According to Singh et al. (1968) the exponential law of the attenuation of $B(L)$ is not valid in lower layers of a soy-bean stand.

Experiments performed by Impens and Lemeur (1969ab) have shown that in stands with predominantly vertical leaves (sunflower, beans) C_B decreases with the increasing solar elevation $h_{\odot}$. This dependence does not exist on a cloudy day. In a maize stand the best agreement with experimental data is obtained by using an exponential formula in the form

$$B(L) = B_0 \exp(-0.622L \pm 0.0533L^2). \qquad \text{(II.11.14)}$$

Rosenthal et al. (1977) in a maize stand have used the formula

$$B(L) = B_0 \exp(-0.398L + 148). \qquad \text{(II.11.15)}$$

The space distribution of net radiation inside a plant stand is generally similar to that of total radiation (Allen & Lemon 1972, Impens et al. 1970). The distribution function of net radiation is bimodal in shape (see Figure 49) in upper layers of the stand. In non-closed stands of maize (Denmead et al. 1962) and soybeans (Luxmoore et al. 1970) net radiation varies in a wide range. In shaded areas B may assume negative values, and in sunflecks $B(L)$ may exceed B_0 values. The latter circumstance is due to two causes: first, as shown in Figure 116, B_I increases with the depth L, and second, according to Figure 101 the total radiation in sunflecks Q° may exceed the total radiation above the stand. According to measurements by Rudnev (1977) in a 15-year old stand of oak, 6 m high, and a pine stand, 15 m high, the distribution function of B below the canopy in the solar elevation range 30 to 60° has a single maximum and is asymmetrical towards lower values.

As estimated by Utina and Shechter (1977) watering increases net radiation in oases by up to 60%. The primary factors which contribute to this increase are as follows: the decrease in the albedo ($\triangle A_Q = 0.04$) and the increase in net long-wave radiation B_I of up to 40% due to the decrease in ground surface temperature. Endeavours to reduce net radiation in droughty areas by spraying a white kaolin suspension on field plots (Lemeur & Rosenberg 1975, Baradas et al. 1976, Fuchs et al. 1976) have not been successful.

Among various types of plant stands forests, as a rule, have the greatest amount of net radiation. That is due to the low albedo and the relatively low ground surface temperature. According to data of Stanhill et al. (1973) the annual total of B_0 in a scrub forest is 11% higher than in a pasture. In the absence of clouds this difference increases up to 15%. According to Rudnev (1977) in summer the daily totals of net radiation in a pine stand are 50% greater than in a dry meadow and 30% greater than in a stand of oak trees.

Conclusion

The principal theoretical concept for the development of a quantitative theory of the radiation regime in plant stands has been to model the stand as a horizontally uniform anisotropic turbid plate medium. This concept assumes a random distribution of plant organs and uses the Poisson distribution. This allows using the mean values of radiation characteristics alone.

Another concept was to divide the radiation field inside the stand into two areas: that of gaps and that of shade, screened by the overlying plant's parts from the parallel rays of the Sun. No assumptions as to the dimensions or number of gaps have been made, only their relative part was taken into account and a random distribution in the horizontal plane was assumed.

However, as shown by our experimental data on the penetration of direct solar radiation in crops, as also by direct measurements of the distribution of phytoelements in stands conducted by Warren Wilson (1961, 1965a) using the inclined point quadrat method, the concept of a random distribution of phytoelements and, consequently, the assumption of a horizontal uniformity of stands are limited and not always valid. According to data of Warren Wilson, deviations from a random distribution are observed, both towards a semiregular and towards a clumped distribution. Our data show that in maize and sorghum crops the distribution of phytoelements, as a rule, deviates towards a clumped one, though in the upper layer and early in the growing season deviations towards a semiregular distribution can be observed. Whatever the seeding technique, crops always show a clumped distribution of phytoelements before the closing stage though within the volume of a single plant the distribution of phytoelements is semiregular, as a rule. It is established in quantitative ecology that the distribution of plants in natural stands tends towards a clumped one. In reality, therefore, a deviation from a random towards a clumped distribution is characteristic of the distribution of plant parts in real stands, and the penetration in real stand types is larger than predicted by theory. As shown by our measurements, these discrepancies are very considerable exceeding 100% in some cases.

The above circumstances call for new ways towards developing a theory of the radiation regime in a horizontally non-uniform stand. A move towards this objective has been made (see Chapter II.4.4) but the problem is very difficult.

The horizontal nonuniformity of a stand is primarily determined by two factors: first, by the peculiarities of the distribution of plants over the ground surface and, second, by those of the distribution of plant parts within the volume of a single plant. It is these factors, we believe, that must be taken as a basis for describing the structure of a nonuniform stand. Much attention is paid in quantitative ecology to the distribution of plants in the stand, but much less to a quantitative description of the distribution of phytoelements within the volume of a single plant, i.e. the habitus of plants. The quantitative description of the distribution of plant parts within the volume of a single plant and development of the theory of the habitus is, therefore, one of the topical problems of phytometry (Ross 1972) associated with the description of a nonuniform stand.

A second topical problem of phytoactinometry is the account of penumbra in the quantitative theory of the radiation regime. This problem remains unresolved notwithstanding the success achieved in this respect by Tanner et al. (1971a, b) and Norman et al. (1971). Penumbras, along with the structural nonuniformities of the stand are the main factors that determine the horizontal inhomogeneity of the radiation field and call for a statistical approach to the radiation regime. The evolution of a statistical theory of the radiative transfer, however, is hampered by major principal and mathematical problems characteristic not only of phytoactinometry but of the statistical theory of transfer in general.

The spatial variability of the radiation field inside a plant stand depends on its peculiar geometrical structure. Therefore, the first stage of investigation should be devoted to collecting new experimental data on the statistical characteristics of the radiation field and the geometrical structure of plant stands and finding empirical or semiempirical relations between them.

A third problem that has to be investigated is the temporal variability of the radiative field inside plant stands. Short-period variations of radiation can be of considerable importance for biological processes in plants, but the available data on such variations are contradictory. The diurnal progress of the Sun's disk over the sky causes considerable changes in the radiation regime in foliage. At a low solar elevation mainly the upperlayers of a stand are irradiated. As solar elevation increases the irradiation of the middle and lower layers is improved. It is not yet clear whether the dynamics of irradiation of individual leaves, which also determines the dynamics of their photosynthetic activity, can be treated as 'the principle of division of labour' in the life of individual plants. After all, both short- and long-period variations of the radiation field in a stand are averaged in time, thus reducing the spatial contrasts of the field within the diurnal cycle. The problem of concurrent influence of the time-spatial variability of the radiation field on the photosynthetic activity of plants has not been studied at all.

A complex approach should be used: theory should cooperate with experiment, research workers in various fields – mathematicians, physicists, biologists, biochemists, agronomists – should be called to assist if the above topical problems are to be solved. The success of a theory is primarily determined by new data from special complex experiments.

In view of the rapid progress in phytoactinometrical research during the past two decades, strengthening contacts between physicists and biologists, new measuring techniques based on the achievements of modern optics and electronics, these topical problems are expected to be solved in the nearest future.

SUPPLEMENT

Description of field experiments

The experimental part of the present book uses the data collected during integrated expeditions on crop stand phytometry, phytoactinometry and photosynthesis. The expeditions were organized by the USSR Academy of Sciences, i.e. its Scientific Council on Photosynthesis, and sent to Estonia, Moldavia and Tadjikistan between 1962–1966. They were led by the Corresponding member of the USSR Academy of Sciences A. A. Nichiporovich; works on phytometry and phytoactinometry were guided by J. K. Ross. Participants in the expeditions were research workers of the Institute of Plant Physiology and Biochemistry of the Moldavian Academy of Sciences, Institute of Plant Biophysics and Physiology of the Tadjik Academy of Sciences, as well as researchers from the Main Geophysical Observatory, and of the Institute of Physics and Astronomy (now the Institute of Astrophysics and Physics of the Atmosphere).

The integrated expeditions were organized with a view to provide an all-round investigation into the stand geometry, radiation and water/heat regimes, photosynthesis and productivity under different climatic conditions of the USSR.

Apart from these, additional phytoactinometric measurements, chiefly in relation to barley stands were undertaken in 1964–1967 and 1969 by the Institute of Physics and Astronomy of the Academy of Sciences of the Estonian Republic in collaboration with the Estonian Academy of Agriculture.

Below is a short description of the experiments, relating to the topics of the present book.

1962

Experiment No. 1. A row crop of horse beans near Tartu, Estonian SSR. The crop density 12.8 plants/m^2 with 60 cm spacing. Measurements were taken on August 17 and September 10 and crop heights of 100 and 130 cm, respectively, were recorded. The phenological stage was that of the formation of inflorescence and flowering.

Experiment No. 2. A row crop of arctic clover in the same locality. Density 117 plants/m^2, height 200 cm. Measurements were taken on August 20 while plants were flowering.

Experiment No. 3. A row crop of maize var. Bukovinsky 3 in an experimental field of Võru region, Estonian SSR. Crop density 16.5 plants/m^2, height 200 cm, leaf area index $L_0 = 8.0$. Measurements were taken in the first half of September, at the tasseling stage.

1963

All the experiments were conducted in the fields of the Jõgeva Selection Station, Estonian SSR.

Experiment No. 4. A row crop of maize var. Bukovinsky 3 in an experimental field with density 30.5 plants/m^2 and 50 cm between rows. Measurements were taken between June and August, from the emergence of seedlings to tasseling. On August 19–20 the crop was beaten down by a strong wind which followed heavy raining.

Experiment No. 5. A square-sown crop of maize var. Bukovinsky 3 in an experimental field with density 31.7 plants/m^2 and the square measuring 50 by 50 cm. Measurements were taken from June to August, the phenological stage covering the emergence of seedlings to tasseling. The crop was beaten down by the same wind and rain as above.

Experiment No. 6. A row crop of maize var. Bukovinsky 3 in a production field with density 27.4 plants/m^2 and 50 cm between rows. In late July the crop suffered considerably from a drought but towards the end of August, the plants recovered. After the plants in the fields allocated for experiments Nos. 4 and 5 were beaten down the expedition transferred to field No. 6. Measurements were taken late in August and early in September at tassel flowering.

Experiment No. 7. A row crop of winter wheat in a production field. Measurements made on August 17 revealed 392 stems/m^2, 15 cm spacing, 99 cm height. The phenological stage was milk ripeness.

Experiment No. 8. A row crop of awnless brome grass (*Bromus inermis*) in a production field with density 391 stems/m^2, height 160 cm and 50 cm spacing. Measurements were carried out on July 13 during the flowering stage.

1964

Experiment No. 9. A row crop of barley var. Domen near the town of Tartu. The density was 380 stems and 450 shoots per m^2, spacing 10 cm. Organic fertilizers were sidedressed at a rate of 300 kg/ha. Measurements were taken from May till August throughout the growing season.

Experiment No. 10. A square-sown crop of maize var. Bukovinsky 3 in an experimental field near the town of Tiraspol. The crop density 5.4 plants/m^2 with the square measuring 60 by 60 cm. Measurements were performed in July at tasseling and tassel flowering.

Experiment No. 11. A square-sown crop of maize var. VIR-42 in Chekolteny, Moldavian SSR. The crop density 3.7 plants/m^2, height 190 cm, $L_0 = 1.7$, the square measuring 60 by 70 cm. Measurements were taken in August during the milk and wax ripeness stage.

1965

Experiment No. 12. A row crop of barley var. Domen near Tartu, Estonian SSR. Density 520 stems and 580 shoots per m^2 with 10 cm spacing. Mineral sidedressing included N_{200}, P_{300}, K_{150}. Measurements were performed from May till August during the whole of the growing season.

Experiment No. 13. A row crop of barley var. Domen near Tartu. Density 450 stems and 410 shoots per m^2. No mineral fertilizers used.

Experiment No. 14. An irrigated row crop of maize var. VIR-338 in Hissar Valley, Tadjikistan. Density 4.5 plants/m^2, spacing 60 cm. Measurements were conducted from May till August (from the emergence of seedlings till milk ripeness).

Experiment No. 15. An irrigated row crop of sorghum in Hissar Valley with density 7.4 plants/m^2, height 360–400 cm, $L_0 = 4.4$, spacing 60 cm. Measurements were carried out from July till August (the tasseling and tassel flowering stage).

Experiment No. 16. An irrigated row crop of sunflower, var. Batyr, in an experimental field in Hissar Valley. Density 4.1 plants/m^2, spacing 60 cm. Time – July and August.

Experiment No. 17. An irrigated row crop of sunflower, var. Batyr, in an individual field in Hissar Valley. Density 8.1 plants/m^2, height 280 cm, spacing 60 cm. As compared with experiment No. 15 this field showed more uniformity and no missing plants. Measurements were taken in July during the inflorescence formation and flowering.

Experiment No. 18. An irrigated row crop of cotton in Hissar Valley with rows spaced 60 cm. Time – August 7–8.

1966

Experiment No. 19. A row crop of maize hybrid var. Bukovinsky 3 in an experimental field of the Jõgeva Selection Station of the Estonian SSR. Density 27.8 plants/m^2 with spacing between rows 60 cm. Extra supply of (N_{180}, P_{80}, K_{120}). Measurements were started during the emergence of seedlings and ended at the ear flowering stage.

Experiment No. 20. A dense row crop of maize var. Bukovinsky 3 in the same locality. Density 44 plants/m^2, spacing 60 cm. Normal supply of N_{90} P_{80} K_{120}. Measurements were conducted starting from seedling emergence till the ear flowering stage.

Experiment No. 21. Hybrid swede var. Kuusiku in the same locality. A row crop with 60 cm spacing thinned at the 5th leaf stage to a square crop of 60 by 60 cm with one plant per square and 2.8 plants/m^2 density. Measurements were started during the emergence of seedlings and ended with the harvesting in the middle of October.

Experiment No. 22. Single potato plants in a field of hybrid swede. Cultivar unknown. The flowering stage.

Experiment No. 23. A row crop of barley, var. Domen, near Tartu, ESSR. Density 550 stems and 350 shoots per m^2 with spacing between rows 10 cm. Fertilizer used was N_{200}, P_{400}, K_{200}. Measurements were conducted from May to August.

Experiment No. 24. A row crop of var. Domen barley in the same locality. Density 480 stems per m^2, no shoots. No mineral nutrition.

Experiment No. 25. A row crop of hybrid maize var. Bukovinsky 3 in the same locality. Density 21.9 plants/m^2, spacing 60 cm. Measurements were conducted from mid-May till mid-August.

Experiment No. 26. A row crop of horse beans var. Jõgeva in the same locality. Density 24.3 plants/m^2, spacing 60 cm. Time – from mid-May till mid-August.

Experiment No. 27. A row crop of hybrid swede var. Kuusiku in the same locality. Density 6.0 plants/m^2 with 60 cm spacing.

Experiment No. 28. A row crop of var. Olev potato in the same locality. Density 5.5 plants/m^2 with 60 cm spacing.

Experiment No. 29. A row crop of barley var. Domen in the same locality.

References

Abashina, E. V. 1968. On the methods of observing and estimating the attenuation and absorption of integral radiation by barley crops. Trans.Inst. Exp. Meteorol., 4: 86–90 (Russian).

Abashina, E. V. & Gorbachev, V. A. 1977. Leaf orientation in barley crops. Trans. Inst. Exp. Meteorol., 8(67): 97–104. (Russian).

Abashina, E. V. & Kozinets, L. V. 1968. Peculiarities of the geometrical structure of barley crops. Trans. Inst. Exp. Meteorol., 4: 91–100. (Russian).

Abashina, E. V. & Sirotenko, O. D. 1973. Calculation of attenuation coeffients of total radiation by barley crops. Trans. Inst. Exp. Meteorol., 3(40): 57–65. (Russian).

Acock, B., Thornley, J. H. M. & Warren Wilson, J. 1970. Spatial variation of light in the canopy. Proc. IBP/PP Techn. Meet., Třebon. Pudoc, Wageningen: 91–102.

Alekseyev, V. A. 1963a. A few problems of the optical properties of the forest. In: Problems of ecology and physiology of forest plants: Leningrad, Publ. Hous of LTA, 1: 47–80. (Russian).

Alekseyev, V. A. 1969. On the light conditions in forest. II. Reflectivity of forest stands. In: Experimental study of biocoenoses of the taiga: 5–13. Nauka Publishers, L. (Russian).

Alekseyev, V. A. 1970. The coming of photosynthetically active radiation to the plants in the lower layers through 'windows' in the forest canopy. In: Phytocoenology and biogeocoenology of dark conifers of the taiga: 19–31. Nauka Publishers, L. (Russian).

Alekseyev, V. A. 1975. The forest light regime. Nauka Publishers, Leningrad.

Alekseyenko, L. N. 1958a. The structure of the plant cover of perennial herbs and their yield. Rep. VASHNIL, 6: 14–18. (Russian).

Alekseyenko, L. N. 1958b. Concerning the biology and ecology of perennial herbs. Sov. Bot. J., 43, 11: 1582–1588. (Russian).

Alekseyenko, L. N. 1960. The structure of the plant cover of pernnial herbs in pure and mixed cultures. In: Problems of haymaking and pasture farming: 115–135. Ministery of Agriculture Press, M. (Russian).

Alekseyenko, L. N. 1964. Some problems of studying the structure of the plant of herbaceous phytocoenoses. Sov. Bot. J., 49, 1: 65–74. (Russian).

Alekseyenko, L. N. 1965a. The weighing method of determining the leaf surfaces of meadow plants and communities. Sov. Bot. J., 50, 2: 205–208. (Russian).

Alekseyenko, L. N. 1965b. Interrelationships between the structure of the plant cover, phytoclimatic conditions and some physiological processes in the plant cover of meadow grasses. Bull. Moscow Naturalists' Soc., 70, 6: 92–98. (Russian).

Alekseyenko, L. N. 1967. Dependence of the productivity of meadow plants on environmental conditions. Leningr. State Univ. Press, L. (Russian).

Alekseyenko, L. N. 1968. The size of leaf surfaces of meadow coenoses and their productivity. Bull. Moscow Naturalists' Soc., 73, 5: 75–88. (Russian).

Allen, L. H., Jr. 1974. Model of light penetration into a widerow crop. Agron. J., 66, 1: 41–47.
Allen, L. H., Jr. & Brown, K. W. 1965. Shortwave radiation in a corn crop. Agron. J., 57, 6: 575–580.
Allen, W. A., Gale, T. V. & Richardson, A. J. 1970. Plant-canopy irradiance specified by the Duntley equations. J. Opt. Soc. Amer., 60, 3: 372–376.
Allen, W. A., Gausman, H. W. & Richardson, A. J. 1970. Mean effective optical constants of cotton leaves. J. Opt. Soc. Amer., 60, 4: 542–547.
Allen, W. A., Gausman, H. W., Richardson, A. J. & Thomas, J. R. 1969. Interaction of isotropic light with a compact plant leaf. J. Opt. Soc. Amer., 59, 10: 1376–1379.
Allen, L. H. & Lemon, E. R. 1972. Net radiation frequency distribution in a corn crop. Boundary-Layer Meteorol. 3, 2: 246–254.
Allen, W. A. & Richardson, A. J. 1968. Interaction of light with a plant canopy. J. Opt. Soc. Amer., 58, 8: 1023–1028.
Anderson, M. C. 1964. Studies of the woodland light climate. I. The photographic computation of light conditions. J. Ecol., 52: 27–41.
Anderson, M. C. 1966a. Stand structure and light penetration. II. A theoretical analysis. J. Appl. Ecol., 3. 1: 41–54.
Anderson, M. C. 1966b. Some problems of simple characterization of the light climate in plant communities. Light as an Ecological Factor. In: Evans, G. C. (Ed.) Brit. Ecol. Soc. Symp., Oxford: 77–90.
Anderson, M. C. 1969. A comparison of two theories of scattering of radiation in crops. Agric. Meteorol., 6: 399–405.
Anderson, M. C. 1970. Radiation climate, crop architecture and photosynthesis. In: Šetlík, J. (Ed.). Prediction and Measurement of Photosynthetic Productivity. Pudoc, Wageningen: 71–78.
Anderson, M. C. & Denmead, O. T. 1969. Shortwave radiation on inclined surfaces in model plant communities. Agron. J., 61: 867–872.
Anderson, M. C. & Miller, E. E. 1974. Forest cover as a solar camera: penumbral effects in plant canopies. J. Appl. Ecol., 11, 2: 691–697.
Anderson, V. 1967. Comparative analysis of the methods for measuring the leaf area of the hybrid swede. In: Abstracts of papers of a sci. conf. on mechanization and electrification of agriculture, Tartu: 27–29. (Russian).
Angus, J. F., Jones, R. & Wilson, J. H. 1972. A comparison of barley cultivars with different leaf inclinations. Austral. J. Agr. Res., 23, 6: 945–957.
Araki, M. 1972a. The studies on the specific leaf area of forest trees. (II). The effects of RLJ, season and shading on the specific leaf area in young birch *(Betula platyphilla Sukatchev var. Japonica (Mig)* stand. J. Jap. Forest Soc., 54, 6: 184–191.
Araki, M. 1972b. Analysis of some characters in foliage in a youg congested birch *(Betula platyphylla Sukatchev var. Japonica (Mig)* stand (A preliminary report). J. Jap. Forest Soc., 54, 6: 192–198.
Baradas, M'. W., Blad, B. L. & Rosenberg, N. J. 1976. Reflectant induced modification of soybean canopy radiation balance, IV. Leaf and canopy temperature. Agron. J., 68, 6: 843–848.
Barashkova, E. P. 1961a. Radiation regime of the territory of the USSR. Hydrometeorol. Press, L. (Russian).
Barashkova, E. P. 1961b. Net radiation of Karadag. In: Shifrin, K. S. (Ed.), Actinometry and Atmospheric Optics: 42–47. Hydrometeorol. Press, L. (Russian).
Barashkova, E. P., Gayevsky, V. L., Dachenko, L. N., Lugina, K. M. & Pivovarova, Z. I. 1961. Radiation regime on the territory of USSR. Hydrometeorol. Press, L. (Russian).
Baumgartner, A. 1953. Das Eindringen des Lichtes in den Boden. Forst-Wissenschaftliches Central-blatt, 72, 516: 172–184.
Bedenko, V. P. & Fedyushin, A. A. 1971. On the determination of incoming photosynthetically active radiation in conditions of vertical zonality in the Zailiisk Alatau. Bull. Acad. Sci. Kasakh. SSR. 2, 310: 68–70. (Russian).
Bell, L. N. 1955. A photoelectric device for measuring photosynthetically active radiation (a photoelectric photoactinometer). Trans. Inst. Plant Physiol. USSR Acad. Sci., 10: 257–264. (Russian).

Bell, L. N. 1956. On the application of luxmeters and selenium photocells for measuring physiological radiation. Sov. Plant Physiol., 3: 176–178. (Russian).

Berger-Landefeldt, U. 1964. Über den Strahlungshaushalt verschiedenen Pflanzenbestände. Berichte der Deutschen botanischen Gesellschaft, 77, 2: 27–48.

Berlyand, T. G. & Mukhenberg, V. V. 1963. Role of absorbed radiation in the formation of net radiation. Trans. Main Geophys. Obs., 139: 3–15. (Russian).

Billings, W. D. & Morris, R. J. 1951. Reflection of visible and intrared radiation from leaves of different ecological groups. Amer. J. Bot., 38, 5: 327–331.

Birkebak, R. & Birkebak, R. 1964. Solar radiation characteristics of tree leaves. Ecology, 45, 3: 646–649.

Blad, B. L. & Baker, D. G. 1972. Orientation and distribution of leaves within soybean canopies. Agron. J., 64, 1: 26–29.

Bohren, C. F. & Thorud, D. B. 1973. Two theoretical models of radiation heat transfer between forest trees and snowpacks. Agric. Meteorol., 11, 1: 3–16.

Bonhomme, R. 1969. Spectral distribution of radiation from sun and sky, an energy available to plants. In: 5th International Biometeorological Congress, Montreux, Switzerland: 1–4.

Bonhomme, R. 1970. Surface relative des taches de soleil dans la végétation. Techn. étude facteurs phys. biosphere: 99–104. Paris.

Bonhomme, R. 1974. Determination des profils d'indice foliaire et de rayonnement dans un couvert vegetal a l'aide de photographies hemispheriques in situ. Thes. Doct. Diss. Univ. Marseille.

Bonhomme, R. & Chartier, P. 1972. The interpretation and automatic measurement of hemispherical photographs to obtain sunlit foliage area and gap frequency. Isr. J. Agric. Res., 22, 2: 53–61.

Bonhomme, R. Chartier, P. & Varlet-Grancher, C. 1971. Assimilation nette, utilisation de l'eau et microclimatologie d'un champ de mais. II. Eclairement d'un plant horizontal au dessus et aux differents niveaux de la culture. Ann. Agron., 22, 4: 383–396.

Brandt, A. B. & Tageyeva, S. V. 1967. Optical parameters of plant organisms. Nauka Publishers, M. (Russian).

Breece, H. T., III & Holmes, R. A. 1971. Bidirectional scattering characteristics of healthy green soybean and corn leaves in vivo. Appl. Opt., 10, 1: 119–127.

Britton, C. M. & Dodd, J. D. 1976. Relationships of photosynthetically active radiation and shortwave irradiance. Agric. Meteorol., 17, 1: 1–7.

Brown, K. W. & Covey, W. 1966. The energy-budget evaluation of the micrometeorological transfer processes within a cornfield. Agric. Meteorol., 3: 73–96.

Budagovsky, A. I. 1964. Evaporation of soil moisture. Nauka Publishers, M. (Russian).

Budagovsky, A. I., Ross, J. & Tooming, H. 1968. Vertical distribution of long-wave radiation fluxes and of radiation balance in the plant cover. In: Actinometry and atmospheric optics: 299–307. Valgus Publishers, Tallinn. (Russian).

Carlson, R. E., Yarger, D. N. & Shaw, R. H. 1971. Factors affecting the spectral properties of leaves with special emphasis on leaf water status. Agron. J., 63: 486–489.

Chance, J. E. & LeMaster, E. W. 1977. Suits reflectance models for wheat and cotton: theoretical and experimental tests. Appl. Opt., 16, 2: 407–412.

Chance, J. E. & LeMaster, E. W. 1978. Plant canopy light absorption model with application to wheat. Appl. Opt., 17, 16: 2629–2636.

Charles-Edwards, D. A. & Thornley, J. H. M. 1973. Light interpretation by an isolated plant, a simple model. Ann. Bot., 37, 152: 919–928.

Charles-Edwards, D. A. & Thorpe, M. R. 1976. Interception of diffuse and direct-beam radiation by a hedgerow apple orchard. Ann. Bot., 40, 167: 603–613.

Chartier, P. 1966. Etude du microclimat lumineaux dans la végétation. Ann. Agron., 17, 5: 571–602.

Coombe, D. E. & Evans, G. C. 1960. Hemispherical photography in studies of plant. Med. Biol. Illustr., 10: 68–75.

Cowan, I. R. 1968. The interception and absorption of radiation in plant stands. J. Appl. Ecol., 5: 367–379.

Cowan, I. Rl 1971. Light in plant stands with horizontal foliage. J. Appl. Ecol., 8, 2: 579–580.

Dadykin, V. P., Potayevich, E. V. & Greshevsky, B. N. 1964. On the spectral composition of light under crowns of trees. Sov. Forest J., 1: 13–16. (Russian).
Davies, J. A. & Buttimor, P. H. 1969. Reflection coeffients, heating coeffients and net radiation at Simcoe, Southern Ontario, Agric. Meteorol., 6: 373–386.
Denmead, O. T. 1976. Temperate cereals. In: Monteith, J. L. (Ed.), Vegetation and Atmosphere, Vol. 2. Case Studies: 1–31. Acad. Press, London.
Denmead, O. T., Fritschen, L. J. & Shaw, R. H. 1962. Spatial distribution of net radiation in a corn field. Agron. J., 54, 6: 505–510.
Dolidze, D. Sh. 1974. Photosynthetically active radiation (PAR) on slopes in conditions of Anassuli. Subtropical Crops, 5: 148–151. (Russian).
Dolph, G. E. 1977. The effect of different calculational techniques on the estimation of leaf area and the construction of leaf size distributions. Bull. Torrey Bot. Club, 104, 3: 264–269.
Ducrey, M. 1975a. Utilisation des photographies hémispheriques pour le calcul de la perméabilité des couverts forestiers au rayonnement solaire. I. Analyse theorique de l'interception. Ann. sci. forest., 32, 2: 73–92.
Ducrey, M. 1975 (1976)b. Utilisation des photographies hémisheriques pour le calcul de la perméabilité des couverts forestiers au rayonnement solaire. 2. Etude experimentale Ann. sci. forest. 32, 4: 205–221.
Duda, M. & Kolek, J. 1970. Vertical biomass distribution during vegetation in the Dvoran variety of barley. Research Project BAB (IBP) Progress report I. Bratislava. Bot. Inst. Slov. Acad. Sci.: 111–117.
Duncan, W. G. 1971. Leaf angles, leaf area and canopy photosynthesis. Crop Sci., 11, 4: 482–485.
Duncan, W. G., Loomis, R. S., Williams, W. A. & Hanau, R. 1967. A model for simulating photosynthesis in plant communities. Hilgardia, 38, 4: 181–205.
Dyatchenko, L. N. 1958. Relationship between net and total solar radiation. Meteorol. i Gidrol. 8: 29–33. (Russian).
Ekern, P. C. 1965. The fraction of sunlight retained as net radiation in Hawaii. J. Geophys. Res., 70, 4: 785–793.
Eiyubov, A. D. & Ragimov, H. Sh. 1973. On the distribution of photosynthetically active radiation on the territory of the Azerbaijan SSR. Sci. pap. Azerbaijan Univ. Ser. geol.-geogr. sci., 6: 52–57.
Evans, G. C. 1956. An area survey method of investigating the distribution of light intensity in woodlands, with particular reference to sunflecks. J. Ecol., 44: 391–428.
Evans, G. C. & Coombe, D. E. 1959. Hemispherical and woodland canopy photography and the light climate. J. Ecol., 47: 103–113.
Evans, G. C., Whitmore, T. C. & Wong, Y. K. 1960. The distribution of light reaching the ground vegetation in a tropical rain forest. J. Ecol., 48: 193–204.
Farrar, J. F. & Mapunda, O. P. 1977, Optical properties of the leaves of some African crop plants. Appl. Opt., 16, 1: 248–251.
Fedyushin, A. A. & Bedenko, V. P. 1974. Investigation of factors for conversion from integral radiation to PAR in South-East Kazakhstan. In: Radiation Processes in the Atmosphere and on the Earth Surface: 455–458. Hydrometeorol. Press, L. (Russian).
Felmery, L. 1974. Amount of photosynthetically active radiation in growth period. Idöjárás, 78, 4: 235–239. (Hung.).
Fleischer, R. 1953. Der Jahrgang der Strahlungsbilanz und ihrer Komponenten. Ann. Meteorol., ser. B. 6: 357–364.
Ford., E. D. 1976. The canopy of a scots pine forest: description of a surface complex roughness. Agric. Meteorol., 17: 9–32.
Fritschen, L. 1967. Net and solar radiations over irrigated field crops. Agric. Meteorol., 4: 55–62.
Fuchs, M., Stanhill G. & Moreshet, S. 1976. Effect of increasing foliage and soil reflectivity on the solar radiation balance of wide-row grain sorghum. Agron. J., 68, 6: 865–871.
Fukai, S., Davison, L. 1977. Estimation of radiation environments in row-planted communities by a leaf geometry model. In: 2nd Austral. Conf. Heat and Mass Transfer, Sydney: 21–28.

Fukai, S. & Loomis, R. S. 1976. Leaf display and light environments in row-planted cottom communities. Agric. Meteorol., 17: 353–379.
Fulga, I. G. 1965. Determination of the leaf area of fruit-trees. Sov. Plant Physiol., 12, 6: 1104–1107. (Russian).
Gagua, V. V. 1968. Methods of indirect determination of hourly totals of net radiation. Trans. Main Geophys. Obs., 233: 49–57. (Russian).
Galaux, A. & Grulois, J. 1968. Echange radiatifs et convectifs en phase vernale. Sta. Rech. Eaux Forêts, Groenendael, Trav. 13, Ser. A: 1–51.
Galperin, B. M. 1938. The albedo of agricultural crops. Socialist Agriculture, 5: 179-192. (Russian).
Gates, D. M. 1968. Transpiration and leaf temperature. Annual Review of Plant Physiology, 19: 211–238.
Gates, D. M., Keegan, H. J., Schleter, J. C. & Weidner, V. R. 1965. Spectral properties of plants. Appl. Opt., 4, 1: 11–20.
Gates, D. M. & Tantraporn, W. 1952. The reflectivity of deciduous trees and herbaceous plants in the infrared to 25 microns. Science, 995, 2997, 6: 613–616.
Gausman, H. W. & Allen, W. A. 1973. Optical parameters of leaves of 30 plant species. Plant Physiol., 52, 1: 57–62.
Gausman, H. W., Allen, W. A., Gardenas, R. & Richardson, A. J. 1972. Effects of leaf age for four growth stages of cotton and corn plants on leaf reflectance, structure, thickness, water and chlorophyll concentrations and selection of wave length for crop discrimination. In: Sharokhy, F. (Ed.), Remote Sensing of Earth Resources Vol. 1: 25–51. The University of Tennessee, Tullahoma, Tennessee. Space Institute.
Gausman, H. W., Allen, W. A., Wiegand, C. L., Escobar, D. E. & Rodriguez, R. R. 1971. Leaf light reflectance, transmittance, absorptance, and optical and geometrical parameters for eleven plant genera with different leaf mesophyll arrangements. In: Proc. Seventh Int. Symp. Rem. Sens. Environ. 17–21 May. Univ. Mich. Ann Arbor, Michigan. Vol. III: 1599–1626.
Gausman, H. W., Allen, W. A., Wiegand, C. L., Escobar, D. E., Rodriguez, R. R. & Richardson, A. J. 1973. The leaf mesophylls of twenty crops, their light spectra and optical and geometrical parameters. Techn. Bull. U.S. Dep. Agr., 1465: 1–59.
Gavrilov, N. I. & Yeremenko, L. L. 1959. A device for measuring the leaf area. Sov. Plant Physiol., 6, 4: 508–512. (Russian).
Gay, L. W. 1971. The regression of net radiation upon solar radiation. Arch. Met. Geoph. Bioklium. Ser. B., 19, 1: 1–14.
Glover, J. 1972. Net radiation over tall and short grasses. Agric. Meteorol., 10, 6: 455–459.
Goisa, N. I., Gatsenko, R. V. & Perelyot, N. A. 1971. Vertical structure of the radiation field in the winter wheat crop under conditions of water stress. Trans. Ukrainian Hydrometeorol. Res. Inst., 108: 21–31. (Russian).
Goisa, N. I. & Perelyot, N. A. (Compilers). 1977. Methodological instructions for the calculation of photosynthetically active radiation (PAR). Ukrainian Hydrometeorol. Res. Inst., Kiev: 1–26. (Russian).
Goisa, N. I. & Rogachenko, A. D. 1969. On the methods of determining phenometric and radiation parameters of maize crops. Transact. Ukrainian Hydrometeorol. Res. Inst., 84: 83–93. (Russian).
Gorska, K. & Slomka, J. 1976. Inflow of the photosynthetically active energy in solar radiation at Belsk and Pulawy in the years 1973–1974. Publ. Inst. Geophys. Pol. Acad. Sci., D, 3: 49–55.
Goudriaan, J. 1977. Crop micrometeorology: a simulation study: 1–249. Pudoc, Wageningen.
Greig-Smith, P. 1964. Quantitative Plant Ecology. Vol. 12. 2nd Ed: 1–256. Butterworths Scientific Publication, London.
Grin, A. M., Rauner, J. L. & Utekhin, V. D. 1970. Efficiency of using radiation and moisture in forest steppe ecosystems. Proc. USSR Acad. Sci. Ser. geogr., 4: 10–23. (Russian).
Groen, P. 1973. On the heat exchange by infrared radiation in a horizontally homogeneous plant stand. Arch. Meteorol. Geophys. Bioklimatol., A 22, 4: 393–407.
Grulois, J. 1967. Extinction du rayonnement global tropismes et paramètres foliares. Bull. Soc. Roy. Bot. Belgique, 100: 315–334.

Gulyayev, B. I. 1965. On the methods of measuring photosynthetically active radiation. In: Photosynthesis and productivity of plants: 176–194. Naukova Dumka Publishers, Kiev. (Russian).
Gulyayev, B. I., Lavrentovich, D. I., Mitrofanov, B. A. & Manuilsky, V. D. 1967. Radiation regime and photosynthesis of crops. In: Ways of Raising the Intensity and Productivity of Photosynthesis: 82–96. Naukova Dumka Publishers, Kiev. (Russian).
Gulyayev, B. I., Mitrofanov, B. A. & Manuilsky, V. D. 1969. Peculiarities of the distribution of photosynthetically active radiation in the winter wheat crop. In: Ways fo Raising the Intensity and Productivity of Photosynthesis, 3: 87–95. Naukova Dumka Publishers, Kiev. (Russian).
Guminetsky, S. G. & Rvachev, V. P. 1966. Determination of the absorption spectra of individual cells of plant leaves. Sov. J. Appl. Spectroscopy, 5, 5: 674–680. (Russian).
Häckel, H. 1975. Modellrechnungen zum nächtlichen Strahlenhaushalt von Pflanzen. Meteorol. Runschau, 28, 5: 148–154.
Hamidov, A. H. 1972. Vertical distribution of the leaf area in the cotton canopy. Trans. Central Asian Regional Res. Inst. Hydrometeorol., 4, 85: 33–36. (Russian).
Härtel, O. 1954. Über spektrale Veränderungen des Lichtfeldes durch die Vegetation. Wetter und Leben, 6, 516: 118–122.
Heinicke, D. R. 1963a. Note on estimation of leaf area and leaf distribution in fruit trees. Can. J. Plant Sci., 43, 4: 597–598.
Heinicke, D. R. 1963b. The micro-climate of fruit trees. II. Foliage and light distribution patterns in apple trees. Amer. Soc. Hort. Sci., 83: 1–11.
Heslehurst, M. R. & Wilson, G. L. 1971. Studies on the productivity of tropical pasture plants. III. Stand structure, light penetration and photosynthesis in field swards of Setaria and green leaf Desmodium. Austral. J. Agr. Res., 22, 6: 865–878.
Hogetsu, K., Osima, Y., Midorikawa, B., Tezuka, Y., Sakamotu, M., Mototani, I. & Kimura, M. 1960. Growth analytical studies on the artificial communities of *Helianthus tuberosus* with different densities. Jap. J. Bot., 17, 2: 278–305.
Honda, H. 1971. Description of the form of trees by the parameters of the tree-like body: effects of the branching angle and the branch length on the shape of the tree-like body. J. Theor. Biol., 31, 2: 331–338.
Horie, T. 1966. Preliminary report of a method for estimating sunlit leaf area within a corn canopy. J. Agric. Meteorol., 22, 2: 45–49.
Horie, T. 1969. Vertical distribution of photosynthetic intensity within a sunflower community. Photosynthesis and Utilization of Solar Energy. Level III Experiments: 85–90. Tokyo.
Horie, T. & Udagawa, T. 1970. Canopy architecture and radiation environment within sunflower communities. Photosynthesis and Utilization of Solar Energy. Level III Experiments: 9–15. Tokyo.
Horn, H. S. 1971. The Adaptive Geometry of Trees. Princeton Univ. Press, New Jersey.
Hozumi, K., Koyama, H. & Kira, T. 1955. Intraspecific competition among higher plants. IV. A preliminary account on the interaction between adjacent individuals. J. Inst. Polytech. Osaka City Univ., Ser. D, 6: 121–130.
Hozumi, K., Shinozaki, K. & Kira, T. 1962. Effect of light intensity and planting density on the growth of *Hibiscus Moscheutos Linn.* III. Analysis of leaf growth based on logistic theory. Physiol. Ecol., 11, 1: 62–77. (Jap.)
Hutchison, B. A. 1975. Photographic assesment of deciduous forest radiation regimes. Phil. Doct. Dissert. Yale Univ.: 1–164.
Huxley, P. A. 1971. Leaf volume: a simple method for measurement and some notes on its use in studies of leaf growth. J. Appl. Ecol., 8, 1: 147–153.
Ichimura, S. 1954. Ecological studies on the plankton in paddy fields. I. Seasonal fluctuations in the standing crop and productivity of plankton. Jap. J. Bot., 14: 269–279.
Idso, S. B. 1968. An analysis of the heating coeffient concept. J. Appl. Meteorol., 7, 4: 716–717.
Idso, S. B. & Baker, D. G. 1967. Method for calculating the photosynthetic response of a crop to light intensity and leaf temperature by an energy flow analysis of the meteorological parameters. Agron. J., 59, 1: 13–21.

Idso, S. B., Baker, D. G. & Blad, B. L. 1969. Relations of radiations fluxes over natural surfaces. Quart. J. Roy. Meteorol. Soc., 95, 404: 244–257.
Idso, S. B., Jackson, R. D., Ehrler, W. L. & Mitchell S. T. 1969. A method for determination of infrared emittance of leaves. Ecology, 50, 5: 899–902.
Idso, S. B. & de Wit, C. T. 1970. Light relations in plant canopies. Appl. Opt., 9: 177–184.
Ilyina, A. A. 1947. Absorption and reflection spectra of plant leaves, Sov. J. Phys. Chem., 21, 2: 145–159. (Russian).
Impens, I. & Lemeur, R. 1969a. The radiation balance of several field crops. Arch. Met. Geoph. Biokl. Ser. B, 17, 2–3: 261–268.
Impens, I. & Lemeur, R. 1969b. Extinction of net radiation in different crop canopies. Arch. Met. Geoph. Biokl. Ser. B, 17: 403–412.
Impens, I., Lemeur, R. & Moermans, R. 1970. Spatial and temporal variation of net radiation in crop canopies. Agric. Meteorol., 7: 335–337.
Instructions to hydrometeorological stations and observation posts. Agrometeorological observations at stations and at observation posts. Part 1. Fundamental agrometeorological observation, Fasc. 11. Hydrometeorol. Press, L. 1963.
Isobe, S. 1962a. Preliminary studies on physical properties of plant communities. Bull. Nat. Inst. Agric. Sci. (Jap.), Ser. A, 9: 29–66.
Isobe, S. 1962b. An analytical approach to the expression of light intensity in plant communities. J. Agric. Meteorol., 17, 4: 143–150. (Jap.).
Isobe, S. 1962c. An analytical approach to the expression of light intensity in plant communities. J. Agric. Meteorol., 18, 1: 19–21. (Jap.).
Isobe, S. 1969. Theory of the light distribution and photosynthesis in canopies of randomly dispersed foliage area. Bull. Nat. Inst. Agric. Sci., (Jap.), Ser. A. 16: 1–25.
Ito, A. 1969. Geometrical structure of rice canopy and penetration of direct solar radiation. Proc. Crop Sci. Soc. Jap., 38, 3: 355–363. (Jap.).
Ito, A. & Udagawa, T. 1971. Phytometrical studies of crop canopies. I. Geometrical structure of soybean canopy and sun light penetration. J. Agric. Meteorol., 26, 4: 187–195. (Jap.).
Ito, A., Udagawa, T. & Uchijima, Z. 1973. Phytometrical studies of crop canopies. II. Canopy structure of rice crops in relation to varieties and growing stage. Proc. Crop Sci. Soc, Jap., 42, 3: 334–342. (Jap.).
Ivaki, H. 1959. Ecological studies on interspecific competition in a plant community. I. An analysis of growth of competing plants in mixed stands of buckwheat and green grams. Jap. J. Bot., 17, 1: 120–138.
Ivanov, L. A. 1932. On the regularity of the distribution of light in forest communities. Sov. Bot. J., 17, 4: 339–351. (Russian).
Ivanov, L. A. 1946. Light and moisture in the life of our tree species. 5th Timiryazev Readings. USSR Acad. Sci. Press, M. (Russian).
Jahnke, L. S. & Lawrence, D. B. 1965. Influence of photosynthetic crown structure on potential productivity of vegetations, based primarily on mathematical models. Ecology, 46, 3: 319–326.
Jain, T. C. & Misra, D. K. 1966. Leaf area estimation by linear measurements in Ricinus communis. Nature, 212, 5063: 741–742.
Jarvis, P. G., James, J. B. & Landsberg, J. J. 1976. Coniferous forest. In: Monteith, J. L. (Ed.), Vegetation and Atmosphere, Vol. 2. Case Studies: 171–240. Acad. Press, London.
Jordan, C. F. 1969. Derivation of leaf-area index from quality of light on the forest floor. Ecology, 50, 4: 663–666.
Kalma, J. D. 1972. The radiation balance of a tropical pasture. II. Net all-wave radiation. Agric. Meteorol., 10, 4–5: 261–275.
Kalma, J. D. & Stanhill, G. 1969. The radiation climate of an irrigated orange plantation. Solar Energy, 12: 491–508.
Kanemasu, E. T., Stone, L. R. & Powers, W. L. 1976. Evapotranspiration model tested for soybean and sorghum. Agron. J., 68: 569–572.

Kasanaga, H. & Monsi, M. 1954. On the light-transmission of leaves, and its meaning for the production of matter in plant communities. Jap. J. Bot., 14: 304–324.

Kastrov, V. G. 1955. On the diurnal course of the albedo of the Earth's surface. Trans. Centr. Aerol. Obs., 14: 12–22. (Russian).

Khosrowshahi, M. 1970. The breeding of ideal types of cereal species. Meld. Norges Landbrukshogskole, 49, 6: 8.

Kimura, K. 1974. Effect of light on leaf inclination of Triticum aestivum. I. Monochromatic light. Ber. Ohara Inst. Landwirt. Biol. Okayama Univ., 16, 2: 47–56.

Kimura, K. 1975a. Effect of light on leaf inclination of Triticum aestivum. II. Intensity of blue light. Ber. Ohara Inst. Landwirt. Biol. Okayama Univ., 16, 3: 129–133.

Kimura, K. 1975b. Effect of light on leaf inclination of Triticum aestivum. III. Seedling age and photosensitive region. Ber. Ohara Insti. Landwirt. Biol. Okayama Univ. 16. 3: 135–146.

Kinerson, R., Jr. & Fritschen, L. J. 1971. Modelling a coniferous forest canopy. Agric. Meteorol., 8, 6: 439–445.

Kleschnin, A. F. 1960. Die Pflanze und das Licht. Akademie-Verlag, Berlin.

Kondratyev, K. Ya. 1954. The radiant energy of the sun. Hydrometeorol. Press. L. (Russian).

Kondratyev, K. Ya. 1969. Radiation in the atmosphere: 1–912. Acad. Press, N. Y.

Kornher, A. 1971. Untersuchungen zur Stoffproduktion von Futterpflanzenbeständen. II. Über die Lichtverhältnisse in Beständen von Wiesenschwingel *(Festuca pratensis Huds.)* und Wiesenliechschgras *(Phleum pratense L)* in Bezuhung zur Bestandmorphologie und Produktivität. Acta agr. scand., 21, 4: 249–263.

Kornher, A. & Rodskjer, N. 1967. Über die Bestimmung der Globalstrahlung in Pflanzenbeständen. Flora, B 157, 2: 149–164.

Koyama, H. & Kira, T. 1956. Intraspecific competition among higher plants. VIII. Frequency distribution of individual plant weight as affected by the interaction between plants. J. Inst. Polytech. Osaka City Univ. Ser. D, 7: 73–94.

Krinov, E. L. 1947. Spectral reflectivity of natural formations. USSR Acad. Sci. Press, M. (Russian).

Kudryavtseva, A. A. 1940. Reflection, absorption and penetration of solar radiation in the plant cover of agricultural crops. Rep. VASHNIL, 2. (Russian).

Kumar, R. & Silva, L. 1973. Light ray tracing through a leaf cross section. Appl. Opt., 12, 12: 2950–2954.

Kupchenko, G. S. 1968. Some biometrical characteristics of winter crops. Trans. Main Geophys. Obs., 229: 7–22. (Russian).

Kuroiwa, S. 1968a. A new calculation method for total photosynthesis of a plant community under illumination consisting of direct and diffused light. In: Eckardt, F.E. (Ed.), Funct. Terrestr. Ecosyst. Primary Prod. Level: 391–398. Unesco, Paris.

Kuroiwa, S. 1968b. Theoretical analysis of light factor and photosynthesis in plant communities. 3. Total photosynthesis of a foliage under parallel light in comparison with that under isotropic light condition. J. Agric. Meteorol., 24, 2: 23–38. Tokyo.

Kuroiwa, S. 1970. Total photosynthesis of a foliage in relation to inclination of leaves. In: Šetlik, I. (Ed.), Prediction and Measurement of Photosynthetic Productivity: 78–89. Pudoc, Wageningen.

Květ, J., Svoboda, J. & Fiala, K. 1967. A simple device for measuring leaf inclinations. Photosynthetica, 1, 1–2: 127–128.

Laisk, A. 1964. Spectrophotometer for detailed research of radiation field within plant cover. In: Investig. Atm. Phys., Inst. Phys. and Astron., Acad. Sci. ESSR. Tartu, 6: 99–109. (Russian).

Laisk, A. 1965a. An improved photoelectric planimeter and a device for determining the orientation of leaves. In: Questions on the Radiation Regime of Plant Cover: 102–113. Inst. Phys. and Astron. Acad. Sci. ESSR, Tartu. (Russian).

Laisk, A. 1965b. Experimental investigations of a statistic character of the radiation field in maize crop. Proc. Acad. Sci. Estonian SSR., ser. phys.-math. and techn., 14, 1: 107–119.

Laisk, A. 1965c. Statistical analyzer for bioactinometric measurements. In: Questions on the Radiation Regime of Plant Cover: 114–126. Inst. Phys. and Astron. Acad. Sci. ESSR, Tartu. (Russian).

Laisk, A. 1968. Statistical character of light extinction in plant communities. In: Solar Radiation Regime in Plant Canopy: 81–111. Inst. Phys. and Astron. Acad. Sci. ESSR, Tartu. (Russian).

Laisk, A. 1969. Measurement of plant cover transparency. In: Photosynthetic Productivity of Plant Stand: 174–185. Inst. Phys. and Astron. Acad. Sci. ESSR, Tartu. (Russian).

Lakin, G. F. 1968. Biometrics. Higher School Publ., M.

Lakso, A. N. 1976. Characterizing apple tree canopies by fisheye photography. Hort Sci., 11, 4: 404–405.

Lang, A. R. G. 1973. Leaf orientation of a cotton plant. Agric. Meteorol., 11, 1: 37–51.

Lang, A. R. G. & Shell, G. S. G. 1976. Sunlit areas and angular distribution of sunflower leaves for plants in single and multiple rows. Agric. Meteorol., 16, 1: 5–15.

Ledent, J.-F. 1974. Une formule générale pour le calcul de la surface foliare soumise au rayonnement solaire direct. Biometrie-Praximetrie XIV, 3–4: 69–75.

Ledent, J.-F. 1976. Beam light interception by twisted leaf surfaces. Agr. Meteorol., 17, 4: 271–280.

Lemeur, R. 1971. Mathematical model for the study of parallel, isotropic and scattered radiation in theoretical and real canopies. Ph. D. Thesis, Chair of Plant Ecology, University of Ghent. 1–102. (Dutch).

Lemeur, R. 1973a. Effects of spatial leaf distribution on penetration and interception of direct radiation. Plant Response to Climatic Factors. Proc. Uppsala Symp. 1970 (Ecology and Conservation): 349–356. Unesco, Paris.

Lemeur, R. 1973b. A method for simulating the direct solar radiation regime in sunflower, Jerusalem artichoke, corn and soybean canopies using actual stand structure data. Agric. Meteorol., 12, 2: 229–247.

Lemeur, R. & Blad, B. L., 1974. A critical review of light models for estimating the shortwave radiation regime of plant canopies. Agr. Meteorol., 14, 1–2: 255–286.

Lemeur, R. & Rosenberg, N. J. 1975. Reflectant induced modification of the radiation balance for increased crop water use efficiency. Heat and Mass Transf. Biosphere. Part 1. Trans. Processes Plant Environ.: 479–488. Washington D.C.

Lemeur, R. & Rosenberg, N. J. 1976. Reflectant induced modification of soybean canopy radiation balance. III A comparison of the effectiveness of celite and kaolinite reflectants. Agron. J., 68, 1: 30–35.

Levy, E. B. & Madden, E. A. 1933. The point method of pasture analysis. New Zeal. J. Agric., 46: 267–279.

Lingova, St. 1970. On the distribution of photosynthetically active irradiation in Bulgaria. Bull. Methodi Popoff Inst. Plant Physiol., 16: 155–166. (Bulg.).

Lingova, St. & Stanev, V. 1969. Photosynthetically active radiation and the development of the maize crop depending on its density. Hidrol. i meteorol., 18, 4: 47–56. (Bulg.).

Loomis, R. S. & Williams, W. A. 1969. Productivity and morphology of crop stands: Patterns with leaves. In: Eastin, J. D., Haskins, F. F., Sullivan, C. Y. & van Bavel, C. H. M. (Eds.), Physiological Aspects of Crop Yield: 27–47. Am. Soc. Agron., Madison, Wisc.

Loomis, R. S., Williams, W. A. & Duncan, W. G. 1967. Community architecture and the productivity of terrestrial plant communities. In: San Pietro, A., Greerr, F. & Army, T. J. (Eds.), Harvesting the Sun: 291–308. Acad. Press.

Loomis, R. S., Williams, W. A., Duncan, W. G., Dovrat, A. & Nunez, F. 1968. F. Quantitative descriptions of foliage display and light interception in field communities of corn. Crop Sci., 8, 3: 352–356.

Lopukhin, E. A. 1948a. On the spectral absorption of radiation by cotton plantations. Rep. Acad. Sci., Uzbek SSR, 9: 13–17. (Russian).

Lopukhin, E. A. 1948b. Concerning the role of radiation in the life of a cotton plant. Trans. Tashkent Geophys. Obs., 3: 14–37. (Russian).

Lopukhin, E. A. 1950. On the radiation regime of cotton. Proc. Acad. Sci., USSR, ser. geogr. and geophys., 14, 3: 260–266. (Russian).

Lopukhin, E. A. 1951a. Method of determining shortwave radiation absorbed by leaves of a cotton plant. Proc. Acad. Sci., Uzbek SSR, 4: 32–39. (Russian).

Lopukhin, E. A. 1951b. An approximate method of determining the distribution of total radiation in cotton plantations. Trans. Tashkent Geophys. Obs., 5: 79–81. (Russian).

Luzganov, A. G. & Solodko, A. S. 1968. Shape of the tree crown and scattered light. Sov. Forest J., 6: 29–33. (Russian).

Luxmoore, R. J., Millington, R. J. & Peters, D. B. 1970. Symap and net radiation in a soybean canopy. Agron. J., 62, 6: 830–831.

Madgwick, H. A. & Brumfield, G. L. 1969. The use of hemispherical photographs to assess light climate in the forest. J. Ecol., 57, 2: 537–542.

Maggs, D. H. 1956. A photometer for estimating the area of detached leaves. In: Milthorpe, F. L. (Ed.), The Growth of Leaves: 200–201. Butterworths Sci. Publ., London.

Mägi, H. & Ross, J. 1969a. Phytometric characteristic and photosynthetic productivity of the barley stand. I. Agrometeorological conditions and geometrical structure of the stand. In: Photosynthetic Productivity of Plant Stand: 102–143. Acad. Sci. ESSR, Inst. Phys. Astron., Tartu. (Russian).

Mägi, H. & Ross, J. 1969b. Phytometric characteristics and photosynthetic productivity of the barley stand. II. Growth dynamics of the assimilation area and increase of the dry-matter content. In: Photosynthetic Productivity of Plant Stand: 144–173. Acad. Sci. ESSR, Inst. Phys. Astron., Tartu. (Russian).

Marchuk, G. I. 1961. Methods of calculating nuclear reactors. State Atom Publishers. M.

Marshall, J. K. 1968. Methods for leaf area measurement of large and small leaf samples. Photosynthetica, 2, 1: 41–47.

McChaughey, J. H. & Davies, J. A. 1974. Diurnal variation in net radiation depletion within a corn crop. Boundary-Layer Meteorol., 5, 4: 505–511.

McCree, J. K. 1966. A solarimeter for measuring photosynthetically active radiation. Agric. Meteorol., 3, 516: 535–366.

McCree, K. J. 1972a. The action spectrum, absorptance and quantum yield of photosynthesis in crop plants. Agric. Meteorol., 9, 3–4: 191–216.

McCree, K. J. 1972b. Test of current definitions of photosynthetically active radiation against leaf photosynthesis data. Agric. Meteorol., 10, 6: 443–453.

McCree, K. J. 1973. The measurement of photosynthetically active radiation. Solar Energy, 15, 1: 83–87.

McCree, K. J. 1976. A comparison of experimental and theoretical spectra for photosynthetically active radiation at various atmospheric turbidities. Agric. Meteorol., 16, 3: 405–412.

McKee, G. W. 1964. A coefficient for computing leaf area in hybrid corn. Agron. J., 56, 2: 240–241.

Midorikawa, B. 1957. The distribution of CO_2 concentration of the air in some plant communities. Jap. J. Ecol., 7: 72–76.

Miller, P. C. 1967 (1968). Leaf temperatures, leaf orientation and energy exchange in quaking aspen *(Populus tremuloides)* and Gambell's oak *(Quercus gambellii)* in central Colorado. Oecologia plantarum, 2, 3: 241–270.

Miller, E. E. & Norman, J. M. 1971a. A sunfleck theory for plant canopies. I. Lenghts of sunlit segments along a transect. Agron. J., 63, 5: 735–738.

Miller, E. E. & Norman, J. M. 1971b. A sunfleck theory for plant canopies. II. Penumbra effect: intensity distribution along sunfleck segments. Agron. J., 63, 5: 739–743.

Milthorpe, F. L. (Ed.) 1956. The Growth of Leaves. Butterworths Scientific Publ., London.

Molchanov, A. G. & Khazanov, V. S. 1975. Measurement and calculation of PAR absorption by pine shoots. Forest Management, 2: 75–79. (Russian).

Moldau, H. 1964. Spectrogoniograph for measuring the indicatrixes of reflection and transmission of leaves. In: Investig. Atm. Phys., Acad. Sci. ESSR, Inst. Phys. Astron., Tartu, 6: 92–98. (Russian).

Moldau, H. 1965. Vegetative course of the luminance factor of leaves of some plants. In: Questions on Radiation Regime of Plant Stand: 89–95. Acad. Sci. ESSR, Inst. Phys. Astron., Tartu. (Russian).

Moldau, H. 1967. Optical model of plant leaf. In: Phytoactinometric Investigations of Plant Canopy: 89–109. Valgus Publishers, Tallinn. (Russian).

Moldau, H., Ross, J., Tooming, H. & Undla, I. 1963. Geopgraphic distributions of photosynthetically active radiation (PAR) over the European part of the USSR. In: Photosyn. and Problems of Plant Produc.: 149–158. USSR Ac. of Sci. Press, M. (Russian).
Monsi, M. & Saeki, T. 1953. Über den Lichtfaktor in den Pflanzengesellschaften und seine Bedeutung für die Stoffproduktion. Jap. J. Bot., 14, 1: 22–52.
Moteith, J. L. 1965a. Light distribution and photosynthesis in field crops. Ann. Bot., 29, 113: 17–37.
Monteith, J. L. 1965b. Radiation and crops. Exp. Agric. Rev., 1: 241–251.
Monteith, J. L. 1969. Light interception and radiative exchange in crop stands. In: Eastin, J. D. et al. (Ed.), Physiological Aspects of Crop Yield: 89–115. Amer. Soc. Agron & Crop Sci. Soc. Amer., Madison, Wisc.
Monteith, J. L. & Szeicz, G. 1961. The radiation balance of bare soil and vegetation. Quart. J. Roy. Meteorol. Soc., 87: 159–170.
Monteith, J. L. & Szeicz, G. 1962. Radiative temperature in the heat balance of natural surface. Quart. J. Roy. Meteorol. Soc., 88, 378: 496–507.
Monteni, B. & Gosse, G. 1978. Variations du rayonnement photosynthétiquement actif en région tropical humide. Arch. Met. Geoph. Bioklim. Ser. B. 25, 4: 375–382.
Moon, P. & Spencer, D. E. 1942. Illumination from a non-uniform sky. Trans. Illum. Engng Soc. N. Y., 37: 707–712.
Morozov, V. K. 1963. Is heliotropism observed in a blossoming sunflower head? Sov. Bot. J., 48, 6: 885–888. (Russian).
Mototani, I. 1968. Horizontal distribution of light intensity in plant communities. In: Photosynthesis and Utilization of Solar Energy. Level III Experiments 1966–1967: 25–28, Tokyo.
Nichiporovich, A. A. 1955. On methods of determining and studying photosynthesis as a factor of productivity. Trans. Inst. Plant Physiol., 10: 210–249. (Russian).
Nichiporovich, A. A. 1958. Photosynthesis and the theory of obtaining high yields. In: 15th Timiryazev Readings. USSR Ac. of Sci. Press, M. (Russian).
Nichiporovich, A. A. 1960. Conference on measurement of visible radiation in plant physiology. Sov. Plant Physiol., 7, 6: 744–747. (Russian).
Nichiporovich, A. A. 1961. On properties of plants as an optical system. Sov. Plant Physiol., 8, 5: 536–546. (Russian).
Nichiporovich, A. A. 1963. On ways of raising productivity of photosynthesis of plants in crops. In: Photosyn. and Problems of Plant Produc.: 5–36, USSR Ac. of Sci. Press, M. (Russian).
Nichiporovich, A. A. 1966. Tasks of scientific papers on the study of photosynthetic activity of plants as a factor of productivity. In: Photosynthesizing Systems of High Productivity: 7–50, Nauka Publishers, M. (Russian).
Nichiporovich, A. A. & Chmora, S. N. 1958. On the utilization of solar radiation in the photosynthesis of potato crops. Sov. Plant Physiol., 5, 4.: 320. (Russian).
Nichiporovich, A. A. Strogonova, L. E., Chmora, S. N. & Vlasova, M. P. 1961. Photosynthetic activity of plants in crops. USSR Ac. of Sci. Press., M. (Russian).
Niilisk, H. J. 1960. On computing atmospheric infra-red radiative fluxes. In: Investig. Atm. Phys., 2: 67–114. Acad. Sci. ESSR, Inst. Phys. Astron., Tartu. (Russian).
Niilisk, H. 1962. Simplified spectrophotometer for measurements of spectral fluxes of diffuse radiation. In: Investig. Atm. Phys., 3: 150–159. Acad. Sci. ESSR, Inst. Phys. Astron., Tartu. (Russian).
Niilisk, H. 1963. Photometrical rod for determination of direct solar radiation in crops and plant cover. In: Investig. Atm. Phys., 4: 120–123. Acad. Sci. ESSR, Inst. Phys. Astron., Tartu. (Russian).
Niilisk, H. 1964. Spectral radiation regime of a maize crop and calculation of the photosynthetically active radiation (PAR). Proc. Acad. Sci. ESSR, ser. physico-math. and techn., 13, 3: 177–191. (Russian).
Niilisk, H. 1965. Spectrophotometer for measuring spectral radiation fluxes in plant cover. Proc. Acad. Sci. ESSR, ser. physico-math. and techn., 14, 4: 528–553. (Russian).

Niilisk, H., Nilson, T. & Ross, J. 1970. Radiation in plant canopies and its measurement. In: Šetlik, I. (Ed.), Prediction and Measurement of Photosynthetic Productivity: 165–177. Centre for Agricultural Publishing and Documentation, Wageningen.
Niilisk, H. & Ross, J. 1969. On some features of solar radiation in vegetation. In: Basic Problems of Biol. Produc.: 155–160, Nauka Publishers, L. (Russian).
Nilson, T. 1968a. The calculation of spectral fluxes of shortwave radiation in plant communities. In: Solar Radiation Regime in Plant Stand: 55–80. Acad, Sci. ESSR, Inst. Phys. Astron., Tartu. (Russian).
Nilson, T. 1968b. On the optimum geometrical arrangement of foliage in the plant cover. In: Solar Radiation Regime in Plant Stand: 112–146. Acad. Sci. ESSR, Inst. Phys. Astron., Tartu. (Russian).
Nilson, T. 1971a. A theoretical analysis of the frequency of gaps in plant stands. Agric. Meteorol., 8: 25–38.
Nilson, T. 1971b. On the character of dispersion of foliage elements in crop stands. In: Quantitative Methods for Analysis of Vegetation: 209–215, Riga. (Russian).
Nilson, T. 1972. Radiation conditions in plant canopy. In: Theoretical Foundations of the Photosynthetic Productivity: 420–424. Nauka Publishers, M. (Russian).
Nilson, T. 1977a. A theory of radiation penetration into non-homogeneous plant canopy. In: The Penetration of Solar Radiation into Plant Canopy: 5–70. Acad. Sci. ESSR, Inst. Phys. Astron., Tartu. (Russian).
Nilson, T. 1977b. On estimating radiation characteristics of plant stands from hemispherical photographs. In: Phytoactinometrical Investigations in Mountain Forests: 45–60. Acad. Sci. USSR, Inst. Georgr. Pacif. Oc., Vladivostok. (Russian).
Nilson, T. & Ross, V. 1971. The determination of canopy density by means of a photographic method. In: Biological Productivity of Spruce Forests: 129–132. Estonian Rep. Comm. for IBP, Tartu. (Russian).
Norman, J. M. & Jarvis, P. G. 1974. Photosynthesis in Sitka spruce. (Picea sitchensis (Bong.) (Carr.). III. Measurements of canopy structure and interception of radiation. J. Appl. Ecol., 11: 375–398.
Norman, J. M. & Jarvis, P. G. 1975. Photosynthesis in Sitka spruce *(Picea sitchensis (Bong.) Carr.)*. V. Radiation penetration theory and a test case. J. Appl, Ecol., 12, 3: 839–878.
Norman, J. M., Miller, E. E. & Tanner, C. B. 1971. Light intensity and sunfleck size distributions in plant canopies. Agron. J., 63, 5: 743–748.
Nosatovsky, A. I. 1947. On leaf position relative sun beams. Trans. Krasnodar Inst. Food Industry, 2: 35–44. (Russian).
Obeid, M., Machin, D. & Harper, J. L. 1967. Influence of density to plant variation in fiber flax, Linum ussitatissimum L. Crop Sci., 7, 5: 471–473.
Oikawa, T. 1977a. Light regime in relation to plant population geometry. II. Light penetration in a square-planted population. Bot. Mag. Tokyo, 90, 1017: 11–22.
Oikawa, T. 1977b. Light regime in relation to plant population geometry. III. Ecological implications of a square-planted population from the viewpoint of utilization efficiency of solar energy. Bot. Mag. Tokyo, 90, 1020: 301–311.
Oikawa, T. & Saeki, T. 1972. Light regime in relation to population structure. An experimental approach based on the Monte Carlo simulation model. Rep. 1971 JIBP/PP-Photosynthesis Level III Group: 107–116. Tokyo.
Oikawa, T. & Saeki, T. 1977. Light regime in relation to plant population geometry. I. A Monte Carlo simulation of light microclimates within a random distribution foliage. Bot. Mag. Tokyo: 90, 1017: 1–10.
Ondok, J. P. 1968. Measurement of leaf area in Phragmites communis Trin. Photosynthetica, 2, 1: 25–30.
Ondok, J. P. 1075. Photosynthetically active radiation in a stand of *Phragmites communis Trin.* IV. Stochastic model. Photosynthetica, 9, 2: 201–210.
Orth, R. 1938. Strahlung, Lichtfeld und Pflanze (Sammelreferat). Bioklim. Beibl. d. Meteorol. Zeitschrift, 5, 2: 68–75.

Pak Sen Su. 1957. Determination of the coefficients of heat radiation of different materials at ordinary temperatures (23–30 °C). Proc. Acad. Sci. Uzbek SSR, ser. Phys.-math., 2: 93–102. (Russian).

Paltridge, G. W. 1973. On the shape of trees. J. Theor. Biol., 38, 1: 111–137.

Pavlov, A. V. 1962. Calculation method for the determination of net radiation using total radiation and albedo. Proc. Acad. Sci. USSR, geogr. ser., 6: 94–100. (Russian).

Pearce, R. B., Brown, R. H. & Blaser, R. E. 1967. Photosynthesis in plant communities as influenced by leaf angle. Crop Sci., 7, 4: 321–324.

Pendleton, J. W., Smith, G. E., Winter, S. R. & Johnston, T. J. 1968. Field investigations of the relationships of leaf angle in corn *(Zea mays L)* to grain yield and apparent photosynthesis. Agron. J., 60, 4: 422–424.

Perelyot, N. A. 1970. Dependence of the interrelationship between photosynthetically active and integral radiation on the transparency of the atmosphere and on cloudiness. Trans. Ukr. Hydrometeorol. Res. Inst., 94: 46–51. (Russian).

Perelyot, N. A. 1971. Distribution of PAR on the territory of the Ukraine. Trans. Ukr. Hydrometeorol. Res. Inst., 102: 3–12. (Russian).

Philip, J. R. 1965a. The distribution of foliage density with foliage angle estimated from inclined point quadrat observations. Austral. J. Bot., 13, 2: 357–366.

Philip, J. R. 1965b. The distribution of foliage density of single plants. Austral. J. Bot., 13, 3: 411–418.

Philip, J. R. 1966a. Some integral equations in geometrical probability. Biometrika, 53, 3–4: 365–374.

Philip, J. R. 1966b. The use of point quadrats, with special reference to stem-like organs. Austral. J. Bot., 14, 1: 105–125.

Pleshko, S. I. 1948. On cases of absence of heliotropism in sunflower. Trans. Acad. Sci. USSR, Tadjik Branch, 8. (Russian).

Põldmaa, V. 1963. On distribution of diffuse radiation over the sky. In: Investig. Atm. Phys., Acad. Sci. ESSR, Inst. Phys. Astron., 4: 111–119. (Russian).

Polykarpov, G. G. 1954. Does a blossoming sunflower head turn to the Sun? Sov. Nat., 5: 116–117. (Russian).

Proctor, J. T. A., Kyle, W. J. & Davies, J. A. 1972. The radiation balance of an apple tree. Can. J. Bot., 50, 8: 1731–1740.

Pugachev, V. S. 1968. Introduction to the theory of probability. Nauka Publishers, M. (Russian).

Rabideau, G. S., French, C. S. & Holt, A. S. 1946. The absorption and reflection spectra of leaves, chloroplast suspensions, and chloroplast fragments as measured in a Ulbricht sphere. Amer. J. Bot., 33, 10: 769–777.

Rabinowitch, E. I. 1945. Photosynthesis and Related Processes. Vol. I. Interscience Publishers, N. Y.

Rabotnov, T. A. 1950. Some problems of studying the structure of meadow herb stands. Bull. Moscow Naturalists Soc., 55, 2: 51–71. (Russian).

Rademacher, B. 1950. Über die Lichtverhältnisse in Kulturpflanzenbeständen mit besondere in Hinblick auf dem Unterwuchs. Z. Acker- u. Pflanzenbau, 29: 129.

Rauner, J. L. 1966. Investigation into the heat balance of th active forest layer. In: Contemporary problems of climatology: 83–93. Hydrometeorol. Press, L. (Russian).

Rauner, J. L. 1968. Radiation regime and biometric characteristics of the forest vegetation. In: Actinometry and Atmospheric Optics: 335–342, Valgus Publishers, Tallinn. (Russian).

Rauner, J. L. 1972. Heat balance of the plant cover: 1–209, Hydrometeorol. Press, L. (Russian).

Rauner, J. L. 1976. Deciduous forests. In: Monteith, J. L. (Ed.), Vegetation and Atmosphere, Vol. II: 241–264. Acad. Press, London.

Rauner, J. L. & Ananyeva, L. M. 1967. Net radiation and the structure of evaporation of meadow steppe vegetation. Proc. Acad. Sci. USSR, geogr. ser., 4: 104–112. (Russian).

Reemann, J. 1963. Simplified electronic integrator for actinometric measurements. In: Investig. Atm. Phys., 4: 124–134. Acad. Sci. ESSR, Inst. Phys. Astron., Tartu. (Russian).

Rodskjer, N. & Kornher, A. 1974. An optical leaf area planimeter based on the potentiometer principle. Swed. J. Agric. Res., 4: 3–9.

Rosenthal, W. D., Kanemasu, E. T., Raney, R. J. & Stone L. R. 1977. Evaluation of an evapotranspiration model for corn. Agron. J., 69, 3: 461–464.
Ross, J. 1957. About shortwave radiation regime on the surface occupied by vegetation. Sci. Pap. Inst. Geol. and Geogr., Acad. Sci. Lith. SSR, Vilnius, 6: 41–60. (Russian).
Ross, J. 1962. Concerning the theory of vegetable cover albedo. Sci. Pap. Inst. Geol. and Geogr., Acad. Sci. Lith. SSR, Vilnius, 13: 151–165. (Russian).
Ross, J. 1964. Photosynthetically active radiation (PAR) in plant stands and its mathematical modelling. In: Actinometry and Atmospheric Optics: 251–256. Nauka Publishers, M. (Russian).
Ross, J. 1967. The role of solar radiation in the photosynthesis of crops. In: Nichiporovich A. A. (Ed.) Photosynthesis of Productive Systems: 44–52. Israel Progr. for Sci. Transl., Jerusalem.
Ross, J. 1968. The present state of research into the radiation regime of the plant cover. In: Actinometry and Atmospheric Optics: 255–260. Valgus Publishers, Tallinn. (Russian).
Ross, J. 1970. Structural organization of crops and coenoses from the standpoint of maximal utilization of solar radiation. In: Fndamental Problems of Photosynthesis in Plant Husbandry: 38–51. Kolos Publishers, M. (Russian).
Ross, J. 1972. Topic of problems in phytoactinometry. In: Proc. of Intern. Radiation Symp. in Sendai, Japan, May 26-June 2: 158–163.
Ross, J. & Bichele, Z. 1968. Calculation of photosynthesis of leaf canopies. I. In: Photosynthesis and Productivity of Plant Stand.: 75–110. Acad. Sci. ESSR, Inst. Phys. Astron., Tartu. (Russian).
Ross, J. & Bichele, Z. 1969. Calculation of photosynthesis of leaf canopies. II. In: Photosynthetic Productivity of Plant Stand.: 5–43. Acad. Sci. ESSR, Inst. Phys. Astron., Tartu. (Russian).
Ross, J. & Mägi, H. 1971. A calculation method for determinating leaf area and its vertical distribution in the barley crop. In: Estonian Contributions to IBP. III.: 101–112. Tartu.
Ross, J. & Nilson, T. 1963. concerning the theory of plant cover radiation regime. In: Investig. Atm. Phys., Acad. Sci. ESSR, 4: 42–64. Inst. Phys. Astron., Tartu. (Russian).
Ross, J. & Nilson, T. 1965. The extinction of direct radiation in crops. In: Questions on Radiation Regime of Plant Stand: 25–64. Acad. Sci. ESSR, Inst. Phys. Astron., Tartu. (Russian).
Ross, J. & Nilson, T. 1967a. The vertical distribution of biomass in crop stands. In: Nichiporovich, A. A. (Ed.), Photosynthesis of Productive Systems: 75–85. Israel Progr. for Sci. Transl., Jerusalem.
Ross, J. & Nilson, T. 1967b. The spatial orientation of leaves in crop stands and its determination. In: Nichiporovich, A. A. (Ed.), Photosynthesis of Productive Systems, 86–99 Israel Progr. for Sci. Transl., Jerusalem,
Ross, J. & Nilson, T. 1967c. The radiation regime of plant communities with horizontal leaves. In: Phytoactinometric Investigations of Plant Stand: 5–34. Valgus Publishers. Tallinn. (Russian).
Ross, J. & Nilson, T. 1968a. A mathematical model of radiation regime of plant cover. In: Actinometry and Atmospheric Optics: 263–281. Valgus Publishers, Tallinn. (Russian).
Ross, J. & Nilson, T. 1968b. The calculation of photosynthetically active radiation in plant communities. In: Solar Radiation Regime in Plant Canopy: 5–54. Acad. Sci. ESSR, Inst. Phys. Astron., Tartu. (Russian).
Ross, J. & Ross, V. 1969a. Vertical distribution of leaf area in crop stands. In: Photosynthetic Productivity of Plant Stand: 44–59. Acad. Sci. ESSR, Inst. Phys. Astron., Tartu. (Russian).
Ross, J. & Ross, V. 1969b. Spatial orientation of leaves in crop stands. In: Photosynthetic Productivity of Plant Stand: 60–82. Acad. Sci. ESSR, Inst. Phys. Astron., Tartu. (Russian).
Ross, J. & Tooming, H. G. 1968. Extinction of direct solar and total radiation within crops and semiempirical formulae for describing this phenomenon. In: Actinometry and Atmospheric Optics: 283–288. Valgus Publishers, Tallinn. (Russian).
Ross, J. & Vlasova, M. P. 1967. Biometric characteristics of maize stand and the dynamics of its development. In: Nichiporovich, A. A. (Ed.), Photosynthesis of Productive Systems: 60–74. Israel Progr. for Sci. Transl., Jerusalem.
Ross, V. 1967. On the determination of plant leaf area. In: Phytoactinometric Investigations of Plant Canopy: 150–162. Valgus Publishers, Tallinn. (Russian).

Ross, V. & Ross, J. 1969. Biometric measurements in agricultural crops. In: Methodological Instructions for Determinating and Checking Basic Indices of Photsynthetic Activity of Plants in Crops: 25–34. VASHNIL, M. (Russian).
Roussel, L. 1962. Etat actuel de la photologie forestiere, Reveu Forestiere Française, 1: 1–13.
Rozhnyatovsky, I. I. 1954. Determination of leaf area with the help of light sensitive paper. Sov. Bot. J., 39, 3: 419–420. (Russian).
Rudnev, N. I. 1977. Forest net radiation: 1–126, Nauka Publishers, M. (Russian).
Rvachev, V. P. 1966. Introduction to biophysical photometry. Lvov University Press. (Russian).
Rvachev, V. P., Guminetsky, S. G. & Sakhanovsky, M. J. 1965. Optical parameters of plant leaves within the spectral interval of 235 to 1000 μm. Sov. Biophysics, 10, 4: 658–664. (Russian).
Rvachev, V. P. & Yatsenko, G. I. 1964. A photoelectrical photoplanimeter. Abstracts of Reports at 20th Sci. Session, section of phys. sci.: 51, Chernovtsy. (Russian).
Rybicki, G. B. 1965. A new differential equation approach to transfer problems. Research in Space Science, Special Report No. 174, Cambridge, Massachusetts, USA.
Saeki, T. 1960a. Leaf growth as influenced by dry matter production. Bot. Mag., Tokyo, 74, 872: 70–78.
Saeki, T. 1960b. Interrelationships between leaf amount, light distribution and total photosynthesis in a plant community. Bot. Mag., Tokyo, 73, 860: 55–63.
Saeki, T. 1961. Analytical studies on the development of foliage of a plant community. Bot. Mag., Tokyo, 74, 877–878: 342–348.
Saeki, T. & Kuroiwa, S. 1959. On the establishment of the vertical distribution of photosynthetic system in a plant community. Bot. Mag., Tokyo, 72, 848: 27–35.
Saito, T. 1964. Method of measurement of transmissivity of atmospheric radiation and calculation of net long wave radiation within plant communities. J. Agric. Meteorol., 20, 1: 7–10.
Samarina, N. N. 1965. Distribution values of net radiation in plant stand. Proc. Acad. Sci. USSR, ser. geogr., 4: 92–97. (Russian).
Samarina, N. N. 1969. Methods and some results of the study of biometric structure of agricultural crops. Proc. Acad. Sci. USSR, ser. geogr., 1: 106–114. (Russian).
Santhirasegaram, K. & Black, J. N. 1968. The distribution of leaf area and light intensity within wheat crops differing in row direction, row spacing and rate of sowing. A contribution to the study of undersowing pasture with cereals. J. Brit. Grassl. Soc., 23, 1: 1–12.
Sauberer, F. 1937. Zur Kentnis der Strahlungsverhältnisse in Pflanzenbeständen. Bioklim. Beibl. d. Meteorol. Zeitschrift., 4: 145–155.
Sauberer, F. 1948. Über die Strahlungseigenschaften der Pflanzen in Infrarot. Wetter und Leben, 1: 231–234.
Sauberer, F. & Härtel, O. 1959. Pflanze und Strahlung. Leipzig.
Sauberer, F. & Trapp, E. 1947. Bioklimatische Strahlungsuntersuchungen, Jahrbuch Hochschule Bodenkultur, Wien, 1: 67.
Saugier, B. 1976. Sunflower. In: Monteith, J. L. (Ed.), Vegetation and the Atmosphere. Vol 2, Case Studies: 87–119. Acad. Press, London, N. Y., San-Francisco.
Savikovsky, I. A. & Kasperovich, J. I. 1973. Calculation of photosynthetically active radiation for the territory of Byelorussia. Collect. Pap. Byelorussian Territ. Hydrometeorol. Centre, 2: 92–101. (Russian).
Savina, S. S. 1968. Absorption of direct radiation by maize crops depending on the relative leaf area and solar elevation. In: Actinometry and Atmospheric Optics: 288–289. Valgus Publishers, Tállinn. (Russian).
Schnelle, F. 1955. Pflanzen-Phänologie. Akademische Verlag Geest und Portig K.-G. Leipzig.
Scott, D. & Wells, J. 1969. Leaf orientation in barley lupin and lucerne stands. New Zeeland J. Bot., 7, 4: 372–388.
Seybold, A. 1933. Über die optischen Eigenschaften der Laubblätter III. Planta (Berl.), 20: 577.
Seybold, A. & Weissweiler, A. 1942. Spektrophotometrische Messungen an grünen Pflanzen und an Chlorophyllösungen. Bot. Arch., 43, 3: 252.

Shaw, R. H. & Weber, C. R. 1967. Effects of canopy arrangements on light interception and yield of soybeans. Agron. J., 59: 155–159.
Shell, G. S. G., Lang, A. R. G. & Sale, P. J. M. 1974. Quantitative measures of leaf orientation and heliotropic response in sunflower, bean, pepper, and cucumber. Agric. Meteorol., 13, 1: 25–37.
Shell, G. S. G. & Lang, A. R. G. 1975. Description of leaf orientation and heliotropic response of sunflower using directional statistics. Agric. Meteorol., 15: 33–48.
Shell, G. S. G. & Lang, A. R. G. 1976. Movements of sunflower leaves over a 24-H period. Agric. Meteorol., 16, 2: 161–170.
Shifrin, K. S. 1953. Concerning the theory of albedo. Trans. Main Geophys. Obs., 39, 101: 244–257. (Russian).
Shiman, L. M. 1967. Determination of the orientation of plant leaves in space. Sov. Plant Physiol., 14, 2: 381–383. (Russian).
Shinozaki, K. & Kira, T. 1977. Canopy structure and light utilization. Jap. IBP Synthesis. Vol. 16. Primary productivity of Japanese forests: 75–86. Tokyo.
Shinozaki, K., Yoda, K., Hozumi, K. & Kira, T. 1964. A quantitative analysis of plant form – the pipe model theory. II. Further evidence of the theory and its application in forest ecology. Jap. J. Ecol., 14, 4: 133–139.
Shulgin, I. A. 1960. On the optical characterization of xeromorphy and succulency of plant leaves. Rep. Acad. Sci. USSR, 134, 4: 972–975. (Russian).
Shulgin, I. A. 1963. Morphophysiological adaptation of plant to light: 1–74. Moscow Univ. Press, M. (Russian).
Shulgin, I. A. 1967. On the adaptation of green plants to the spectral composition of solar radiation. Sov. Plant Physiol., 14, 4: 592–602. (Russian).
Shulgin, I. A. 1973. the plant and the Sun. Hydrometeorol. Press, L. (Russian).
Sinclair, T. R., Schreiber, M. M. & Hoffer, R. M. 1973. Diffuse reflectance hypothesis for the pathway of solar radiation through leaves. Agron. J., 65, 2: 276–283.
Singh, M., Peters, D. B. & Pendleton, J. W. 1968. Net and solar radiation in soybean canopies. Agron. J., 60: 542–545.
Sinnot, E. W. 1960. Plant morphogenesis. McGraw-Hill Book Comp. Inc., N. Y., Toronto, London.
Slatyer, R. O. 1967. Plant-water relationships. Acad. Press, London, N. Y.
Slomka, J. & Slomka, K. 1973. Calculation of active solar radiation energy in photosynthesis (PhAR). Wiad. ekol., 19, 2: 186–193. (Polish).
Smith, J. A. & Oliver, R. E. 1974. Effects of changing canopy directional reflectance of feature selection. Appl. Opt., 13, 7: 1599–1604.
Smith, S. & Scoular, K. 1975. A data structure analogue for modelling tree crowns. Can. J. Forest Res., 5, 4: 574–579.
Smurov, A. V. 1975. Statistical methods in the investigations of the spatial patterns of organisms. In: Methods of Soil-Zoological Investigations: 217–240. Nauka Publishers, M. (Russian).
Sobolev, V. V. 1956. Transfer of radiant energy in the atmosphere of stars and planets. GITTL, M. (Russian).
Stanhill, G. Cox, J. T. H. & Moreshet, S. 1968. The effect of crop and climatic factors on the radiation balance of an irrigated maize crop. J. Appl. Ecol., 5: 707–8720.
Stanhill, G. & Fuchs, M. 1977. The relative flux density of photosynthetically active radiation. J. Appl. Ecol., 14, 1: 317–322.
Stanhill, G. Hofstede, G. J. & Kalma, J. D. 1966. Radiation balance of natural and agricultural vegetation. Quart. J. Roy. Meteorol. Soc. 92, 391: 128–140,
Stanhill, G., Israeli, M. & Rosenzweig, D. 1973. The solar radiation balance of shrub forest and pasture on the Carmel Mountain. Ecology, 54, 4: 819–828.
Stephens, G. R. 1969. Productivity of red pine. 1. Foliage distribution in tree crown and stand canopy. Agric. Meteorol., 6, 4: 275–282.
Suits, G. H. 1972a. The cause of azimuthal variations in directional reflectance of vegetative canopies. Rem. Sens. Envir., 2, 3: 175–187.

Suits, G. H. 1972b. The calculation of the directional reflectance of a vegetative canopy. Rem. Sens. envir., 2, 2: 117–125.

Szeicz, G. 1966. Field measurements of energy in the 0.4–0.7 micron range. Light as an Ecological Factor: 41–51. Blackwell Sci. Publishers, Oxford.

Szwarcbaum, I. & Shaviv, G. 1976. A Monte-Carlo model for the radiation field in plant canopies. Agric. Meteorol., 17, 5: 333–352.

Tajchman, S. J. 1972. The radiation and energy balances of coniferous and deciduous forests. J. Appl. Ecol., 9, 2: 359–375.

Tamme, E. & Sõrmus, T. 1967. On the solution of anisotropic scattering transport equation. Proc. Acad. Sci. ESSR, ser, phys.-math. and techn., 16, 4: 412–418. (Russian).

Tanaka, S. 1969. Estimation of sunlit leaf area in tobacco plant community by the Monte-Carlo method. Estimation on direct sunlight. In: Photosynthesis and Utilization of Solar Energy. Level III Experiments, 1968: 76–79. Tokyo.

Tanaka, S. 1970. Geometrical distribution of leaves in tobacco plant community. In: Photosynthesis and Utilization of Solar Energy. Level III Experiments, 1969: 16–19. Tokyo.

Thomas, J. R., Namken, L. N., Oerther, Y. F. & Brown, R. G. 1971. Estimation of leaf water content by reflectance measurements. Agron. J., 63: 845–847.

Timoshenko, G. L. 1966. On the phytometric characterization of the plant cover of winter wheat. Trans. Ukrainian Hydrometeorol. Res. Inst., 90: 63–77. (Russian).

Tinney, F. W., Aamodt, O. S. & Ahlgren, H. L. 1937. Preliminary report on a study of methods used in botanical analysis of pasture swards. J. Amer. Soc. Agron., 29: 835–840.

Tooming, H. 1960. Daily and seasonal variations of albedo on some natural surfaces in the Estonian SSR. In: Investig. Atm. Phys., Inst. Phys. and Astron. Acad. Sci. ESSR, Tartu, 2: 115–163. (Russian).

Tooming, H. 1967. An approximate method for determining the attenuation and reflection of PAR and of near infrared radiation in a maize stand from the measurements of total radiation. In: Nichiporovich, A. A. (Ed.), Photosynthesis of Productive Systems: 100–113. Isr. Progr. for Sci. Transl., Jerusalem.

Tooming, H. 1968a. Adaptation of plant communities to light intensity and its mathematical modelling. Sov. J. General Biol., 19, 5: 549–563. (Russian).

Tooming, H. 1968b. Influence of foliage geometrical structure on the radiation regime, photosynthesis and growth of plants in geographical connection. Trans. Tartu Univ., 211, Papers on Bot., 8: 30–58. (Russian).

Tooming, H. 1969. Determination of absorbed radiation and the efficiency of solar energy conversion in special cases. In: Methodological Instructions for Determining and Checking Basic Indices of Photosynthetic Activity of Plant in Crops: 50–57. VASHNIL, M. (Russian).

Tooming, H. 1977. Solar radiation and yield production. Hydrometeorol. Press, L. (Russian).

Tooming, H. & Gulyayev, B. I. 1967. Methods of measurement of photosynthetically active radiation. Nauka Publishers, M. (Russian).

Tooming, H. & Niilisk, H. 1967. Coefficients for transition from integral to photosynthetically active radiation under field conditions. In: Phytoactin. Investig. of Plant Canopy: 140–149. Valgus Publishers, Tallinn. (Russian).

Tooming, H. & Nilson, T. 1967. The energetics of light adaptation of plant communities. In: Phytoactin. Investig. of Plant Canopy: 35–63. Valgus Publishers, Tallinn. (Russian).

Tooming, H. & Ross, J. 1964. The regime of short-wave radiation on different levels in a maize field and its approximation formulae. In: Investig. Atm. Phys., Inst. Phys. and Astron., Acad. Sci. ESSR, 6: 63–80. (Russian).

Tooming, H. & Ross, J. 1965. The extinction of radiation in various maize crops. In: Problems of Radiation Regime in Plant Canopy: 65–72. Acad. Sci. ESSR, Inst. Phys. and Astron., Tartu (Russian).

Tooming, H. & Ross, J. 1969. Determination of absorption and utilization coefficients of PAR for agricultural crops. In: Methodological Instructions for Determining and Checking Basic Indices of Photosynthetic Activity of Plants in Crops: 35–49. VASHNIL, M. (Russian).

Trenbath, B. R. & Angus, J. F. 1975. Leaf inclination and crop production. Field Crop Abst., 5: 231–244.

Tsunoda, S. 1959. A developmental analysis of yielding ability in varieties of field crops. I. Leaf area per Plant and leaf area ratio. Jap. J. Breed. 9, 2–3: 161–168.

Tucker, C. J. & Garrat, M. W. 1977. Leaf optical system modelled as a stochastic process. Appl. Opt., 16, 3: 635–642.

Tung Liang & Wing Ming Chan. 1977. Estimating solar energy absorption potential for Macadamia nut orchard design: a theoretical approach. Trans. ASAE, 20, 6: 1045–1049.

Uchijima, Z. 1961. On characteristics of heat balance of water layer under paddy plant cover. Bull. Nation. Inst. Agric. Sci., A. 8: 243–263.

Uchijima, Z. 1976a. Maize and Rice. In: Monteith, J. L. (Ed.), Vegetation and the Atmosphere. Vol. 2. Case Studies: 33–64. Acad. Press, London, N.Y., San-Francisco.

Uchijima, Z. 1976b. Microclimate of the rice crop. In: Climate and Rice. Intern. Rice Res. Inst. Los Baños. Philippines: 115–140.

Udagawa, T., Horie, T. & Uchijima, Z. 1970. Density effect on the geometrical structure of crop canopies. In: Photosynthesis and Utilization of Solar Energy. Level III Experiments, 1969: 6–8. Tokyo.

Udagawa, T., Uchijima, Z. & Horie, T. 1969. The penetration of direct solar radiation into a corn canopy and the intensity of direct solar radiation on the foliage surface. In: Photosynthesis and Utilization of Solar Energy. Level III Experiments. 1968: 80–84. Tokyo.

Udagawa, T., Uchijima, Z., Horie, T. & Kobayashi, K. 1968. Canopy structure of corn plant. In: Photosynthesis and Utilization of Solar Energy. Level III Experiments, 1966–1967: 20–24. Tokyo.

Utekhin, V. D. 1977. Primary biological productivity of forest ecosystem: 1–146. Nauka Publishers, M. (Russian).

Utina, Z. M., Shechter, F. H. 1977. The influence of irrigation on the ground surface net radiation and its components. Meteorol. i. Gidrol. 6: 17–23. (Russian).

Utkin, A. I. & Dylis, N. V. 1966. Study of the vertical distribution of phytomass in forest biogeocoenoses. Bull. Moscow Naturalists' Soc., 71, 6: 79–91. (Russian).

Varlet-Grancher, C. & Bonhomme, R. 1971. Assimilation nette, utilisation de l'eau et microclimatologie d'un champ de mais. III Composition spectrale de la lumiere dans la culture. Ann. Agron., 22, 5: 515–525.

Verhagen, A. M. W. & Wilson, J. H. 1969. The propagation and effectiveness of light in leaf canopies with horizontal foliage. Ann. Bot., 33, 132: 711–727.

Verhagen, A. M. W., Wilson, J. H. & Britten, E. J. 1963. Plant production in relation to foliage illumination. Ann. Bot., 27, 108: 627–640.

Vidovič, J. 1973. Relationship of leaf angle in corn (*Zea mays L.*) to grain yield. Report to the 7th Meeting of Eucarpia, Maize and Sorhum Section, Zagreb-Stubičke Toplice: 1–12.

Vygodskaya, N. N. 1967. Radiation regime of a thirty-year-old oak stand in diurnal and seasonal dynamics. In: Light Regime, Photosynthesis and Productivity of the Forests: 77–94. Nauka Publishers, M. (Russian).

Warren Wilson, J. 1959a. Analysis of the spatial distribution of foliage by two-dimensional point quadrats. New Phytologist, 58: 92–101.

Warren Wilson, J. 1959b. Analysis of the distribution of foliage area a in grassland. In: Ivins, J. D. (Ed.), The Measurement of Grassland Productivity: 51–61. London.

Warren Wilson, J. 1960. Inclined point quadrats. New Phytologist, 59, 1: 1–8.

Warren Wilson, J. 1961. Influence of spatial arrangement of foliage area on light interception and pasture growth. Proc. 8th Intern. Grassland Congress: 275–279.

Warren Wilson, J. 1963a. Errors resulting from thickness of point quadrats. Austral. J. Bot., 11, 2: 178–188.

Warren Wilson, J. 1963b. Estimation of foliage denseness and foliage angle by inclined point quadrats. Austral. J. Bot., 11, 1: 95–105.

Warren Wilson, J. 1965a. Stand structure and light penetration. I. Analysis by point quadrats. J. Appl. Ecol., 2, 2: 383–390.

Warren Wilson, J. 1965b. Point quadrat analysis of foliage distribution for plants growing singly or in rows. Austral. J. Bot., 13, 3: 405–409.

Warren Wilson, J. 1967. Stand structure and light penetration. III. Sunlit foliage area. J. Appl. Ecol., 4: 159–165.

Watson, D. J. 1937. The estimation of leaf area in field crops. J. Agric. Sci., 27, 3: 474–483.

Wien, H. C. & Wallace, D. H. 1973. Light-induced leaflet orientation in Phaseolus vulgaris L. Crop Sci., 13, 6: 721–724.

Wijk, W. R. van (Ed.). 1963. Physics of plant environment. Amsterdam.

Williams, J. G. 1976. Small variation in the photosynthetically active fraction of solar radiation on clear days. Arch. Met. Geoph. Biokl., ser. B, 24: 209–217.

Williams, W. A., Loomis, R. S., Duncan, W. G., Dovrat, A. & Nunez, F. 1968. Canopy architecture at various population densities and the growth and grain yield of corn. Crop Sci., 8, 3: 303–308.

Willstätter, A. & Stoll, K. 1918. Untersuchungen über die Assimilation der Kohlensäure. Verlag-Springer, Berlin.

Winter, S. R. & Ohlrogge, A. J. 1973. Leaf angle, leaf area and corn *(Zea mays L.)* yield. Agron. J., 65: 395–397.

Wit, C. T., de. 1965. Photosynthesis of leaf canopies. Agric. Res. Report No. 663, Wageningen: 1–57.

Woolley, J. T. 1971. Reflectance and transmittance of light by leaves. Plant Physiol., 47, 5: 656–662.

Yanishevsky, Y. D. 1940. Methodological instructions for performing phytoactinometric observations and investigations. Manual of Agrometeorology. TsIP, M. (Russian).

Yanishevsky, Y. D. 1941. Methods of radiation measurements for ecological purposes. Trans. Bot. Inst., ser. IV, 5: 22–45. (Russian).

Yefimova, N. A. 1965. Distribution of photosynthetically active radiation on the territory of the Soviet Union. Trans. Main Geophys. Obs., 179: 118–130. (Russian).

Yefimova, N. A. 1968. Peculiarities of the formation of phytoclimate of agricultural crops. Trans. Main Geophys. Obs., 233: 127–138. (Russian).

Yefimova, N. A. 1969. Geographical distribution of the totals of photosynthetically active radiation. In: General Theoretical Problems of Biological Productivity: 160–164. Nauka Publishers, L. (Russian).

Yefimova, N. A. 1970. Influence of the type of plant cover on the formation of phytoclimate of agricultural crops. In: Agroclimatic Resources of the Natural Zones of the USSR and their Utilization: 141–152. Hydrometeorol. Press, L. (Russian).

Yin, H. Z., Wang, T. D., Li, Y. Z., Qin, G. X., Yang, S. Y. & Shen, G. M. 1959. Community structure and light utilization of rice field. Acta Biologiae Experimentalis Sinica, 6, 3: 243–261.

Zelniker, Y. L. 1969. Radiation regime under the forest canopy. Nauka Publishers, M. (Russian).

Author index

E

F

G

H

I

J

K

S

T

U

V

W

Y

Z

Subject index

S

T

U

V

W

MIX
Papier aus verantwortungsvollen Quellen
Paper from responsible sources
FSC® C105338

If you have any concerns about our products,
you can contact us on
ProductSafety@springernature.com

In case Publisher is established outside the EU,
the EU authorized representative is:
Springer Nature Customer Service Center GmbH
Europaplatz 3, 69115 Heidelberg, Germany

Printed by Libri Plureos GmbH
in Hamburg, Germany